STRUCTURAL STEELWORK

FOR BUILDING AND ARCHITECTURAL STUDENTS

TREFOR J. REYNOLDS
B.Sc., A.M.I.C.E., M.I.Struct.E., M.Sth.W.Inst.E.
*Formerly Lecturer-in-Charge of Structural Engineering,
L.C.C. School of Building, Brixton*

and

LEWIS E. KENT
O.B.E., B.Sc., F.I.C.E., F.I.Struct.E.
*Consulting Structural Engineer,
Formerly Lecturer in Structural Engineering,
L.C.C. School of Building, Brixton*

with

DAVID W. LAZENBY
D.I.C., M.I.C.E., F.I.Struct.E.
Consulting Engineer

HODDER AND STOUGHTON
London Sydney Auckland Toronto

by the same authors

Introduction to Structural Mechanics
for Building and Architectural
Students

048 1132 3

ISBN 0 340 16132 9 (Boards Edition)
ISBN 0 340 16133 7 (Paperback Edition)
First printed 1936
Editions 1939; 1942; 1943; 1945 (2); 1946; 1947; 1948; 1952
Eleventh Edition 1955, Second Impression 1957
Twelfth Edition 1961, Second Impression 1965
Thirteenth Edition 1967
Fourteenth Edition 1976

Copyright © 1976 D. W. Lazenby
All rights reserved. No part of this publication may be
reproduced or transmitted in any form or by any means,
electronic or mechanical, including photocopy, recording,
or any information storage and retrieval system, without
permission in writing from the publisher.
Printed in Great Britain for Hodder and Stoughton Educational,
a division of Hodder and Stoughton Ltd,
by Hazell Watson and Viney Ltd, Aylesbury, Bucks

D
624.1821
REY

Preface to Fourteenth Edition

Since the publication of the thirteenth edition of this book, the relevant British Standards have been amended, and most importantly the building industry has 'gone metric'.

In the foreseeable future further major developments can be expected, particularly with regard to Limit State design. As new design concepts are introduced, the practice of structural steelwork will no doubt continue to develop.

The change to a metric system is not merely a case of simple conversion from imperial to metric measure, but it involves the acceptance of some entirely different units for the evaluation of qualities and quantities. Moreover, simple conversion would lead to some arithmetic values which could be difficult to handle, and a certain amount of 'rounding off' is necessary. In practice the SI metric system has been generally adopted, and it is so in this book. A table of conversion factors has been included immediately before the main text.

Whilst revising the contents to accord with the major change of metrication outlined above, the opportunity has been taken of amending sections of the text to generally reflect current practice. For example, Chapter 4, previously dealing with riveted and bolted connections, has now been amended to reflect the small use of riveting, and extended to cover high strength friction grip bolts, and welded connections.

Many of the photographs are newly introduced, whilst a number of the original ones have now been omitted as no longer being representative of modern methods. Acknowledgements for the photographs are noted with their titles.

The standard section tables have again been included, as currently issued by the B.C.S.A., and CONSTRADO. References to and extracts from British Standards are included in this publication by arrangement with the British Standards Institution.

D.W.L.

Preface to First Edition

During recent years a good deal of research work has been carried out with a view to improving the quality of the materials used in building construction and, in this connection, structural steel has received considerable attention. Improved quality in the materials of construction naturally leads to investigation into the possibility of their more economical use and, in the case of structural steelwork, much thought has been given to the question of better methods of design.

Building regulations have had to take into account the progress which has been made both in the quality of structural steel and in the knowledge of its most effective employment in steel-framed buildings.

As the regulations contained in the Building Acts of the London County Council are acknowledged throughout the country to represent a high standard of building construction, it will be useful to state the present position with regard to the regulations affecting steelwork construction in London.

In 1932 the Council issued a 'Code of Practice' for the use of structural steel and other materials in buildings, approved by the Council as a basis of consideration of applications for relief from the Third Schedule of the London Building Act, 1930. By the London Building Act (Amendment) Act, 1935 the Council has obtained powers to make by-laws with respect to a number of matters, including those affecting steelwork (and reinforced concrete) construction.

In writing this book the authors have endeavoured to show, in as simple a manner as possible, the relationship between the established principles of structural mechanics and modern methods of steelwork calculations—as exemplified in structures of not too difficult a character. It is written as a text-book for the student, whether he be a full-time student in a technical college or a young assistant in an office supplementing his practical experience by private, or part-time evening, study.

Preface to First Edition v

Throughout the book theoretical demonstration has been immediately followed by practical illustration in the form of a worked numerical example.

The mathematics employed has been of the simplest possible character consistent with effective demonstration and should present little difficulty to students in advanced building courses. On the few occasions in which theoretical investigation has involved the use of the methods of Calculus, the results have been clearly set out in simple language, so that their employment does not demand a knowledge of this branch of mathematics.

Students preparing for the examinations of the Institute of Builders, the Institution of Structural Engineers, and the Royal Institute of British Architects will, it is hoped, find the book of material assistance. Candidates for the Inter. R.I.B.A. examination should read Chapters 1 to 7, the part of Chapter 9 dealing with maximum bending moments, and the more elementary portions of Chapter 11. For the final examination, architectural students will require to read Chapter 11 more closely and also to take up those parts of the book which deal with practical design.

Building students preparing for the National Diploma or Certificate in Building should find the groundwork covered in the theory and design of structural steelwork.

The authors wish to acknowledge freely the many sources of the theoretical principles which, together with the results of practical experience, constitute the text of the book. The practical value of the book has been considerably enhanced by the assistance received from a number of well-known constructional firms who have supplied diagrams, photographs, and other practical data. The authors' thanks are especially due to:

Messrs. Dorman, Long & Co., Ltd.; Messrs. Redpath, Brown & Co., Ltd.; Messrs. R. A. Skelton & Co. Steel & Engineering, Ltd.; Messrs. Dawnays, Ltd.; Messrs. The Kleine Company, Ltd.; Messrs. Caxton Floors, Ltd.; Messrs. The Quasi-Arc Company, Ltd.

The British Standards Institution kindly granted permission for extracts to be made from recent B.S. and this, together with the permission of the London County Council to quote from building regulations, notably the 'Code of Practice', has made it possible for the authors to refer frequently to the practical considerations which influence purely theoretical results.

Acknowledgement is also due to the British Steelwork Association for its courtesy in permitting the publication of certain property tables which will be found in the text, and to the Institution of Structural Engineers for permission to quote from the valuable report which it issued on the metallic arc welding of structural steelwork.

The authors wish to record their thanks to Mr. F. E. Drury, M.Sc., M.I.Struct.E., Principal of the L.C.C. School of Building, Brixton, for helpful advice given on this occasion, as on other occasions, whenever sought.

Finally it is desired to express appreciation of the interest taken by Dr. H. H. Burness in the preparation and production of the book.

1936   T.J.R.
                                     L.E.K.

Contents

Preface to Fourteenth Edition	iii
Preface to First Edition	iv
Table of Conversion Factors	x
1. Stress, Strain, and Elasticity	1
2. Strength, Factor of Safety, and Working Stress	17
3. Fabrication of Steelwork	38
4. Practical Design of Riveted and Bolted Connections	50
5. Theory of Beam Design	82
6. Design of Simple Beams. Moment of Resistance	103
7. Properties of Compound Beam Sections	145
8. Deflection of Beams. Theory and Practice	161
9. Shear and its Applications	185
10. Fixed and Continuous Beams	202
11. Practical Design of Compression Members	227
12. Stanchion and Column Bases	281
13. Encasement of Steelwork. Concrete Floors	299
14. Introductory Principles of the Welding of Steelwork	307
15. Plate and Lattice Girders. Theory and Practical Design	326
Appendix 1 Design of a Steel Frame for a small Warehouse Building, with Typical Details	346
Appendix 2 British Standards and Codes of Practice	379

Appendix 3 Spacing of Holes in Standard Sections	384
Appendix 4 Geometrical Properties of a Parabola	386
Answers to Numerical Questions	388
Index	391

Illustrations

Plates   Facing page

1	Denison Universal Testing Machine	20
2	Steel Plate Rolling Mill	21
3	Bolting by the use of a Torque-controlled Spanner	52
4	Bolting with Load-indicating Washers by the use of a Power Tool	53
5	Welded Stanchion Base	282
6	(a) Section of Fillet Weld	283
	(b) Section of Butt Weld	283
7	Automatic Submerged-arc Welding Machine	314
8	Examples of Welded Construction	
	(a) Tubular Grid Structure	314
	(b) Portal Framed Building	314
9	Detail of Welded Plate Construction	315
10	Lattice Girder Construction	346

Folding Charts

Charts

1	Plate Girder	332
2	Lattice Girder	340
3	Roof Trusses	352
4	Steel Plans and Stanchion Schedule	368

Table of Conversion Factors

A **newton** is the force required to accelerate a mass of **1 kilogram** by **1 metre per second per second**.

1 N	= 0·224 81 lbf
1 kN	= 224·81 lbf
1 lbf	= 4·448 22 N
1 tonf	= 9964 N
1 inch	= 25·4 mm = 0·0254 m
1 foot	= 304·8 mm = 0·3048 m
1 m	= 3·280 84 ft = 39·37 in
1 lbf/ft	= 14·593 86 N/m
1 tonf/ft	= 32 690·2464 N/m = 32·69 kN/m
1 lbf/in^2	= 0·006 894 76 N/mm^2 = 6894·76 N/m^2
1 tonf/in^2	= 15·44 N/mm^2
1 lbf/ft^2	= 47·8803 N/m^2
1 tonf/ft^2	= 107 251·872 N/m^2 = 107·252 kN/m^2
1 lbf/ft^3	= 157·088 N/m^3

1

Stress, Strain, and Elasticity

Nature of structural steel

Steel is not a simple element. It is mainly composed of iron, but the iron is alloyed, or associated with, various other materials. It is upon the nature and relative amounts of these special ingredients that the physical properties of the steel depend. For example, if the metal chromium is introduced into the composition, the resulting steel is able to exhibit, among other useful properties, a pronounced resistance to rusting and is given the name *stainless steel*. The element manganese, on the other hand, gives good wearing properties to steel, making it suitable for use in the manufacture of rails. There are, therefore, various types of steels, known respectively as *chromium steels*, *manganese steels*, and so on, according to the alloying elements which give the steels their characteristic properties.

A substance which plays an important part in the type of steel used in building construction is the element *carbon*. The percentage of carbon in steel directly influences its essential structural properties. An increase in carbon content results in an increase in strength, but this is accompanied by a marked decrease in ductility. Ductility, or absence of brittleness, is one of the important requisites of a structural steel. It promotes equalisation of load between the steel fibres of a member. Of such importance is this property of ductility that, in the commercial testing of structural steel, an upper limit of strength is prescribed for the steel in addition to a definite minimum value for the percentage elongation.

Steel which is to be used in general building construction is subject to a number of standardised requirements. The standards of quality required are laid down in one of the standards issued by the British Standards Institution. These standards are known as BS (British Standards), and that relating to structural steel for general building work is BS 4360.

2 Structural Steelwork

Table 1.1 Typical analyses of structural steels.

	Typical composition %	
	A lower-grade steel	A higher-yield-stress steel
Carbon	0·2	0·2 max.
Manganese	0·5	0·7–1·0
Chromium	—	0·5
Copper	—	0·25–0·5
Silicon	0·04	0·2 max.
Sulphur	0·04	0·05 max.
Phosphorus	0·04	0·05 max.
Iron	99·18	balance

In the case of a higher-yield-stress steel, a typical analysis would indicate a maximum carbon content of 0·2 per cent. The mechanical and physical properties depend, not simply upon the carbon-content, but upon certain other alloying elements such as copper, nickel, chromium, etc. It is usual in a specification to lay down maximum limits for certain of the component materials. The reader should consult the British Standard for further requirements specified in t‚   ‚-
facture of these steels.

Stress

To understand the provisions of BS 4360 and similar specifications, it is necessary to study the subject of *stress*.

Fig. 1.1 shows a tension member AB subjected to a pull of L kN at

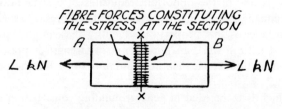

Fig. 1.1 *Diagram illustrating the nature of stress.*

each end. Considering a typical section XX, we see that the load L kN, at the end A, is trying to detach the portion AX of the member from the portion XB. It is unable to do so, because of the numerous little pulls which the fibres of the material exert, and which are shown to the right of the section plane XX. Similarly the pull L kN at end B cannot effect separation at the section, because there are fibre pulls acting there, to the left. Such a system of actions and reactions, acting over the cross-sectional area of a member, constitutes what is termed a *stress* at the section.

If the cross-sectional area of the member at XX were A mm^2, and if the load L kN were distributed uniformly over the section, the ratio L kN $\div A$ mm^2 would give the *intensity of stress* at the section. Assuming the load to be 100 kN, and the sectional area to be 2000 mm^2, the *stress* at the section (the word *intensity* is always omitted in practice) would be $\dfrac{100 \text{ kN}}{2000 \text{ mm}^2} = 0{\cdot}05 \text{ kN/mm}^2$.

Varying stress

The distribution of load over the section of a member may be of a non-uniform character. For example, in a beam section, not only is there variation of load value, but there is a complete reversal from *tension* (pulling) to *compression* (pushing). In such a case, the 'stress' cannot be obtained by the simple formula load $\div$ area: the stress will vary in value from point to point in the beam section. But, however its value may be obtained, a stress has always the nature of *load intensity per unit area*. When a phrase such as '*extreme fibre stress equals* $0{\cdot}15$ kN/mm^2' is used, what is meant is simply that, were the fibres over one square mm to be subjected to this given stress value, the total load carried would be $0{\cdot}15$ kN. The structural designer does not think in terms of average stress values, but in terms of 'maximum stress intensity' in any single fibre.

Forms of stress

There are three forms of simple stress: *tension, compression*, and *shear*.

Tensile stress

This is the kind of stress induced in a member when it is subjected to a pull.

$$\text{Tensile stress at section XX} = \frac{\text{load}}{\text{area}}$$

$$= \frac{L \text{ kN}}{A \text{ mm}^2} = \frac{L}{A} \text{ kN/mm}^2$$

4 *Structural Steelwork*

(Any convenient units of load and area may be used in stress calculations, but, as stress values in structural steelwork are nearly always expressed in kN/mm^2, this unit has been adopted.)

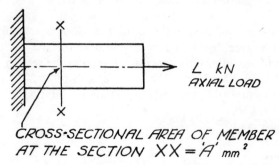

Fig. 1.2 *Tensile stress.*

Example 1 A solid circular steel tie-rod 20 mm diameter carries an axial load of 40 kN. Calculate the stress in the material of the rod.

$$\text{Tensile stress} = \frac{\text{load}}{\text{area}}$$

$$\text{Sectional area of tie rod} = \frac{\pi d^2}{4} = \frac{\pi \times 20^2}{4} \text{ mm}^2$$

$$= 314 \text{ mm}^2$$

$$\text{Tensile stress} = \frac{40 \text{ kN}}{314 \text{ mm}^2}$$

$$= 0\cdot127 \text{ kN/mm}^2$$

$$\text{or } 127\cdot0 \text{ N/mm}^2$$

Example 2 Find the maximum safe axial load for a mild steel tie-bar, 20 mm wide by 5 mm thick, if the tensile stress is not to exceed 0·12 kN/mm^2.

$$\text{Sectional area of tie bar} = 20 \text{ mm} \times 5 \text{ mm} = 100 \text{ mm}^2$$

$$\text{Stress} = \frac{\text{load}}{\text{area}}$$

$$0.12 \text{ kN/mm}^2 = \frac{L \text{ kN}}{100 \text{ mm}^2}$$

$$L = 0.12 \times 100 \text{ kN} = 12 \text{ kN}$$

Example 3 Calculate the necessary thickness of a tie-bar 100 mm wide, if it has to carry an axial load of 60 kN without the maximum stress exceeding 120 N/mm².

Sectional area of tie-bar = 100 mm × t mm = 100t mm²

$$0.12 \text{ kN/mm}^2 = \frac{60 \text{ kN}}{100t \text{ mm}^2}$$

$$\therefore 100t \times 0.12 = 60$$

$$t = \frac{60}{12} = 5.0 \text{ mm}$$

Compressive stress

Columns and *struts* (i.e. members which support thrusts) have their fibres in this condition of stress.

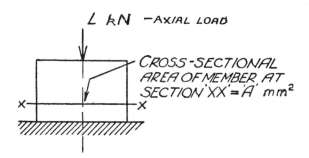

Fig. 1.3 *Illustration of compressive stress.*

Compressive stress at section XX = $\dfrac{\text{load}}{\text{area}}$

$$= \frac{L \text{ kN}}{A \text{ mm}^2} = \frac{L}{A} \text{ kN/mm}^2$$

Slender compression members are liable to failure by side bending or 'buckling', in addition to direct crushing. This type of member is fully dealt with later in this volume (Chapter 11).

Example A solid circular steel column supports 20 m² of floor area, for which the total inclusive load is 10 kN/m². Assuming it to be necessary to limit the maximum compressive stress in the column to 0·05 kN/mm², obtain the minimum permissible diameter for the column.

Total load on column (assumed axial) $= 10 \text{ kN/m}^2 \times 20 \text{ m}^2$

$\hspace{6cm} = 200 \text{ kN}$

Let d mm = diameter of column.

$$\text{Sectional area of column} = \frac{\pi d^2}{4} \text{ mm}^2$$

$$\therefore \quad 0{\cdot}05 \text{ kN/mm}^2 = \frac{200 \text{ kN}}{\pi d^2/4 \text{ mm}^2}$$

$$\therefore \quad 0{\cdot}05 \times \pi d^2/4 = 200$$

$$d^2 = \frac{16\,000}{\pi}$$

$$\therefore \quad d = 71{\cdot}3 \text{ or, say, } 75 \text{ mm}.$$

Shear stress

When one portion of a member tends to slide over another portion at a given section, the fibres at the section are said to be in *shear stress*.

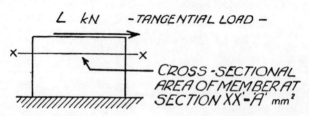

Fig. 1.4 Member subjected to shear stress.

$$\text{Shear stress at section XX} = \frac{\text{shear load}}{\text{area under shear}}$$

Stress, Strain, and Elasticity 7

$$\text{Shear stress (XX)} = \frac{L \text{ kN}}{A \text{ mm}^2} = \frac{L}{A} \text{ kN/mm}^2$$

The rivets or black bolts, in a simple joint, are examples of structural units subjected to this form of stress.

Example Assuming the load in fig. 1.5 to be 20 kN, and the rivet diameter to be 15 mm, find the shear stress in the rivet.

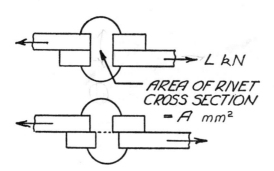

Fig. 1.5   *Shear stress in a rivet.*

$$\text{Sectional area of rivet} = \frac{\pi d^2}{4} = \frac{\pi \times 15^2}{4} \text{ mm}^2$$

$$= 176 \cdot 7 \text{ mm}^2$$

$$\text{Shear stress} = \frac{\text{load}}{\text{area}} = \frac{20 \text{ kN}}{176 \cdot 7 \text{ mm}^2}$$

$$= 0 \cdot 113 \text{ kN/mm}^2$$

$$\text{or } 113 \text{ N/mm}^2$$

The subject of rivet strength is treated, in detail, in Chapter 4.

Strain

If we apply a load to a member, not only do we induce in the fibres a state of stress, but in some respect we alter the size, or shape, of the member.

The subject of *strain* is concerned with these geometrical alterations. Each of the stresses which has been referred to is accompanied by its corresponding strain.

Tensile strain

The tensile strain in the member in fig. 1.6 is not measured by the extension x in itself, but by the *ratio of this extension to the original length* of the member.

$$\text{Tensile strain} = \frac{\text{extension}}{\text{original length}}$$

$$= \frac{x \text{ mm}}{l \text{ mm}} = \frac{x}{l} \quad \text{(simply a number)}$$

The reader should carefully note that the value of the strain is not expressed in any dimensional unit. The two length measurements concerned in the computation may be in any units, provided the same unit is employed for both. In the practical employment of a steel member, the extension x is extremely small, and is not discernible by the naked eye.

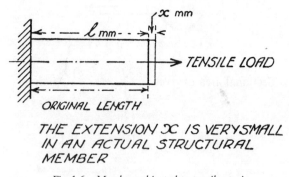

Fig. 1.6 *Member subjected to tensile strain.*

Example 1   A tie member 4 m long is subjected to an axial load which stretches it 0·2 mm. Calculate the strain in the material of the member.

$$\text{Tensile strain} = \frac{\text{extension}}{\text{original length}}$$

$$= \frac{0 \cdot 2 \text{ mm}}{4 \times 1000 \text{ mm}} = \frac{0 \cdot 2}{4000} = 0 \cdot 00005 \quad \text{(a number)}$$

Example 2   A tensile-test specimen undergoes a strain of 0·0004. Find the actual extension on a measured gauge length of 200 mm.

Stress, Strain, and Elasticity 9

$$\text{Tensile strain} = \frac{\text{extension}}{\text{original length}}$$

$$0.0004 = \frac{\text{extension in mm}}{200 \text{ mm}}$$

$\therefore$ extension $= (200 \times 0.0004) \text{ mm} = 0.08 \text{ mm}$

Compressive strain

$$\text{Compressive strain} = \frac{\text{shortening in length}}{\text{original length}} = \frac{x \text{ mm}}{l \text{ mm}} = \frac{x}{l}.$$

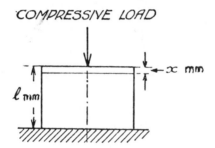

THE SHORTENING IN LENGTH x IS VERY SMALL IN A STEEL STRUT UNDER WORKING CONDITIONS

Fig. 1.7 Member subjected to compressive strain.

Example A column, loaded axially, shortened by 0.5 mm. If the resulting strain were 0.0002, find the original length of the column.

$$\text{Compressive strain} = \frac{\text{shortening}}{\text{original length}}$$

$$0.0002 = \frac{0.5 \text{ mm}}{\text{length in mm}}$$

$$\therefore \text{length} = \frac{0.5 \text{ mm}}{0.0002} = 2500 \text{ mm} = 2.5 \text{ m}$$

Shear strain

The two strains already discussed involve the change in length of a member. Shear strain is concerned with the change of shape, or the distortion, which results from shear stress.

The value of the shear strain is given by the ratio $\dfrac{x \text{ mm}}{l \text{ mm}} = \dfrac{x}{l}$.

It is more important at the present stage for the reader to have a correct appreciation of the nature of shear strain, than to possess a knowledge of its exact determination. It is the type of strain induced in a workshop shaft which is transmitting a twisting moment, or in a key when we are attempting to turn it in a stiff lock.

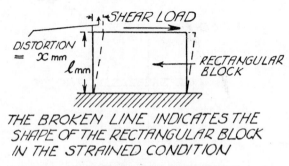

Fig. 1.8   Nature of shear strain.

Relationship between stress and strain

It will now be clear that the terms 'stress' and 'strain' refer, respectively, to two quite different physical conditions of a loaded member. Within certain limits, however, there is a definite, and simple, relationship between the corresponding values of these quantities. The relationship is more clearly defined in some building materials than in others. Steel possesses the property of *elasticity* in a high degree, and obeys the *elastic law* very closely.

Elasticity

A piece of material is said to be *elastic* if, having been deformed by an applied force, it regains its original size and shape when the deforming force is removed. An experimenter, named Hooke, discovered (about the year 1676) that an elastic body would always stretch by an amount which was directly proportional to the applied tensile load, provided the experiment were not conducted beyond a certain maximum limit of stretching. This law of relationship between 'load' and 'extension' is exemplified in the tension spring-balance scales used in shops. The graduation marks on such scales are all equidistant, indicating, for example, that the spring stretches twice as much for a 20 N weight as it does for a 10 N weight. A steel member may be regarded as being, within limits, a very strong and accurate spring, whose extensions (or compressions) are so small that special instruments are required to measure them.

We have seen that the stress in a member is directly proportional to the applied load $\left(\text{stress} = \dfrac{\text{load}}{\text{area}}\right)$ and that the strain is directly proportional to the extension $\left(\text{strain} = \dfrac{\text{extension}}{\text{original length}}\right)$.

Hooke's load–extension law may therefore be expressed in terms of 'stress' and 'strain'. This is the form in which it is usually remembered and quoted, and the law is expressed by the statement that '*stress varies as strain*'. The law will equally apply in tension and compression for all steel members, up to a stress value known as the *elastic limit* for the particular steel.

Young's modulus of elasticity

The reader will, perhaps, more readily understand this very important physical property, if a comparison be made between it and another physical property with which he is already familiar.

If we took a number of different pieces of any given material and weighed them, a graph could be drawn showing the variation of 'weight' with 'volume'. The graph would be a straight line (fig. 1.9), as weight

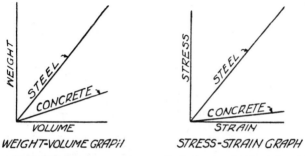

Figs 1.9 and 1.10   Analogy between the two physical constants 'density' and 'Young's modulus'.

would increase uniformly with volume. The value of the ratio $\frac{\text{weight}}{\text{volume}}$ would be the same for any pair of corresponding values of weight and volume, taken at any point on the graph. The ratio would give us the *physical constant* for the material known as its *density*. But stresses and strains follow the same type of law as weight and volumes do (fig. 1.10).

The stress–strain graph will be a straight line, and the ratio of any particular stress to its corresponding strain will be a constant value for all points in the graph. The actual value of this constant does not depend on the size of the member undergoing stress and strain, but simply on the nature of the material of the member, just as *density* is independent of the dimensions of the substance concerned.

The physical constant obtained from the stress–strain ratio is given the name *Young's modulus*, and is denoted, in calculations, by the letter E.

$$E = \frac{\text{stress}}{\text{strain}}$$

Hooke's elastic law holds also in the case of shear stress and shear strain, but the value of the constant $\frac{\text{stress}}{\text{strain}}$ differs in this case from that obtained in tension and compression, and is termed the *shear modulus* of the given material.

Units of Young's modulus As the value of strain is simply expressed as a number, the units of E will be those of stress. If strain = 1, E = stress, so that Young's modulus may, theoretically, be defined as the stress value required to produce unit strain in a tensile specimen of the particular material. Unit strain, however, involves an extension equal to the original length of the specimen. Young's modulus has no significance beyond the elastic limit of the material, which, in the case of steel, represents a strain of the order of 0·001. Although unit strain is impracticable in attainment, the terms of the definition serve to emphasise the nature of the constant E.

The value of Young's modulus for structural steel, in tension or compression, may be taken as 200 kN/mm^2. A lower value is sometimes taken in calculations involving the deflection of steel beams.

The following important facts of elasticity will now be appreciated by the reader:

i) *Stress cannot exist without strain, nor strain without stress.*
ii) *A given elastic stress value is always accompanied, in any particular type of material, by the same value of strain.*

iii) *Young's modulus is the physical constant which enables us to calculate exactly how much strain accompanies a given stress value, and vice versa.*

Example 1 Calculate the value of E from the following results of a steel tensile test.

Sectional area of specimen = 300 mm^2
Measured gauge length on specimen = 200 mm
Applied tensile load = 15 kN
Corresponding elastic extension = 0·05 mm

$$\text{Stress} = \frac{\text{load}}{\text{area}} = \frac{15 \text{ kN}}{300 \text{ mm}^2} = 0{\cdot}05 \text{ kN/mm}^2$$

$$\text{Strain} = \frac{\text{extension}}{\text{original length}} = \frac{0{\cdot}05 \text{ mm}}{200 \text{ mm}} = 0{\cdot}00025$$

$$E = \frac{\text{stress}}{\text{strain}} = \frac{0{\cdot}05 \text{ kN/mm}^2}{0{\cdot}00025} = 200 \text{ kN/mm}^2$$

Example 2 Find the elongation produced in a circular tie-rod, 3 m long and 20 mm diameter, when subjected to an axial load of 30 kN. $E = 200 \text{ kN/mm}^2$.
Let x mm = the extension.

Sectional area of a 20 mm diameter rod = 314 mm^2

$$\text{Stress in rod} = \frac{\text{load}}{\text{area}} = \frac{30 \text{ kN}}{314 \text{ mm}^2}$$

$$= 0{\cdot}096 \text{ kN/mm}^2$$

$$\text{Strain in rod} = \frac{\text{extension}}{\text{original length}}$$

$$= \frac{x \text{ mm}}{(3 \times 1000) \text{ mm}} = \frac{x}{3000}$$

$$E = \frac{\text{stress}}{\text{strain}}$$

$$200 = \frac{0{\cdot}096 \text{ kN/mm}^2}{x/3000}$$

14 *Structural Steelwork*

$$200 \times \frac{x}{3000} = 0.096$$

$$\therefore \quad x = 1.44 \text{ mm}$$

Example 3 In a test to determine the stress induced in a member of a steel frame by the load carried, an instrument was fixed to the member in order to measure the shortening produced in it. Assuming a contraction in length of 0·015 mm to have been measured on a 300 mm gauge length, deduce the stress in the member.

$$\text{Strain in member} = \frac{\text{shortening}}{\text{original length}}$$

$$= \frac{0.015 \text{ mm}}{300 \text{ mm}} = 0.0005$$

$$E = \frac{\text{stress}}{\text{strain}} \quad \therefore \quad \text{stress} = E \times \text{strain}$$

Taking E to be 200 kN/mm².

stress = $200 \times 0.0005 = 0.1$ kN/mm² or 100 N/mm²

The above example illustrates the method employed in research work to ascertain the stress at any part of a loaded specimen. The extremely small alterations in length caused by the application of load are measured by instruments termed 'extensometers'.

The illustration given in plate 1 (facing page 20) shows a typical testing machine. An extensometer is fixed to the test piece in order to find the extension on a known gauge length, i.e. the 'strain' (and hence the 'stress' and 'load') in this member.

Exercises 1

1 A tie-bar in a steel truss carries a load of 75 kN. The section of the bar is rectangular, 75 mm × 10 mm. Calculate the tensile stress in the material of the bar.

2 How many steel suspension bars, 100 mm × 12 mm, would be required to support a load of 700 kN, assuming the load equally divided between the bars? Maximum stress not to exceed 155 N/mm².

3 Find the necessary diameter for a steel column of solid circular section which has to carry an axial load of 800 kN, the maximum allowable stress being 120 N/mm².

Stress, Strain, and Elasticity   15

4 The base of a column is carried on a square concrete slab. The load transmitted to the ground beneath the slab is 600 kN. Assuming a safe bearing pressure on the ground of 400 kN/m², find the minimum dimensions, in plan, for the slab.

5 Find the maximum safe value for *P*, in the gusseted connection given in fig. 1.11, from the point of view of (a) the tension in the tie-bar (60 mm gross width), (b) the tension at section XX in the gusset plate.

(If x mm = width of tie-bar and d mm = diameter of rivet hole, a section taken through a rivet hole will have an 'effective area' of $(x-d)$ mm × 'thickness of tie-bar'. The safe load for the tie-bar must be computed from this 'net' area.)

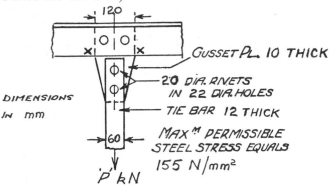

Fig. 1.11.

6 Find the shear stress in the tie-bar bolts of the given connection (fig. 1.12). Find also the necessary thickness of the tie-bar, using a safe

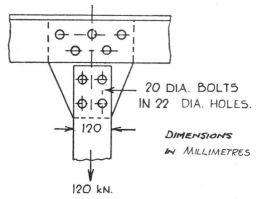

Fig. 1.12.

stress of 215 N/mm². (Two holes must be allowed for in finding the net width of the tie-bar.)

7 Calculate the strain in a column which is shortened by 1·2 mm under the applied axial load, the original length of the column being 4 m. Also determine the contraction in length corresponding to a strain of 0·0004.

8 Obtain Young's modulus from the following results of a practical test:
   Diameter of circular specimen = 20 mm
   Gauge length on specimen = 200 mm
   Applied load = 25 kN
   Corresponding extension (measured on gauge length) = 0·08 mm.

9 Distinguish between the terms 'stress' and 'strain'. Give the three forms of simple stress with examples of typical structural members in which they respectively occur. Name and write down the law which, within certain limits, governs the relative values of these two physical properties, and explain the meaning and nature of 'Young's modulus'.

10 Calculate the extension in a steel tie-rod, 25 mm diameter and 2·5 m long, for an axially applied load of 25 kN. $E = 200$ kN/mm².

11 In a test to determine the 'live load' carried by a member of a steel lattice girder—due to the passage over the girder of a travelling crane—an extensometer, of the direct-reading dial type, was employed. Taking the following test results, determine the live load referred to:
   Gauge length on member = 250 mm
   Value of one dial division on extensometer = $\frac{1}{100}$ mm
   Difference in readings due to passage of load = 20 divisions
   Sectional area of member = 1500 mm²
   Take the usual value for E.

12 An uncased steel member 6 m long was subjected during a fire to a temperature rise of 30 °C. Assuming the ends of the member to have been fixed in such a way as not to allow of any expansion, calculate the stress induced in the steel.
   Coefficient of linear expansion of steel = 0·00001 per °C
   Assume $E = 200$ kN/mm².

2

Strength, Factor of Safety, and Working Stress

Stress–strain graph for tensile test

If a steel specimen were placed in a testing machine and a tensile load applied steadily until the specimen fractured, the stress–strain graph would have the general character given in fig. 2.1.

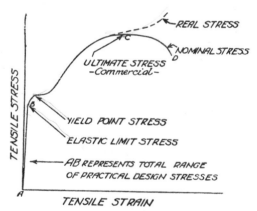

Fig. 2.1 Stress–strain graph for a steel specimen tested to fracture.

The graph may be divided into three parts, indicated in the figure by AB, BC, and CD.

(i) *A to B* Between A and B the graph is a straight line. The point B fixes the upper *limit of proportionality* between stress and strain. The stress value corresponding to this limit is known as the *elastic limit* stress for the steel.* The part AB of the graph is very important from the point

---

*No distinction is made here between E.L. and L. of P., or between an upper and a lower Y.P.

of view of design, as the stresses involved are those corresponding to the elastic strain of the steel—the strain condition of all steel in practical structures. It will be clear that no formula in design, based on the assumption of Hooke's law, will hold for a stress exceeding the 'limit of proportionality' stress.

(ii) *B to C* When the stress has reached a value slightly higher than the elastic limit, a definite yield takes place in the specimen. The strain value increases without corresponding increase in the stress. The stress at this point is known as the *yield-point* stress. The yield point is made apparent in a practical test by the sudden drop of the lever arm of the testing machine, and the temporary refusal of the specimen to take up load. The term *elastic limit* is sometimes used for this stress. After passing the yield point, the stress increases, and the graph takes the form indicated in the figure.

(iii) *C to D* Throughout the test, lateral strain accompanies the longitudinal strain, but extensometers would be required to measure both these strains—inside the elastic limit. Between C and D, however, there is a marked contraction of cross-sectional area, easily visible to the naked eye. Just before fracture the specimen will have the appearance shown in fig. 2.2.

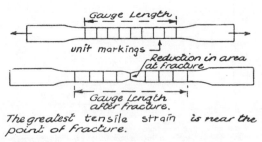

Fig. 2.2 Laboratory tensile-test specimen.

It will be found that to maintain the lever arm *floating* between its stops during this part of the test, load will have to be taken off the specimen. If we calculate stress values on the original sectional area of the specimen, these values will decrease with the decreasing load. This explains the apparent drop in stress before fracture, as represented in the graph. Actually the stress increases up to the point of fracture, and the broken-line graph would be obtained if the reduced cross-sectional area of the specimen were taken into account. Stress calculations, in commercial testing, are based on original dimensions.

Strength

In a practical test to determine the strength of steel, the maximum load carried by the specimen is the important quantity, not the actual load at fracture. This applies equally to 'compression' and 'shear' tests. The strength is obtained by dividing the maximum load during the test by the original sectional area of the specimen.

$$\text{Strength} = \frac{\text{maximum load}}{\text{original sectional area}}$$

Commercial testing of 'structural steel'

BS 4360 gives precise details of the nature of the tests, both on the material and tolerances of the finished product, to be carried out, and of the general procedure which has to be followed. The reader is referred to this specification for fuller particulars than can be given here.

Some of the steel sections, etc., being rolled to an order, are made longer than necessary. The extra lengths are then cut off, as required, for test specimens.

In addition to chemical composition, tests imposed by the BS are (i) a tensile test, (ii) a cold bend test, (iii) a flattening test for hollow sections, and (iv) in some cases an impact test.

Clauses relating to tensile tests

In BS 4360 Clauses 18–21 set out the requirements for the tensile tests, in relation to samples, test pieces, and test results. Tables 7, 9, 10, and 12 of the Standard list the tensile strength, yield stress, and elongation for the various types of steel section, and the various grades of steel.

The minimum yield stress which is required for each grade of steel is an important factor in the derivation of permissible working stresses.

Clauses relating to bend tests

BS 4360 Clauses 22–25 give the requirements for bend tests. Tables 7, 9, and 10 of the Standard list the requirement for each grade of steel, for the various types of steel section—except hollow sections, which are not subject to this form of testing. The requirement is expressed in terms of the maximum bend radius, related to the thickness of the sample.

Clause relating to the flattening test

Clause 26 of BS 4360 gives the procedure for the flattening test to be applied to hollow sections. The minimum requirements for the various grades of steel are set out in Table 12.

Clauses relating to impact tests

BS 4360 clauses 27–30 give the requirements for impact testing, which relate only to the various grade B, C, D, and E steels. This is a most important test, particularly for the higher-strength steels, e.g. 50 C and 55 C, commonly used in structural work.

Laboratory steel testing

In laboratory testing on circular specimens to find the '*percentage contraction in area*' of the specimen at point of fracture, in addition to the other test constants, interesting tests on a range of carbon steels (steels in which the variation of carbon-content provides the characteristic properties) may with advantage be carried out to illustrate how the yield point and tensile strength are increased in value and the ductility constants ('percentage elongation' and 'percentage contraction in area') decreased as the percentage carbon-content increases.

FORMS OF BRITISH STANDARD TENSILE TEST PIECES
(Extracted from BS 18, Methods for Tensile Testing of Materials)

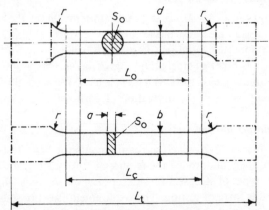

Fig. 1   *Test piece of circular and rectangular cross-section.*

Plate 1   *Denison universal testing machine.*
Reproduced by courtesy of Avery-Denison Ltd.

Plate 2 Steel plate rolling mill.

Reproduced by courtesy of British Steel Corporation, Scunthorpe Group.

TABLE 1. DIMENSIONS OF CIRCULAR SECTION TEST PIECES

Gauge length $L_o = 5.65\sqrt{S_o}$

Cross-sectional area S_o	Diameter d	Gauge length L (See Note 4)	Minimum parallel length $L_c \approx 5.5d$	Minimum transition radius r	Tolerance on diameter (See Note 3)
mm^2	mm	mm	mm	mm	±mm
400	22.56	113	124	23.5	0.13
200	15.96	80	88	15	0.08
150	13.82	69	76	13	0.07
100	11.28	56	62	10	0.06
50	7.98	40	44	8	0.04
25	5.64	28	31	5	0.03
12.5	3.99	20	22	4	0.02

NOTE 1. The gripped ends of the test piece shall be co-axial with the parallel portion within a concentricity tolerance of 0.03 mm.

NOTE 2. The diameter along the parallel length shall not vary by more than 0.03 mm.

NOTE 3. The tolerance on diameter has been determined in relation to the required accuracy of measurement (see 7.1).

NOTE 4. The gauge length is given to the nearest 1 mm and the minimum parallel length is adjusted accordingly (see also 8.1).

NOTE 5. In special cases, test pieces with diameters other than those given in Table 1 may be used provided that the gauge length $L_o = 5.65\sqrt{S_o}$, i.e. approximately equal to 5 diameters.

TABLE 2. DIMENSIONS OF RECTANGULAR
SECTION TEST PIECES (NON-PROPORTIONAL)

Width b	Gauge length L_o	Minimum transition radius r	Approximate total length L_t
mm	mm	mm	mm
40	200	25	450
20	200	25	375
25	100	25	300
12.5	50	25	200
6	24	12	100
3	12	6	50

NOTE 1. Notwithstanding the above, test pieces having a gauge length equal to $5.65\sqrt{S_o}$ are permitted.

NOTE 2. For any width from 3 mm to 25 mm a gauge length of 50 mm may be used, the total length being adjusted accordingly.

NOTE 3. The width of the parallel section shall not vary by more than 0.03 mm for test pieces having a width up to and including 12.5 mm and 0.1 mm for test pieces having a width greater than 12.5 mm.

NOTE 4. The test pieces having a width of 40 mm and 20 mm and a gauge length of 200 mm have been included to correspond to the former inch dimension test pieces having a width of 1½ in and ¾ in and a gauge length of 8 in.

NOTE 5. A straight, parallel test piece without enlarged ends is permissible for any size.

Strength, Factor of Safety, and Working Stress

Example 1 Calculate the usual test constants from the following results of a steel tensile test. State whether the steel represented by this specimen would satisfy the requirements of BS 4360, grade 43a, for round bars.

Diameter of specimen = 19·5 mm
Distance between gauge points = 100 mm
Load at yield point = 85 kN
Maximum load during test = 140 kN
Gauge length (after fracture) = 132 mm
Diameter at fracture = 13·46 mm.

(a) Yield-point stress = $\dfrac{\text{load at Y.P.}}{\text{original area of specimen}}$

Original sectional area of specimen = 300 mm²

$\therefore$ Y.P. stress = $\dfrac{85 \text{ kN}}{300 \text{ mm}^2}$ = 283 N/mm²

(b) Tensile strength = $\dfrac{\text{maximum load during test}}{\text{original area of specimen}}$

$= \dfrac{140 \text{ kN}}{300 \text{ mm}^2}$ = 466 N/mm²

(c) Percentage elongation at fracture = $\dfrac{\text{elongation}}{\text{gauge length}} \times 100$

$= \dfrac{(132 - 100)}{100} \times 100$

$= 32\%$

(d) Percentage contraction in area at fracture

$= \dfrac{\text{contraction in area}}{\text{original area}} \times 100$

Area at fracture = $\dfrac{\pi \times 13\cdot 46^2}{4}$ mm²

$= 142\cdot 3$ mm²

$\therefore$ contraction in area = $(300 - 142\cdot 3) = 157\cdot 7$ mm²

$$\therefore \text{ percentage contraction} = \frac{157 \cdot 7}{300} \times 100$$

$$= 52 \cdot 6\%.$$

Results from (a), (b), and (c) are within the limits of the BS requirements, hence the steel would pass the tensile test.

Example 2 For the specimen given in the last example, evaluate the maximum and minimum test loads, respectively, which would have been permissible for the specimen to pass the BS test. Also, determine the minimum elongation on 100 mm gauge length.

Sectional area of specimen = 300 mm^2
Minimum Y.P. load = 255 N/mm$^2 \times$ 300 mm^2
= 76·5 kN
Minimum elongation = 22% of 100 mm = 22 mm

Factor of safety and working stress

It will be clear that the stress to be used in the actual dimensioning of structural members will have to be somewhat less than the corresponding strength of the material. Some of the principles governing the suitable margin of safety to be allowed in any particular case are indicated later. The stress used in practical design is termed the *safe working stress* or simply *working stress*. This stress value is obtained from the appropriate strength by dividing it by a selected number known as the *factor of safety*.

$$\text{Working stress} = \frac{\text{strength}}{\text{factor of safety}}.$$

Thus, if we take 450 N/mm^2 as the tensile strength for a structural steel, and adopt a factor of safety of 3, the working stress will be

$$\frac{\text{tensile strength}}{\text{factor of safety}} = \frac{450 \text{ N/mm}^2}{3} = 150 \text{ N/mm}^2$$

Choice of a factor of safety

The value of the factor of safety will be influenced by such considerations as the following.

(i) The tensile strength of a material is not always the best stress upon which to base a working-stress value. The limit of useful stress is given by the *elastic limit stress*. Formulae and the elastic method of design are definitely based on the assumption of Hooke's elastic law. In the case of structural steel, this fact alone requires a factor of safety of about 1·7.

(ii) Materials such as timber and cast iron are not so consistent as steel, and are less likely to exhibit, throughout any considerable quantity, the standard of quality represented by the tested specimens. From this point of view, therefore, some materials will require a higher factor of safety than others.

(iii) Structural members may be temporarily overloaded during abnormal conditions. The margin of safe stress is useful then.

(iv) It is not always possible to calculate the actual load a member will have to carry. The load application may be of a doubtful character, and, possibly, the theoretical principles involved may be based on assumptions not wholly justifiable.

(v) Design calculations are made on the basis of a high standard of workmanship in the fabrication of the members. Also (as in the case of reinforced concrete) it is assumed that the details given on the working drawings will be strictly adhered to in the assembly of the members. The factor of safety makes some allowance for the human element in this respect.

Load factor

If W_1 is the safe load for a structure and W_2 is the corresponding load which would just render it unsuitable for its purpose, the ratio $\dfrac{W_2}{W_1}$ is termed the *load factor*.

Working stresses

The designer has, in practice, little to do with the 'factor of safety'. He is supplied with a list of permissible stresses directly. Lists of working stresses are issued, the most important being those of the British Standards Institution, which are adopted generally throughout Great Britain.

BS 449 is concerned with 'The use of structural steel in building', and the reader is strongly advised to make himself acquainted with its contents. With the permission of the British Standards Institution, extracts will be given, where appropriate, from this standard. The GLC Byelaws, and the Building Regulations, accept BS 449 as a 'deemed to satisfy' standard.

The Institution of Structural Engineers, as the results of research work carried out become established, issue reports and recommendations from time to time. BS 449 gives the following maximum working stresses for members in bending, as an example. (The extracts are taken from chapter 4 of the British Standard, by permission of the BSI.)

Clause 19 deals with bending stresses in beams other than plate girders. The following extracts are given here to introduce the present method of dealing with bending stresses.*

BENDING STRESSES (BEAMS OTHER THAN PLATE GIRDERS)

19. a. Rolled I-beams, broad flange beams, Universal beams and columns, castellated beams† with lateral support, rolled channels, rolled Z-beams and compound beams having equal flanges of uniform cross-section throughout or having flanges each of uniform cross-section throughout but where the moment of inertia of the compression flange about the Y–Y axis of the beam exceeds that of the tension flange about the same axis. The bending stress in the extreme fibres, calculated on the effective section, shall not exceed:

1. *for parts in tension.* The appropriate value of p_{bt} in Table 2.

2. *for parts in compression,* for grade 43 steel the lesser of the values of p_{bc} given in Tables 2 and 3a, as appropriate; or for grade 50 steel the lesser of the values given in Tables 2 and 3b as appropriate; or for grade 55 steel the lesser of the values given in Tables 2 and 3c as appropriate.

In Tables 3a, 3b, and 3c:

l = effective length of the compression flanges
r_y = radius of gyration of the beam section about its axis lying in the plane of bending
D = overall depth of beam
T = mean thickness of flange
  = area of horizontal portion of flange, divided by width

For rolled sections, the mean thickness is that given in the appropriate British Standards or other reference book.

---
* BS 449 stresses are normally used in the text.
† Castellated beams *without* lateral support shall be treated as plate girders.

TABLE 2. ALLOWABLE STRESS p_{bc} OR p_{bt} IN BENDING
(See also Clauses 19 and 20, and Tables 3a, 3b and 3c)

Form	Grade	Thickness of material	p_{bc} or p_{bt}
		mm	N/mm^2
Rolled I-beams and channels	43	All	165
Compound girders composed of rolled I-beams or channels plated, with thickness of plate	43	Up to and including 40 Over 40	165 150
Plates, flats, rounds, squares, angles, tees and any sections other than above	43	Up to and including 40 Over 40	165 150
Plate girders with single or multiple webs	43	Up to and including 40 Over 40	155 140
Universal beams and columns	43	Up to and including 40 Over 40	165 150
Plates, flats, rounds, squares and other similar sections, rolled I-beams, double channels forming a symmetrical I-section which acts as an integral unit, compound beams composed of rolled I-beams or channels plated, single channels, angles and tees	50 55	Up to and including 65 Over 65 Up to and including 40 Over 40	230 $Y_s/1.52$ 280 260
Plate girders with single or multiple webs	50 55	Up to and including 65 Over 65 Up to and including 40 Over 40	215 $Y_s/1.63$ 265 245
Hot rolled hollow sections ,, ,, ,, ,, ,, ,, ,, ,,	43 50 55	All All All	165 230 280
Slab bases	All steels		185

where Y_s = yield stress agreed with manufacturer, with a maximum value of 350 N/mm^2.

TABLE 3a. ALLOWABLE STRESS p_{bc} IN BENDING (N/mm²) FOR BEAMS OF GRADE 43 STEEL

l/r_y	D/T							
	10	15	20	25	30	35	40	50
90	165	165	165	165	165	165	165	165
95	165	165	165	163	163	163	163	163
100	165	165	165	157	157	157	157	157
105	165	165	160	152	152	152	152	152
110	165	165	156	147	147	147	147	147
115	165	165	152	141	141	141	141	141
120	165	162	148	136	136	136	136	136
130	165	155	139	126	126	126	126	126
140	165	149	130	115	115	115	115	115
150	165	143	122	104	104	104	104	104
160	163	136	113	95	94	94	94	94
170	159	130	104	91	85	82	82	82
180	155	124	96	87	80	76	72	71
190	151	118	93	83	77	72	68	62
200	147	111	89	80	73	68	64	59
210	143	105	87	77	70	65	61	55
220	139	99	84	74	67	62	58	52
230	134	95	81	71	64	59	55	49
240	130	92	78	69	61	56	52	47
250	126	90	76	66	59	54	50	44
260	122	88	74	64	57	52	48	42
270	118	86	72	62	55	50	46	40
280	114	84	70	60	53	48	44	39
290	110	82	68	58	51	46	42	37
300	106	80	66	56	49	44	41	36

Intermediate values may be obtained by linear interpolation.

NOTE. For materials over 40 mm thick the stress shall not exceed 150 N/mm².

TABLE 3b. ALLOWABLE STRESS p_{bc} IN BENDING (N/mm²) FOR BEAMS OF GRADE 50 STEEL

l/r_y	D/T							
	10	15	20	25	30	35	40	50
80	230	230	230	230	230	230	230	230
85	230	230	230	227	227	227	227	227
90	230	230	228	220	220	220	220	220
95	230	230	222	212	212	212	212	212
100	230	230	215	204	204	204	204	204
105	230	226	209	196	196	196	196	196
110	230	221	203	188	188	188	188	188
115	230	216	196	181	181	181	181	181
120	230	211	190	173	173	173	173	173
130	230	202	177	157	157	157	157	157
140	225	193	165	142	142	142	142	142
150	219	183	152	126	126	126	126	126
160	213	174	139	112	110	110	110	110
170	207	165	126	106	97	94	94	94
180	201	155	114	101	91	85	80	77
190	195	146	109	96	86	80	75	68
200	189	136	104	91	82	75	70	64
210	183	127	100	87	77	71	66	60
220	177	118	96	83	74	67	62	56
230	171	112	92	79	70	64	59	53
240	165	108	89	76	67	61	56	50
250	159	105	86	73	64	58	53	47
260	153	102	83	70	62	56	51	45
270	147	99	80	68	59	53	49	43
280	141	96	77	65	57	51	47	41
290	135	93	75	63	55	49	45	39
300	129	90	72	61	53	47	43	37

Intermediate values may be obtained by linear interpolation.

NOTE. For materials over 65 mm thick the stress shall not exceed $\dfrac{y_s}{1\cdot 52}$ N/mm², where y_s = yield stress agreed with manufacturer, with a maximum value of 350 N/mm².

TABLE 3c. ALLOWABLE STRESS p_{bc} IN BENDING (N/mm²) FOR BEAMS OF GRADE 55 STEEL

l/r_y	D/T							
	10	15	20	25	30	35	40	50
75	280	280	280	280	280	280	280	280
80	280	280	280	277	277	277	277	277
85	280	280	277	267	267	267	267	267
90	280	280	269	257	257	257	257	257
95	280	278	261	247	247	247	247	247
100	280	272	253	238	238	238	238	238
105	280	266	245	228	228	228	228	228
110	280	260	237	218	218	218	218	218
115	280	254	228	208	208	208	208	208
120	280	248	220	198	198	198	198	198
130	273	236	204	178	178	178	178	178
140	266	224	188	158	158	158	158	158
150	258	212	171	138	138	138	138	138
160	250	200	155	121	119	119	119	119
170	243	188	139	114	103	99	99	99
180	235	176	123	107	97	89	84	80
190	227	164	117	101	91	83	78	71
200	219	152	111	96	86	78	73	66
210	212	140	106	91	81	74	69	62
220	204	128	102	87	77	70	65	58
230	196	120	97	83	73	66	61	54
240	189	116	94	79	70	63	58	51
250	181	112	90	76	67	60	55	49
260	173	108	87	73	64	57	53	46
270	165	105	84	70	61	55	50	44
280	158	101	81	68	59	53	48	42
290	150	98	78	65	57	51	46	40
300	142	95	76	63	55	49	44	38

Intermediate values may be obtained by linear interpolation.

NOTE. For materials over 40 mm thick the stress shall not exceed 260 N/mm

The application of the stresses given in Clause 19 are explained in later portions of the book. Clauses 20 and 21 deal with maximum bending stresses in plate girders and cased beams, respectively.

Clause 22 deals with bearing stress and states that the calculated bearing stress on the net projected area of contact shall not exceed 190 N/mm², 260 N/mm², or 320 N/mm² for grades 43, 50, or 55 steel.

Clause 23 (a) deals with the maximum shear stress.

Clause 23 (b) deals with the average shear stress in webs of beams and girders and the following extracts are given.

TABLE 11. ALLOWABLE AVERAGE SHEAR STRESS p_q' IN UNSTIFFENED WEBS

(For stiffened webs see also b above and Tables 12a, 12b and 12c below)

Grade	Thickness	p_q'
	mm	N/mm²
43	Up to and including 40	100
	Over 40	90
50	Up to and including 65	140
55	Up to and including 40	170
	Over 40	160

b. **Average shear stress in webs of I-beams and plate girders.** The average shear stress f_q' on the gross section of the web shall not exceed the values of p_q' given in Table 11 for unstiffened webs or, for stiffened webs, the values given in Tables 12a, 12b, or 12c, as appropriate.

The gross section of the web shall be taken as

For rolled I-beams and channels   The depth of the beam multiplied by the web thickness.

For plate girders   The depth of the web plate multiplied by its thickness.

Compliance with this sub-clause shall be deemed to satisfy the requirements of a above.

Clause 26 deals with the effective lengths of compression flanges of beams and girders.

Clause 30 deals with axial stresses in struts and is dealt with in Chapter 11.

Clause 41 deals with axial stresses in tension and states that the direct stress in axial tension on the net area of section shall not exceed the values given in Table 19, from which the following extract is taken.

TABLE 19. ALLOWABLE STRESS p_t IN AXIAL TENSION

Form	Grade	Thickness or diameter	p_t
		mm	N/mm²
Rolled I-beams and channels	43	All	155
Universal beams and columns	43	Up to and including 40	155
		Over 40	140
Plates, bars and sections other than above	43	Up to and including 40	155
		Over 40	140
Plates, bars and sections	50	Up to and including 65	215
		Over 65	* $Y_s/1.63$
,, ,, ,, ,,	55	Up to and including 40	265
		Over 40	245
Hot rolled hollow sections	43	All	155
,, ,, ,, ,,	50	All	215
,, ,, ,, ,,	55	All	265

* Where Y_s = yield stress agreed with the manufacturer with a maximum value of 350 N/mm².

Clause 50 deals with the allowable stresses in rivets and bolts. The calculated stresses in rivets and bolts shall not exceed the values given in Table 20.

TABLE 20. ALLOWABLE STRESSES IN RIVETS AND BOLTS

Description	Bolts of strength grade designation 4.6	Rivets of material having a yield stress in N/mm² of		
		250	350	430
	N/mm²	N/mm²	N/mm²	N/mm²
In tension:				
Axial stress on gross area* of rivets and on net area* of bolts and tension rods				
Rivets		100	140	170
Bolts	130			
In shear†:				
Shear stress on gross area* of rivets and bolts				
Power-driven shop rivets		110	155	190
Power-driven field rivets		100	140	170
Hand-driven rivets		90	125	155
Close tolerance turned bolts	95			
Black bolts	80			
In bearing (double shear)‡:				
Power-driven shop rivets		315	440	540
Power-driven field rivets		290	405	500
Hand-driven rivets		265	370	455
Close tolerance turned bolts	300			
Black bolts	200			

For bolts of other strength grade designations, the allowable stresses shall be those for grade 4.6 bolts varied in the ratio of the specified stress under proof load§ in kgf/mm² to 22·6.

For rivets of other strength grades, the allowable stresses shall be those in the first column for rivets varied in the ratio of the specified yield stress in N/mm² to 250.

Where parts are connected together by fastenings of a higher grade of material, the local bearing stress in the connected parts shall not exceed $3p_t$ for riveted joints or $2\frac{1}{2}p_t$ for bolted joints where p_t is the allowable stress in axial tension as given in Table 19 for the material of the connected part.

* For gross and net areas see Subclauses 17*b* and *c*.
† For rivets and bolts in double shear, the area to be assumed shall be twice the area defined.
‡ Where the rivets or bolts are in single shear, the allowable bearing stress shall be reduced by 20 %.
§ The stress under proof load is given in Table 18 of BS 4190 and in Table 13 of BS 3692.

NOTE. This does not apply to high strength friction grip bolts, which bolts shall only be used in conformity with BS 4604: The use of high strength friction grip bolts in structural steelwork.

34 Structural Steelwork

Example 1 Taking the tensile strength for grade 43 steel as 460 N/mm², and adopting a factor of safety of 4, calculate the safe axial load for a mild steel tie-bar, 100 mm × 14 mm.

$$\text{Working stress} = \frac{\text{tensile strength}}{\text{factor of safety}}$$

$$= \frac{460 \text{ N/mm}^2}{4} = 115 \text{ N/mm}^2$$

Sectional area of tie-bar = $100 \times 14 = 1400$ mm²
∴ Safe axial load = (1400×115) N = 161 kN.

Example 2 A tie-bar, 50 mm × 10 mm section, is used in a structure to carry 56·5 kN. In a test on the same quality steel, the maximum load carried was 136·4 kN, the test specimen having a sectional area of 280 mm². Find the factor of safety used in the design.

$$\text{Tensile strength for the steel} = \frac{\text{maximum load}}{\text{sectional area}} = \frac{136 \cdot 4 \text{ kN}}{280 \text{ mm}^2}$$

$$= 487 \text{ N/mm}^2$$

$$\text{Actual working stress} = \frac{56 \cdot 5 \times 10^3}{50 \times 10} = 113 \text{ N/mm}^2$$

$$\text{Factor of safety used} = \frac{\text{tensile strength}}{\text{working stress}}$$

$$= \frac{487 \text{ N/mm}^2}{113 \text{ N/mm}^2} = 4 \cdot 30$$

Example 3 Find the safe shear load for one 20 mm diameter rivet in a 22 mm diameter hole, assuming the shearing tendency to be across one section of the rivet, and the rivet to have been power-driven in the 'shop'.

From the table of working stresses given, we find that the stress in such a case can be taken as 110 N/mm².

Sectional area of rivet $= \dfrac{\pi d^2}{4} = \dfrac{\pi \times 22^2}{4}$ mm^2

$= 380 \cdot 1$ mm^2

$\therefore$ Safe shear load $= (380 \cdot 1 \times 110)$N

$= 41 \cdot 8$ kN

(Tables of rivet and bolt strengths, for various working stresses, are given in Chapter 4.)

Example 4 Calculate the safe load, from the point of view of shear in the bolts, for the cleat connection given in fig. 2.3.

Sectional area of one 22 mm diameter bolt $= 380$ mm^2
Working stress for bolts $= 80$ N/mm^2
$\therefore$ Safe load per bolt $= (80 \times 380)$ N $= 30 \cdot 4$ kN
$\therefore$ Safe load for 6 bolts $= 182 \cdot 4$ kN

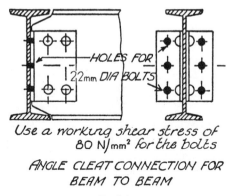

Use a working shear stress of 80 N/mm² for the bolts
ANGLE CLEAT CONNECTION FOR BEAM TO BEAM
Fig. 2.3.

Exercises 2

1 Express the relationship between the following three quantities: tensile strength, working stress, and factor of safety. Assuming a working stress of 125 N/mm^2 to represent a factor of safety of 4, obtain the tensile strength. What would be the working stress in this case for a factor of safety of 5?

36 Structural Steelwork

2 Work out a complete set of test results for the following steel test, and show that the steel would satisfy the requirements of BS 4360, grade 43.

Diameter of specimen (round) = 12 mm
Gauge length = 50 mm
Load at yield point = 40 kN
Maximum load = 73 kN
Gauge length (after fracture) = 65 mm
Diameter at fracture = 8·2 mm.

3 A specimen of steel gave the following calculated results in a tensile test: Y.P. stress = 300 N/mm^2; tensile strength = 500 N/mm^2; percentage elongation at fracture (on 200 mm gauge length) = 30; percentage reduction in area at fracture = 58. The original sectional area of specimen being 400 mm^2, evaluate the experimental results obtained in the test.

4 A piece of steel of rectangular section, 40 mm × 15 mm, fractured at a maximum tensile load of 300 kN. Using a factor of safety of 4, determine the safe working stress and hence find the necessary thickness for a tie-bar of the same quality steel, 75 mm wide, to carry safely an axial pull of 185 kN.

5 A flat bar, 1000 mm^2 in sectional area, has two bolt holes drilled in it 3·7 m apart. Assuming that the load in the bar, when in position in a structure, is 90 kN, and that E for the material of the bar is 200 kN/mm^2, show that the bolt holes will be 1·67 mm out of true under load.

6 A specimen of rivet steel, 250 mm^2 in section, sheared in a test at a load of 110 kN. Adopting a factor of safety of 5, obtain the safe shear load for four 16 mm diameter rivets in 18 mm diameter holes which are resisting shear, as indicated in fig. 1.5.

7 In a tensile test, a specimen of steel of rectangular section, 40 mm × 20 mm, broke at a maximum load of 400 kN. A tie-bar of the same quality steel, and having a rectangular section 100 mm × 16 mm, is used to carry an axial load of 200 kN. What factor of safety does this represent?

8 By an error in printing, the following was given as a problem in 'elasticity':

Sectional area of a steel tensile member = 300 mm^2
Length before application of load = 250 mm

Load applied = 130 kN
E for material = 200 kN/mm^2
Calculate the extension in length produced.

Explain why this problem is not capable of solution. Substitute a possible correct set of values.

3

Fabrication of Steelwork

British Standard sections

Constructional steelwork is built up, or fabricated, from units of standardised shape and dimensions. The British Standards Institution issue BS for the various *sections* employed. BS 4: part 1 is the revised standard specification for hot-rolled structural-steel sections.

The various sections are produced from white-hot steel ingots by passing them through rolls in a *rolling mill*. Plate 2 facing page 21 shows the soft-steel slab ready to be drawn through the rolls. The mill shown in the foreground is the *finishing mill*, and it has reduced the steel to the form of a *plate*. The rolls have grooves cut in them when the sections rolled are of the flanged type. Fig. 3.1 gives the types of rolls used for the various steel sections used in structural work. The diagrams show the rolls used for (i) plates and sheets; (ii) squares and rounds; (iii) flats, angles, etc.; (iv) flanged sections.

Fig. 3.1   *Forms of rolls used in the rolling of steel sections.*
Reproduced by permission and courtesy of Messrs R. A. Skelton & Co., Ltd.

The term 'rolled steel section' is applied to constructional units manufactured in the manner indicated. Smaller 'I' sections are commonly called 'rolled steel joists' (RSJ); larger 'I' sections for use as beams are now rolled as 'Universal Beams' (UB), and those for use as column sections are 'Universal Columns' (UC). A full range of equal and unequal angles is also rolled, together with channels and tees. A few special sections, which are not 'British Standard', are rolled by particular manufacturers for specialist purposes. It is possible with most sections to increase the thickness of certain parts, by spacing the mill rolls farther apart.

Choice of sections

In the practical choice of a section for a particular job in a structure, several factors have to be considered, in addition to the question of strength suitability. The section should be a *standard* one, and, not only so, it should be one fairly frequently rolled. Steelmakers indicate in their lists those sections which are most readily obtainable. They also issue lists of *extras* which have to be paid, for sizes and weights which are outside certain limits in the case of any particular section.

It is not possible, owing to the high temperature of the sections when dealt with—and the usual mode of cutting—to obtain the exact dimensions and weights which might be specified in an order. The reader is referred to BS 4 for full details of the maximum allowable variations. When *exact* lengths are specified, the sections are to be cold-sawn or machined more accurately. An *extra* has to be paid for cutting to exact lengths. The tolerance for weights of flat bars or sections (not stated to be either a maximum or a minimum) is also specified.

Commercial data

The authors are indebted to the British Standards Institution for permission to reproduce details respecting the profiles of the BS sections. Pricing information is based on the published pricing lists of the British Steel Corporation. It should be remembered that commercial data is subject to possible variation from time to time and is given as a guide only, for heavy plain rolled sections.

Universal beams (*UB*)

It will be observed, from an inspection of the profile given in fig. 3.2 that the flange thickness is measured at a point half-way between the extreme edge of the flange and the nearer side of the web.

In naming a section, the *depth* is given first. Thus a $305 \times 127 \times 37$ kg UB is a universal beam, having 305 mm overall depth, 127 mm width of flange, and a mass of 37 kg per metre of length. The smallest UB is a $203 \times 133 \times 25$ kg section, and the largest a $914 \times 419 \times 388$ kg section. In every *nominal* depth of UB, more than one weight can be obtained. The different weights are obtained by an adjustment of the rolls. It must be noted that the *actual* depths and widths of universal beams are almost invariably different from the *nominal* dimensions.

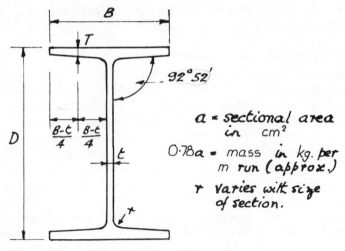

Fig. 3.2 *Universal-beam section.*

The universal beams are rolled by the Grey process which rolls the metal on all faces. Fig. 3.6 illustrates the arrangement of the roll in a universal mill. Fig. 3.7 shows the stages in the production of a universal beam. There are three stages in the rolling, viz. (i) rolling the ingot in a blooming mill into the bloom shape shown in the figure, (ii) passing the bloom through the roughing mill, and (iii) finishing off in the finishing mill. Both the roughing and finishing mills have two sets of rolls, the bloom first passing through the edging roll stand and then through the main roll stand in each mill.

By this process, beams can be rolled with little or no taper on the flanges, the taper on the universal beam being 2° 52′.

Universal columns (UC)

These sections are rolled with parallel flanges. The smallest UC is a $152 \times 152 \times 23$ kg section, and the largest a $356 \times 406 \times 634$ kg section. To illustrate the point that the *actual* depth is different from the *nominal* depth, it will be seen that in the latter section the actual depth is 474·7 mm while the actual width is 424·1 mm.

Small joists with 5° taper flanges

A small range of sections is rolled in this group, the smallest being a $102 \times 64 \times 9·65$ kg section, and the largest a $203 \times 102 \times 25·33$ kg section.

Stock sizes and 'extra' sizes of beams Sections are usually stocked by steel stockholders in even lengths up to 15 m. Universal beams can be obtained in much greater lengths if required. Most beam sections carry an extra in price varying according to the size and characteristics of the section. Extras are also listed for lengths exceeding 15 m or under 6 m. It should be noted that a range of additional sections may also be available.

British Standard equal angles and unequal angles

As indicated in fig. 3.3 an equal angle means one with equal legs, and an unequal angle one in which the legs are of unequal length. An angle

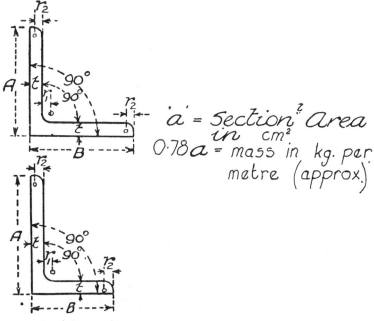

Fig. 3.3 British Standard equal angle and unequal angle sections

section is named by giving the two leg lengths and the thickness thus: $100 \times 100 \times 12$ or $125 \times 75 \times 10$. The angle between the legs is nominally $90°$.

The smallest equal angle is $25 \times 25 \times 3$ and the largest $200 \times 200 \times 24$. Unequal angles are rolled as small as $65 \times 50 \times 5$ and as large as $200 \times 150 \times 18$.

Stock sizes and 'extra' sizes of angles. Larger angle sections are usually obtainable in even lengths up to about 15 m.

Equal angles 300 mm and over combined leg lengths, by 10 mm and 12 mm thick and over, are charged at *basis* price. Extras are charged when the united leg lengths are outside the stated limits, and for all unequal angles. Angles less than 10 mm thick are subject to an extra. The maximum and minimum lengths, without extras, are 15 m and 6 m respectively.

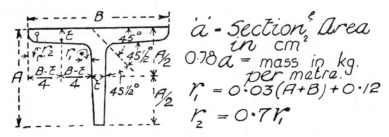

Fig. 3.4 Short-stalk T-bar.

T-bars

In naming a T-section, the flange dimension is given first. Thus a $152 \times 76 \times 11.5$ T-bar would indicate a flange width (B) of 152·4 mm, a web depth (including flange thickness) of 76·2 mm (A), and a mass of 11·5 kg/m.

T-bars can be cut from universal beams or universal columns. They are also rolled in two types of sections called 'rolled short-stalk T-bars' and 'rolled long-stalk T-bars'. The short-stalk T-bars have flange and web tapers of $\frac{1}{2}°$, as shown in fig. 3.4, while the long-stalk T-bars have a flange taper of 8°, but the webs are virtually parallel. Short-stalk T-bars range from $38.1 \times 38.1 \times 6.35$ while long-stalk T-bars range from 25×76 to 127×254. The thicknesses of flanges and web are not the same.

Stock sizes and 'extra' sizes of tees Short-stalk tees are a 'special' item. Long-stalk tees are generally obtainable only at a very high basis price.

Additional sections, not in accordance with BS 4, may be available.

Channels

The profile given in fig. 3.5 indicates that the flanges have a taper of 5°. Standard flange thickness is measured half-way between the extreme

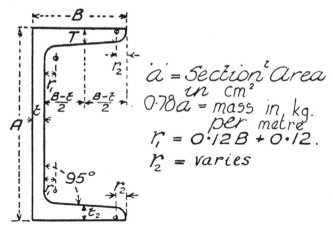

Fig. 3.5 British Standard channel section.

edge of flange and the nearer side of web. In describing a channel, the web depth is given first, then the flange width and mass per metre. A $178 \times 76 \times 20\cdot82$ kg channel would therefore be 178 mm deep overall, with a flange width of 76 mm, and mass per metre of 20·82 kg.

The smallest BSC is $76 \times 38 \times 6.70$ kg and the largest $432 \times 102 \times 65\cdot54$ kg.

Stock sizes and 'extra' sizes of channels BSCs are usually obtainable from 102 mm depth up to 432 mm depth, in lengths of 15 m to 6 m, at no extra over the channel basis price.

Plates

The thinnest plate rolled in an ordinary plate mill is a 5 mm plate. Thinner plates than these are usually termed *sheets*. Plates are rolled to a maximum area, the area depending on the plate thickness. Correspondingly to each thickness there is also a maximum length and a maximum width. Both the maxima cannot be obtained together, so that the maximum width for any given thickness equals the listed maximum area divided by the length required. Intermediate thicknesses will have correspondingly intermediate values to those given.

Flats

An inspection of the diagrams, indicating the forms of rolls employed (fig. 3.1), will show the difference between the methods employed in

44 Structural Steelwork

rolling plates and flats respectively. In the case of flats, it will be seen that the rolls bear on the edges, and thus have a control on the width. There is a special type of flat known as a *universal plate* or *wide flat*. For long and narrow details, as in plate-girder work, flats are superior to plates.

Stock sizes and extra sizes   Basis price is applicable to 1500 m to 12 m, for flats up to 220 mm wide, for all thicknesses produced. Over 220 mm wide, the basis length is 4·5 m up to 12 m, for all thicknesses produced. Availability should always be checked at the particular time.

Connections

The plates and sections used in steelwork construction are usually connected together by welding, riveting, or bolting. Due to the relatively high cost of riveted fabrication, shop connections are largely formed by bolting or welding, and site connections are usually by bolting. The welding of structural steelwork has become accepted as a standard constructional process, and recognition, in the form of a British Standard (BS 1856), first gave the science a definite place in the technique of steel construction. The welding of structural members is dealt with in Chapter 14.

Fig. 3.6   Rolling universal beams, grey process.

Riveting and bolting form the subject of a number of clauses in BS 449. The BS requires as much as possible of the fabrication to be completed in the works where the steelwork is fabricated, using rivets or close-tolerance bolts. Only high-strength friction-grip bolts shall be used in connections subject to impact or vibration.

If a connection contains rivets and bolts, acting together in carrying the load, only rivets with close-tolerance bolts may be used. In other composite connections, each type of fastening must be able, independently, to carry the total load. (See Clauses 48, 51, and 52 of BS 449.)

Rivets

The type of steel used in the manufacture of rivets is given in the British Standard. The usual form of rivet head employed in structural steelwork is the *snap* head (fig. 3.8). Snap heads and *pan* heads form a projection beyond the plate face, and where this is an objection—as in bearings, where continuity of contact between plate and plate, or between plate and masonry, is necessary—a *countersunk* head is employed. Occasionally snap heads are flattened a little to provide clearance.

BS 153 deals with the requirements of good riveting. It states that rivet holes should preferably be drilled through the solid metal. In

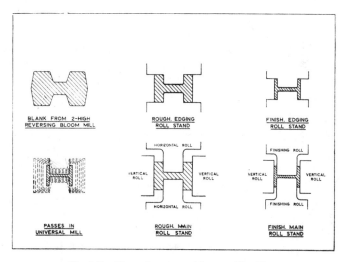

Fig. 3.7   *Stages in universal-beam mill rolling.*
Reproduced by permission and courtesy of Messrs Dorman, Long & Co., Ltd.

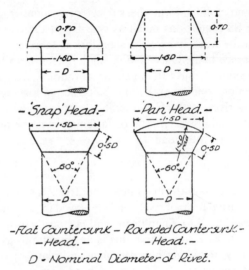

Fig. 3.8 *Form and proportions of rivet heads.*

cases where a compound girder or plate girder is built up of several plates and sections, the parts should be firmly clamped, or tacked together, with temporary bolts, and the holes drilled through in one operation. This procedure ensures correct alignment of holes. Any burrs formed around the holes by drilling should be removed, and the parts re-bolted together in preparation for final riveting-up. The practice of driving in *drifts* (slightly tapered round bars of iron), in order to effect alignment of holes, is forbidden by the BS, but drifts may be used to position the various parts together.

Methods of riveting The rivet, having been rendered soft by heating, is placed in the rivet hole prepared and *closed*, i.e. the second head is formed. The contraction in length on cooling tends to draw the parts connected closer together. The closing of the rivet may be effected in several ways, but the BS recommends some form of machine riveting, preferably of the pressure type. In the hydraulic riveter, pressure is utilised to force a die on to the soft rivet shank, while the other end of the rivet is held firmly by a stationary die. The water, in some forms of pressure riveters, is replaced by compressed air. Such machines have been used in fabrication shops, suitably installed for rapid dealing with the various riveting operations carried out there. The closing of the rivets by pressure leads to the best results, as the rivet hole becomes compactly filled with the metal. Site riveting, and some shop riveting, is carried out by a pistol-shaped compressed-air machine known as the *pom-pom* or *pneumatic hammer*. The head is formed in this case by a rapid succession of blows. The rivet is held tightly in position by one of the riveters (usually known as the 'holder-up'), while the other man plays skilfully upon the projecting shank until he has formed—by repeated blows—the necessary rivet head. This form of riveting gives rise to considerable noise which resembles a machine gun in operation. There is a risk of rivet heads being formed by the 'pom-pom' without the holes being completely filled, and shop pressure riveting is to be preferred.

In order that the hot rivet shall easily be placed in position in the rivet hole, it is necessary to make the diameter of the hole bigger than the nominal rivet diameter. The maximum *clearance* (i.e. difference in diameters) permitted by BS is 2 mm.

Bolts and nuts

The relevant British Standards for bolts are:
BS 3692 ISO metric precision hexagon bolts, screws, and nuts.
BS 4190 ISO metric black hexagon bolts, screws, and nuts.
BS 4320 Metal washers for general engineering purposes. Metric series.
BS 4395 High strength friction grip bolts and associated nuts and washers for structural engineering. Metric series. Part 1—General grade, and Part 2—Higher grade bolts and nuts and general grade washers.
BS 4604 The use of high strength friction grip bolts in structural steelwork. Metric series. Part 1—General grade. Part 2—Higher grade (parallel shank).

Two varieties of bolts, apart from high strength friction grip bolts, are used: (i) *turned or close tolerance* bolts; (ii) *black* bolts.

Turned bolts are carefully turned parallel throughout the length of the barrel, and with the maximum clearance closely limited. It will be observed, from the list of working stresses given in Chapter 2, that a close-tolerance bolt is regarded as being very nearly the equal of a shop rivet.

'Black' bolts are not reduced to precise size, and the hole diameter for such bolts can be made 2 mm bigger than the nominal bolt diameter. Black bolts are not therefore a tight fit in the hole. The allowable shear stress is consequently not so high as for rivets, or for turned bolts. Black bolts are not permissible for all purposes, even with the lower stress value. In shop fabrication, black bolts may be used for the end cleat connections of secondary floor beams. Such bolts may be used on 'site' for roof-truss work and end connections of secondary floor beams; also for some other field connections if the shear forces are otherwise resisted. Bolts, turned and black, must have washers under the nuts of such thickness that the thread is clear of the hole in the plate. The shanks must also project at least one full thread beyond the nuts.

High strength friction grip bolts function by a different principle. The nut is tightened to produce a predetermined tension in the bolt, and the steel members being connected are thus gripped tightly together. The passive friction developed between the adjoining faces of the steel members resists any movement. In order to ensure that the friction resistance is fully developed, the adjoining steel faces must be clean and unpainted, and the shank tension in the bolt must be correct. There are three common ways of checking that the bolt is correctly tightened:

i) the part-turn method, in which the nut is tightened by hand, and then is turned a specified further amount by spanner;

ii) by load-indicating washers or bolts, which have projecting spacer lugs which are gradually flattened by tightening the nut, until the correct gap is achieved, which is checked by a feeler-gauge;

iii) by means of a torque-spanner, which is a special spanner so constructed as to 'break' at a pre-set torque which is sufficient to develop the required shank tension in the bolt.

The photographs in plates 2 and 3, facing pages 52 and 53, illustrate two methods of installing High Strength friction grip bolts.

The following BS are referred to in this chapter: 4, 153, 449, 1856, 3692, 4190, 4320, 4395, and 4604. The extracts have been made by permission of the British Standards Institution, 2 Park Street, London, W1A 2BS, from whom official copies of the specifications may be obtained.

Exercises 3

1 Give a few considerations which influence the choice of a structural section, apart from the question of strength.

2 What are the trade allowances for structural sections in (a) length, and (b) weight?

3 Name the British Standard sections in common use. Which dimension is given first in naming (a) a tee-bar, (b) a beam?

Fabrication of Steelwork 49

4 Which of the standard sections have tapered flanges, and which have not? Give the angle between the component parts in each case, and state where the standard thicknesses are measured.

5 Distinguish between a plate and a sheet, and between a plate and a flat. Give the maximum thickness of plates usually kept in stock.

6 Name and sketch the forms of rivet heads which were adopted in structural steelwork. Indicate which was the most commonly used of these forms and give, for this head, the proportions laid down in the BS.

7 Distinguish between 'black bolts', 'high strength friction grip bolts', and 'close tolerance' bolts. Give any BS regulations you know which affect the use of these bolts.

8 Give values of working stresses in shear which exemplify the superiority of turned bolts over black bolts.

9 What special precautions need to be taken in preparing steelwork for high strength friction grip bolted connections?

4

Practical Design of Riveted and Bolted Connections

Introduction

It is mentioned elsewhere that riveting is no longer in general use in structural steelwork fabrication, but the matters covered in this chapter, and the general principles of design, can be taken to apply, in a corresponding manner, to either riveting or bolting. Black bolts have a limited application (by regulations), even with their reduced working stress. In the normal steelwork connection, a rivet or bolt is called upon to resist shear and *bearing* stresses only. Rivets and bolts may, however, under certain conditions, be designed to resist tension. In the connections dealt with in this chapter, the rivets or bolts are not intentionally subjected to tension. Figs 4.1 and 4.2 (kindly supplied by the British Steel Corporation and the British Constructional Steelwork Association) illustrate a number of types of structural steelwork connections, in which bolts and rivets of various forms are used. The conventional methods adopted to indicate these forms should be carefully noted, particularly those where countersinking is necessary (denoted by an asterisk).

Strength of one rivet or bolt

Shear strength

According to the type of a given joint, the connecting rivets or bolts may be subjected to *single shear* or to *double shear*. Fig. 4.3 illustrates these two shearing tendencies. (This does not apply to high strength friction grip bolts.)

In *single shear*, the shearing action is across one cross-sectional plane of the rivet or bolt; in *double shear*, two such cross-sectional areas are involved.

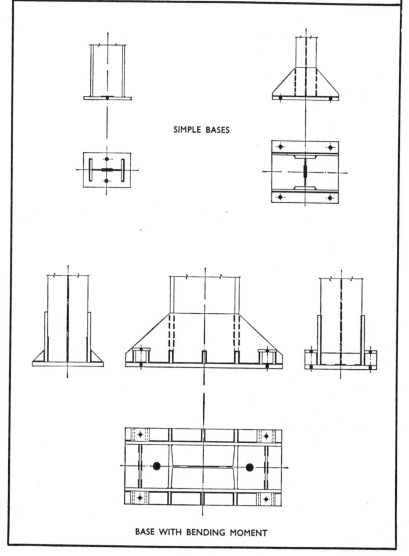

Fig. 4.1(a)

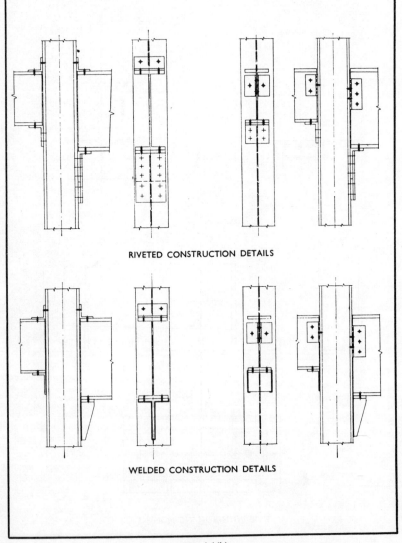

Fig. 4.1(b)

Plates I(a) and I(b) reproduced by courtesy of the British Steel Corporation and the British Constructional Steelwork Association.

Plate 3   Bolting by the use of a torque-controlled spanner.
Reproduced by courtesy of Norbar Torque Tools Ltd.

Plate 4 Bolting with load-indicating washers by the use of a power tool.
Reproduced by courtesy of Messrs. Cooper & Turner Ltd.

TYPICAL ROOF TRUSS CONNECTIONS

RIVETED OR BOLTED

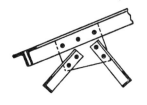

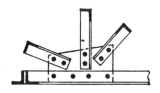

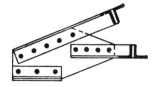
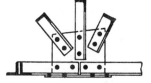

Fig. 4.2(a)

TYPICAL ROOF TRUSS CONNECTIONS

WELDED

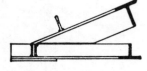

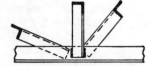
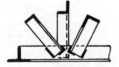

Fig 4.2(b)

Plates II(a) and II(b) reproduced by courtesy of the British Steel Corporation and the British Constructional Steelwork Association.

If d mm = diameter of rivet, the area of metal provided in one cross-sectional area = $\pi d^2/4$ mm². Using the symbol f_s for the working stress in shear (in N/mm²) for the rivet material, the formula for the strength of one rivet in single shear becomes $(\pi d^2/4)f_s$ N. The sectional area of metal provided being twice as much, in the case of double shear, as that in the case of single shear, the corresponding expression for the strength of one rivet in double shear is $2(\pi d^2/4)f_s$ N. The latter value is accepted by steelwork regulations.

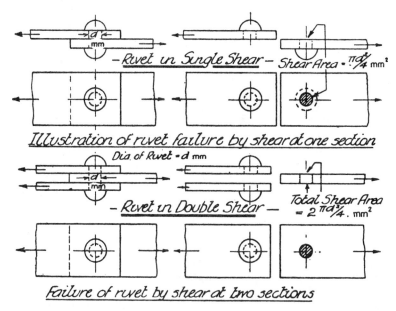

Fig. 4.3 *Single shear and double shear in a rivet or bolt.*

Bearing strength

If we walk on smooth sand, the depth of the impression left depends upon the type of shoe worn. Flat-bottomed sand shoes, which provide a large *bearing* area, would cause a shallow depression, but, if shoes with well-defined heels are worn, the impression is much deeper. In the latter case the *intensity of bearing pressure* is higher, owing to the reduction in bearing area. In the same way, the intensity of bearing stress between a plate and a rivet—for a given applied force in the plate—becomes greater as the bearing area between the two becomes less. With the usual plate thicknesses the bearing strength of a rivet is less than its strength in double shear, but greater than that in single shear (see page 59).

As shown in fig. 4.4, the bearing area is taken as *diameter of rivet* ×

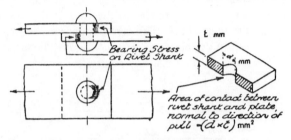

Fig. 4.4   Bearing stress in rivet or bolt.

plate thickness, i.e. $d \times t$ mm². If f_b N/mm² is the working stress in 'bearing' for the rivets, the strength of one rivet in bearing will be given by the formula dtf_b N. We have, therefore, the following three important formulae:

$$\text{Single shear (S.S.) strength of one rivet} = \frac{\pi d^2}{4} f_s \text{ N}$$

$$\text{Double shear (D.S.) strength of one rivet} = \frac{2\pi d^2}{4} f_s \text{ N}$$

$$\text{Bearing strength of one rivet} = dtf_b \text{ N}$$

The actual strength, or value, of one rivet in a joint will be the lesser of its shear and bearing values.

The value of d in the formulae above may be taken to be the diameter of the *finished* rivet, i.e. the actual diameter of the rivet hole.* In the case of countersunk rivets, one half of the depth of the countersink must be omitted in calculating bearing area.

Example 1   Calculate the actual value or worth, in kN, of one rivet in the following circumstances: rivet diameter = 18 mm, hole diameter = 20 mm, plate thickness = 12 mm; the rivets are in single shear and are power-driven in the shop.

The working stresses to be used in this case are $f_s = 110$ N/mm²; $f_b = 0.8 \times 315$ N/mm².

*Rivet-hole diameter always when calculating loss of plate area.

S.S. value of one rivet $= \dfrac{\pi d^2}{4} f_s = \dfrac{\pi 20^2}{4} \times \dfrac{110}{1000}$ kN $= 34\cdot6$ kN

B.V. (bearing value) $= dtf_b = 20 \times 12 \times 0\cdot8 \times 315/1000 = 60\cdot5$ kN

∴ The actual value of one rivet in this case $= 34\cdot6$ kN.

Example 2 Find the maximum safe load, from the point of view of rivet strength, for the joint given in fig. 4.5.

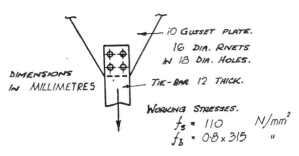

Fig. 4.5.

In this case the rivets are in single shear and bearing in 10 mm plate (the thinner of the two plates concerned).

S.S. value of one rivet $= \dfrac{\pi d^2}{4} f_s = \dfrac{\pi \times 18^2}{4} \times \dfrac{110}{1000}$ kN

$= 28$ kN

B.V. of one rivet $= dtf_b = (18 \times 10 \times 0\cdot8 \times 315/1000)$ kN
$= 45\cdot4$ kN

Actual value $= 28$ kN

Rivet strength of joint $= (4 \times 28)$ kN
$= 112$ kN

Tables of rivet and bolt strengths

The tables on pages 59 to 65 give the shearing and bearing values for rivets and bolts (including black bolts), and are reproduced by permission of the British Constructional Steelwork Association · and CONSTRADO.

Readers will note that where the rivets or bolts are in single shear, the permissible bearing stress must be reduced by 20%.

Enclosed bearing value means the value of one rivet or bolt obtained by using the appropriate stresses.

58 Structural Steelwork

Simple bearing value indicates that the appropriate stresses have been reduced by 20%.

The *nominal* diameter means the diameter of the rivet when cold and before driving in the hole.

The *gross area* is the area of section of a rivet when it fits the area of hole provided.

Rivet or bolt diameter and plate thickness

The choice of a suitable rivet diameter for a given structural connection involves several factors, and it is not possible to lay down any hard and fast rule. As far as possible, rivet diameters are kept constant throughout any particular built-up unit. For example, 18 mm diameter rivets might be used solely throughout a plate girder, and the rivet positions be designed to maintain this uniformity. It is usual to select a rivet size with reference to the plate thicknesses involved.

The size of rivet to be used in standard cases of compound girders etc. will be found tabulated in the section books issued by steel firms. The diameters in common use in building work are 18 mm, 20 mm, and 22 mm. Heavy engineering work sometimes requires 24 mm diameter rivets.

Design of riveting or bolting detail

The positioning of rivets and rivet lines forms the subject of a number of practical regulations and theoretical considerations. The latter will be taken up later in the book. BS 449 part 2: 1969, in Clauses 48 and 50–52, gives recommendations in respect of riveted and bolted connections. The reader should carefully consider these. A brief reference to some of the important matters is given below. The various steel-fabrication regulations are in close agreement in regard to the main essentials.

Minimum pitch of rivets or bolts

The distance between centres of rivets shall not be less than $2\frac{1}{2}$ times the normal diameter of the rivet.

Maximum pitch of rivets or bolts

The straight-line pitch in the direction of stress in riveted girders, columns, or other members shall not exceed the following values:

for parts in tension, 16 times the thickness of the thinnest outside plate or angle, with a maximum pitch of 200 mm;

SHEARING AND BEARING VALUES

IN KILONEWTONS FOR POWER-DRIVEN SHOP RIVETS OF STEEL HAVING A YIELD STRESS OF 250 N/mm²

BASED ON BS 449 1969

Gross Dia. of Rivet after driving in mm	Area in cm²	Shearing Value @ 110 N/mm²		Simple Bearing Value @ 80% of 315 N/mm² and Enclosed Bearing Value @ 315 N/mm² (see footnote) Thickness in mm of plate passed through or of enclosed plate											
		Single Shear	Double Shear	5	6	7	8	9	10	12	15	18	20	22	25
10	0.79	8.64	17.3	**12.6** 15.8	**15.1** 18.9	17.6 22.0	20.2 25.2	22.7							
12	1.13	12.4	24.9	**15.1** 18.9	**18.1** 22.7	**21.2** 26.5	**24.2** 30.2	**27.2** 34.0	30.2	36.3					
14	1.54	16.9	33.9	**17.6** 22.0	**21.2** 26.5	**24.7** 30.9	**28.2** 35.3	**31.8** 39.7	35.3 44.1	42.3	52.9				
16	2.01	22.1	44.2	**20.2** 25.2	**24.2** 30.2	**28.2** 35.3	**32.3** 40.3	**36.3** 45.4	**40.3** 50.4	48.4 60.5	60.5	72.6			
18	2.54	28.0	56.0	**22.7** 28.3	**27.2** 34.0	**31.8** 39.7	**36.3** 45.4	**40.8** 51.0	**45.4** 56.7	**54.4** 68.0	68.0 85.0	81.6	90.7		
20	3.14	34.6	69.1	**25.2** 31.5	**30.2** 37.8	**35.3** 44.1	**40.3** 50.4	**45.4** 56.7	**50.4** 63.0	**60.5** 75.6	75.6 94.5	90.7 113	101		
22	3.80	41.8	83.6	**27.7** 34.6	**33.3** 41.6	**38.8** 48.5	**44.4** 55.4	**49.9** 62.4	**55.4** 69.3	**66.5** 83.2	**83.2** 104	99.8 125	111 139	122	
24	4.52	49.8	99.5	**30.2** 37.8	**36.3** 45.4	**42.3** 52.9	**48.4** 60.5	**54.4** 68.0	**60.5** 75.6	**72.6** 90.7	**90.7** 113	109 136	121 151	133	
27	5.73	63.0	126	**34.0** 42.5	**40.8** 51.0	**47.6** 59.5	**54.4** 68.0	**61.2** 76.5	**68.0** 85.0	**81.6** 102	**102** 128	122 153	136 170	150	170

Upper line Bearing Values for each diameter of rivet are Simple Bearing Values.
Lower line Bearing Values for each diameter of rivet are Enclosed Bearing Values.
For areas to be deducted from a bar for one hole, see table on page 80.
For explanation of table, see Notes on page 131.
1 kilonewton may be taken as 0.102 metric tonne (megagramme) force, but see page 102.

Table reproduced by permission and courtesy of the British Constructional Steelwork Association.

BASED ON
BS 449
1969

SHEARING AND BEARING VALUES

IN KILONEWTONS FOR POWER-DRIVEN FIELD RIVETS OF STEEL HAVING A YIELD STRESS OF 250 N/mm²

Gross Dia. of Rivet after driving in mm	Area in cm²	Shearing Value @ 100 N/mm²		Simple Bearing Value @ 80% of 290 N/mm² and Enclosed Bearing Value @ 290 N/mm² (see footnote)										
				Thickness in mm of plate passed through or of enclosed plate										
		Single Shear	Double Shear	5	6	7	8	9	10	12	15	18	20	22	25
10	0.79	7.85	15.7	11.6 / 14.5	13.9 / 17.4	16.2 / 20.3	18.6 / 23.2	20.9							
12	1.13	11.3	22.6	13.9 / 17.4	16.7 / 20.9	19.5 / 24.4	22.3 / 27.8	25.1 / 31.3	27.8	33.4					
14	1.54	15.4	30.8	16.2 / 20.3	19.5 / 24.4	22.7 / 28.4	26.0 / 32.5	29.2 / 36.5	32.5 / 40.6	39.0	48.7				
16	2.01	20.1	40.2	18.6 / 23.2	22.3 / 27.8	26.0 / 32.5	29.7 / 37.1	33.4 / 41.8	37.1 / 46.4	44.5 / 55.7	55.7	66.8			
18	2.54	25.4	50.9	20.9 / 26.1	25.1 / 31.3	29.2 / 36.5	33.4 / 41.8	37.6 / 47.0	41.8 / 52.2	50.1 / 62.6	62.6 / 78.3	75.2	83.5		
20	3.14	31.4	62.8	23.2 / 29.0	27.8 / 34.8	32.5 / 40.6	37.1 / 46.4	41.8 / 52.2	46.4 / 58.0	55.7 / 69.6	69.6 / 87.0	83.5 / 104	92.8		
22	3.80	38.0	76.0	25.5 / 31.9	30.6 / 38.3	35.7 / 44.7	40.8 / 51.0	45.9 / 57.4	51.0 / 63.8	61.2 / 76.6	76.6 / 95.7	91.9 / 115	102		
24	4.52	45.2	90.5	27.8 / 34.8	33.4 / 41.8	39.0 / 48.7	44.5 / 55.7	50.1 / 62.6	55.7 / 69.6	66.8 / 83.5	83.5 / 104	100 / 125	111 / 139	122	
27	5.73	57.3	115	31.3 / 39.1	37.6 / 47.0	43.8 / 54.8	50.1 / 62.6	56.4 / 70.5	62.6 / 78.3	75.2 / 94.0	94.0 / 117	113 / 141	125 / 157	138	157

Upper line Bearing Values for each diameter of rivet are Simple Bearing Values.
Lower line Bearing Values for each diameter of rivet are Enclosed Bearing Values.
For areas to be deducted from a bar for one hole, see table on page 80.
For explanation of table, see Notes on page 131.
1 kilonewton may be taken as 0.102 metric tonne (megagramme) force, but see page 102.

Tables reproduced by permission and courtesy of

SHEARING AND BEARING VALUES

IN KILONEWTONS FOR HAND-DRIVEN RIVETS OF STEEL HAVING A YIELD STRESS OF 250 N/mm²

BASED ON BS 449 1969

Gross Dia. of Rivet after driving in mm	Area in cm²	Shearing Value @ 90 N/mm²		Simple Bearing Value @ 80% of 265 N/mm² and Enclosed Bearing Value @ 265 N/mm² (see footnote)											
				Thickness in mm of plate passed through or of enclosed plate											
		Single Shear	Double Shear	5	6	7	8	9	10	12	15	18	20	22	25
10	0.79	7.07	14.1	10.6	12.7	14.8	17.0	19.1							
				13.3	15.9	18.5	21.2								
12	1.13	10.2	20.4	12.7	15.3	17.8	20.4	22.9	25.4	30.5					
				15.9	19.1	22.3	25.4	28.6							
14	1.54	13.9	27.7	14.8	17.8	20.8	23.7	26.7	29.7	35.6	44.5				
				18.5	22.3	26.0	29.7	33.4	37.1						
16	2.01	18.1	36.2	17.0	20.4	23.7	27.1	30.5	33.9	40.7	50.9	61.1			
				21.2	25.4	29.7	33.9	38.2	42.4	50.9					
18	2.54	22.9	45.8	19.1	22.9	26.7	30.5	34.3	38.2	45.8	57.2	68.7	76.3		
				23.8	28.6	33.4	38.2	42.9	47.7	57.2	71.5				
20	3.14	28.3	56.5	21.2	25.4	29.7	33.9	38.2	42.4	50.9	63.6	76.3	84.8		
				26.5	31.8	37.1	42.4	47.7	53.0	63.6	79.5	95.4			
22	3.80	34.2	68.4	23.3	28.0	32.6	37.3	42.0	46.6	56.0	70.0	84.0	93.3		
				29.1	35.0	40.8	46.6	52.5	58.3	70.0	87.5	105			
24	4.52	40.7	81.4	25.4	30.5	35.6	40.7	45.8	50.9	61.1	76.3	91.6	102	112	
				31.8	38.2	44.5	50.9	57.2	63.6	76.3	95.4	114	127		
27	5.73	51.5	103	28.6	34.3	40.1	45.8	51.5	57.2	68.7	85.9	103	114	126	143
				35.8	42.9	50.1	57.2	64.4	71.5	85.9	107	129	143		

Upper line Bearing Values for each diameter of rivet are Simple Bearing Values.
Lower line Bearing Values for each diameter of rivet are Enclosed Bearing Values.
For areas to be deducted from a bar for one hole, see table on page 80.
For explanation of table, see Notes on page 131.
1 kilonewton may be taken as 0.102 metric tonne (megagramme) force, but see page 102.

the British Constructional Steelwork Association.

BASED ON
BS 449
1969

SHEARING AND BEARING VALUES

IN KILONEWTONS FOR CLOSE TOLERANCE AND TURNED BOLTS OF STEEL OF STRENGTH GRADE DESIGNATION 4.6

Dia. of Bolt Shank in mm	Area in cm^2	Shearing Value @ 95 N/mm^2		Simple Bearing Value @ 80% of 300 N/mm^2 and Enclosed Bearing Value @ 300 N/mm^2 (see footnote)										
				Thickness in mm of plate passed through or of enclosed plate										
		Single Shear	Double Shear	5	6	7	8	9	10	12	15	18	20	22	25
10	0.79	7.46	14.9	**12.0**	**14.4**	**16.8**	**19.2**	**21.6**							
				15.0	*18.0*	*21.0*	*24.0*								
12	1.13	10.7	21.5	**14.4**	**17.3**	**20.2**	**23.0**	**25.9**	**28.8**						
				18.0	*21.6*	*25.2*	*28.8*								
14	1.54	14.6	29.2	**16.8**	**20.2**	**23.5**	**26.9**	**30.2**	**33.6**	**40.3**					
				21.0	*25.2*	*29.4*	*33.6*	*37.8*							
16	2.01	19.1	38.2	**19.2**	**23.0**	**26.9**	**30.7**	**34.6**	**38.4**	**46.1**	**57.6**				
				24.0	*28.8*	*33.6*	*38.4*	*43.2*	*48.0*						
18	2.54	24.2	48.3	**21.6**	**25.9**	**30.2**	**34.6**	**38.9**	**43.2**	*51.8*	*64.8*	*77.8*			
				27.0	*32.4*	*37.8*	*43.2*	*48.6*	*54.0*	*64.8*					
20	3.14	29.8	59.7	**24.0**	**28.8**	**33.6**	**38.4**	**43.2**	**48.0**	**57.6**	*72.0*	*86.4*	*96.0*		
				30.0	*36.0*	*42.0*	*48.0*	*54.0*	*60.0*	*72.0*	*90.0*				
22	3.80	36.1	72.2	**26.4**	**31.7**	**37.0**	**42.2**	**47.5**	**52.8**	**63.4**	*79.2*	*95.0*	*106*		
				33.0	*39.6*	*46.2*	*52.8*	*59.4*	*66.0*	*79.2*	*99.0*	*119*			
24	4.52	43.0	86.0	**28.8**	**34.6**	**40.3**	**46.1**	**51.8**	**57.6**	**69.1**	*86.4*	*104*	*115*		
				36.0	*43.2*	*50.4*	*57.6*	*64.8*	*72.0*	*86.4*	*108*	*130*			
27	5.73	54.4	109	**32.4**	**38.9**	**45.4**	**51.8**	**58.3**	**64.8**	**77.8**	**97.2**	*117*	*130*	*143*	
				40.5	*48.6*	*56.7*	*64.8*	*72.9*	*81.0*	*97.2*	*122*	*146*	*162*		

Upper line Bearing Values for each diameter of bolt are Simple Bearing Values.
Lower line Bearing Values for each diameter of bolt are Enclosed Bearing Values.
For areas to be deducted from a bar for one hole, see table on page 80.
For explanation of table, see Notes on page 131.
1 kilonewton may be taken as 0.102 metric tonne (megagramme) force, but see page 102.

Tables reproduced by permission and courtesy of

SHEARING AND BEARING VALUES

IN KILONEWTONS FOR BLACK BOLTS OF STEEL OF STRENGTH GRADE DESIGNATION 4.6

BASED ON BS 449 1969

Dia. of Bolt Shank in mm	Area in cm²	Shearing Value @ 80 N/mm²		Simple Bearing Value @ 80% of 200 N/mm² and Enclosed Bearing Value @ 200 N/mm² (see footnote) Thickness in mm of plate passed through or of enclosed plate											
		Single Shear	Double Shear	5	6	7	8	9	10	12	15	18	20	22	25
10	0.79	6.28	12.6	8.00 10.0	9.60 12.0	11.2 14.0	12.8 16.0	14.4 18.0	16.0						
12	1.13	9.05	18.1	9.60 12.0	11.5 14.4	13.4 16.8	15.4 19.2	17.3 21.6	19.2 24.0	23.0	28.8				
14	1.54	12.3	24.6	11.2 14.0	13.4 16.8	15.7 19.6	17.9 22.4	20.2 25.2	22.4 28.0	26.9 33.6	33.6	40.3			
16	2.01	16.1	32.2	12.8 16.0	15.4 19.2	17.9 22.4	20.5 25.6	23.0 28.8	25.6 32.0	30.7 38.4	38.4 48.0	46.1 57.6	51.2		
18	2.54	20.4	40.7	14.4 18.0	17.3 21.6	20.2 25.2	23.0 28.8	25.9 32.4	28.8 36.0	34.6 43.2	43.2 54.0	51.8 64.8	57.6		
20	3.14	25.1	50.3	16.0 20.0	19.2 24.0	22.4 28.0	25.6 32.0	28.8 36.0	32.0 40.0	38.4 48.0	48.0 60.0	57.6 72.0	64.0 80.0	70.4	
22	3.80	30.4	60.8	17.6 22.0	21.1 26.4	24.6 30.8	28.2 35.2	31.7 39.6	35.2 44.0	42.2 52.8	52.8 66.0	63.4 79.2	70.4 88.0	77.4	
24	4.52	36.2	72.4	19.2 24.0	23.0 28.8	26.9 33.6	30.7 38.4	34.6 43.2	38.4 48.0	46.1 57.6	57.6 72.0	69.1 86.4	76.8 96.0	84.5 106	96.0
27	5.73	45.8	91.6	21.6 27.0	25.9 32.4	30.2 37.8	34.6 43.2	38.9 48.6	43.2 54.0	51.8 64.8	64.8 81.0	77.8 97.2	86.4 108	95.0 119	108

Upper line Bearing Values for each diameter bolt are Simple Bearing Values.
Lower line Bearing Values for each diameter bolt are Enclosed Bearing Values.
For areas to be deducted from a bar for one hole, see table on page 80.
For explanation of table, see Notes on page 131.
1 kilonewton may be taken as 0.102 metric tonne (megagramme) force, but see page 102.

the British Constructional Steelwork Association.

HIGH STRENGTH FRICTION GRIP BOLTS

TO BS 4395 : PART 1 : 1969

GENERAL GRADE

SHEAR VALUES OF BOLTS PER INTERFACE

Diameter of Bolt Shank in mm	Shear Value without wind kilonewtons	Shear Value including wind kilonewtons
12	15.9	18.5
16	29.6	34.5
20	46.3	54.0
22	56.9	66.4
24	66.5	77.6
27	75.2	87.8
30	91.9	107
36	134	157

For explanation of table, see page 132.
1 kilonewton may be taken as 0.102 metric tonne (megagramme) force, but see page 102.

Table reproduced by permission and courtesy of the British Constructional Steelwork Association.

AREAS IN SQUARE CENTIMETRES TO BE DEDUCTED FOR ONE HOLE THROUGH A MEMBER

Dia. of Hole in mm	THICKNESS OF MEMBER AT HOLE—IN MILLIMETRES												
	5	6	8	10	12	15	18	20	22	25	28	30	32	35	40
10	0.50	0.60	0.80	1.00	1.20	1.50	1.80	2.00	2.20	2.50	2.80	3.00	3.20	3.50	4.00
11	0.55	0.66	0.88	1.10	1.32	1.65	1.98	2.20	2.42	2.75	3.08	3.30	3.52	3.85	4.40
12	0.60	0.72	0.96	1.20	1.44	1.80	2.16	2.40	2.64	3.00	3.36	3.60	3.84	4.20	4.80
13	0.65	0.78	1.04	1.30	1.56	1.95	2.34	2.60	2.86	3.25	3.64	3.90	4.16	4.55	5.20
14	0.70	0.84	1.12	1.40	1.68	2.10	2.52	2.80	3.08	3.50	3.92	4.20	4.48	4.90	5.60
15	0.75	0.90	1.20	1.50	1.80	2.25	2.70	3.00	3.30	3.75	4.20	4.50	4.80	5.25	6.00
16	0.80	0.96	1.28	1.60	1.92	2.40	2.88	3.20	3.52	4.00	4.48	4.80	5.12	5.60	6.40
17	0.85	1.02	1.36	1.70	2.04	2.55	3.06	3.40	3.74	4.25	4.76	5.10	5.44	5.95	6.80
18	0.90	1.08	1.44	1.80	2.16	2.70	3.24	3.60	3.96	4.50	5.04	5.40	5.76	6.30	7.20
19	0.95	1.14	1.52	1.90	2.28	2.85	3.42	3.80	4.18	4.75	5.32	5.70	6.08	6.65	7.60
20	1.00	1.20	1.60	2.00	2.40	3.00	3.60	4.00	4.40	5.00	5.60	6.00	6.40	7.00	8.00
21	1.05	1.26	1.68	2.10	2.52	3.15	3.78	4.20	4.62	5.25	5.88	6.30	6.72	7.35	8.40
22	1.10	1.32	1.76	2.20	2.64	3.30	3.96	4.40	4.84	5.50	6.16	6.60	7.04	7.70	8.80
23	1.15	1.38	1.84	2.30	2.76	3.45	4.14	4.60	5.06	5.75	6.44	6.90	7.36	8.05	9.20
24	1.20	1.44	1.92	2.40	2.88	3.60	4.32	4.80	5.28	6.00	6.72	7.20	7.68	8.40	9.60
25	1.25	1.50	2.00	2.50	3.00	3.75	4.50	5.00	5.50	6.25	7.00	7.50	8.00	8.75	10.00
26	1.30	1.56	2.08	2.60	3.12	3.90	4.68	5.20	5.72	6.50	7.28	7.80	8.32	9.10	10.40
27	1.35	1.62	2.16	2.70	3.24	4.05	4.86	5.40	5.94	6.75	7.56	8.10	8.64	9.45	10.80
28	1.40	1.68	2.24	2.80	3.36	4.20	5.04	5.60	6.16	7.00	7.84	8.40	8.96	9.80	11.20
29	1.45	1.74	2.32	2.90	3.48	4.35	5.22	5.80	6.38	7.25	8.12	8.70	9.28	10.15	11.60

Notes on tables of rivets and bolts

Working stresses

Safe-load tables have been evaluated on pp. 59–65 for rivets of material having a yield stress of 250 N/mm² and for bolts of strength grade designation 4.6 in accordance with the allowable stresses given in clause 50 of BS 449: 1969 as follows:

	Allowable stresses in N/mm^2	
Description	Simple shear	Bearing (double shear)
Power-driven shop rivets	110	315
Power-driven field rivets	100	290
Hand-driven rivets	90	265
Close-tolerance turned bolts	95	300
Black bolts	80	200

Multiple shear

For rivets and bolts in double shear, the area to be assumed must be twice the area for single shear. Where the rivets or bolts are in single shear, the permissible stress must be reduced by 20%.

Critical values

Bearing values printed in ordinary type are less than single shear. In these cases, the bearing values are the determining factors. Bearing values printed in bold type are greater than single and less than double shear, so that, in the case of:

a) single shear, the shearing value is the criterion;
b) double shear, the bearing value is the criterion.

Bearing values printed in italic type are equal to or greater than double shear. In these cases, the shearing values are the criterion.

for parts in compression, 12 times the thickness of the thinnest outside plate or angle, with a maximum pitch of 200 mm.

The following variation is commonly accepted: *where two rows of staggered rivets occur in one flange of a single angle (as in* 120×120 *or* 150×150 *L's), the straight-line pitch in the direction of stress shall not exceed* $1\frac{1}{2}$ *times the above.* This will apply to angles in tension or compression. Tacking rivets, i.e. rivets merely used for tacking flange plates together, and not subjected to calculated stress (see Chapter 9) may be spaced farther apart (BS Clause 51e).

Applying these rules to a 20 mm diameter rivet, the minimum pitch equals $2\frac{1}{2} \times 20$ mm = 50 mm. Similarly, for a 22 mm diameter rivet the minimum pitch = $2\frac{1}{2} \times 22$ mm = 55 mm. If a plated member has an outside plate 12 mm thick, or an angle thickness 12 mm, the maximum pitch (for rivets in a single row in one angle flange) in the case of the tension flange = 16×12 mm = 192 mm (the maximum allowed is 200 mm). For the compression flange it would be 12×12 mm = 144 mm (see Clause 51c (ii)).

Edge distance of rivets or bolts

Certain regulations give the minimum distance in the form of a rule: *the distance from the edge of a rivet hole or bolt hole to the edge of a plate, bar, or member shall not be less than the diameter of the rivet or bolt.**
BS 449, Clause 51d, gives a series of edge distances in terms of the nominal rivet diameter and whether the edge is sheared or sawn, etc.

Riveting or bolting for built-up girders

BS 449 gives a few practical regulations respecting the edge distance of rivet lines for flange plates. These regulations are important, as they influence the choice of flange-plate width in plated joists and girders. Fig. 4.6 shows two cases: (A) *where one flange plate is used*, (B) *where two or more flange plates are used.*

The maximum distance x mm permitted is measured in each case from the centre line of the rivets—connecting the flange plates to the web construction—which lie nearest the plate edge. For a single plate the maximum distance is 40 mm + 4 times the plate thickness. For two or more plates the maximum distance is 40 mm + 4 times the thickness of the thinner outside plate, provided that the plates are tack riveted in accordance with Clause 51e. The spacing of such tacking rivets does not

---

*Centre distance = $1\frac{1}{2} d$ mm if d mm = finished rivet diameter.

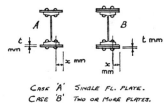

Fig. 4.6 Edge distance of rivets or bolts.

involve the type of calculation given later in the book for angle rivets, but their pitch must not exceed 32 times the thickness of the outside plate or 300 mm, whichever is the lesser, except that, where the plates are exposed to the weather, the pitch in line shall not exceed 16 times the thickness of the outside plate or 200 mm, whichever is the lesser. (See Clause 51e.)

Principles of design of riveted and bolted joints

The principles of design will be illustrated by consideration of some of the common forms of joints used in structural steelwork. Whilst the text deals with many riveted details, it is important to note that the principles and requirements for bolts are as for rivets.

Joint in a tie-bar

The plates may be joined together in one of the three ways illustrated in fig. 4.7. The third method shown is the best, as methods (a) and (b) have a tendency to subject the rivets or bolts to tension.

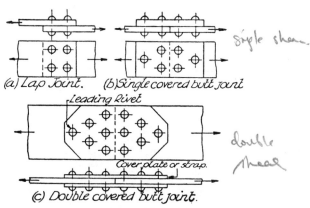

Fig. 4.7 Forms of joints in tension members.

In (a) and (b) the rivets are in single shear and bearing, and in (c) in double shear and bearing.

Tie-bar joints have to be designed to resist liability to failure in the following ways:

i) *By failure of the rivets* If n rivets are provided altogether in a lap joint, or to each side of the butt in a butt joint, and V kN is the value of one rivet, the rivet strength of the joint will be nV kN.

ii) *By the tearing of the tie-bar across a section weakened by one or more rivet holes* Consider section 1 in fig. 4.8; the effective solid width of the tie-bar $= (x-d)$ mm. The net area of metal provided $= (x-d)t$ mm². If f_t N/mm² $=$ the working tensile stress in the plate, the safe load for section 1 $= (x-d)tf_t$ N.

Similarly the net sectional area of plate at section 2 is $(x-2d)t$ mm², giving a strength—from the point of view of tearing only—of $(x-2d)tf_t$ N. But the rivet situated at section 1 acts as a peg, and would have to be got rid of before failure at section 2 could actually take place. The actual strength at section 2 is therefore $\{(x-2d)tf_t + V\}$ N. In the same way, the strength for section 3 would be $\{(x-2d)tf_t + 3V\}$ N. The reason for the adoption of the *leading rivet* form of rivet arrangement will now be apparent. Where a serious deduction of metal is made by rivet holes—as at section 2—compensation is afforded by the strength of one rivet, so that one hole only is made without alternative strength being supplied.

iii) *By failure of the covers (in a butt joint)* If the covers failed at section 3 (see fig. 4.8), the joint would fail without any assistance from

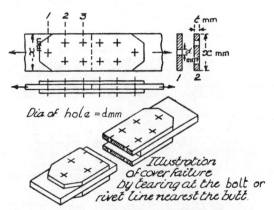

Fig. 4.8 *Double-covered butt joint in tension member.*

the rivets. The cover-plate strength of the joint illustrated is given, therefore, by the expression $\{(x-2d) \times 2Tf_t\}$ N, where T mm is the thickness of one cover. In a double-covered butt joint the thickness of each cover should be about $\frac{5}{8} \times$ the plate thickness.

iv) *By the rivets being placed too near the edge of a plate* The tendency to split, or shear out, the intervening piece of metal between the rivet and plate edge is guarded by the rules already given for minimum edge distance.

Example A tie member in a frame has to transmit an axial dead load of 400 kN. The plate thickness is to be 15 mm and the rivet diameter 20 mm. Design the general joint details and evaluate the percentage efficiency of the connection. Working stresses $f_s = 110$ N/mm², $f_b = 315$ N/mm², $f_t = 155$ N/mm². The joint is to be a double-covered butt joint.

D.S. value of one rivet $= \dfrac{2\pi d^2}{4} f_s = \dfrac{2 \times \pi \times 22^2}{4} \times \dfrac{110}{1000}$ kN

$$= 83.6 \text{ kN}$$

B.V. of one rivet $= dtf_b = (22 \times 15 \times 315/1000)$ kN $= 104$ kN

$$\therefore \quad V = 83.6 \text{ kN}$$

Number of rivets required each side of butt $= \dfrac{400}{83.6} = 5$.

The rivets are arranged as in fig. 4.9.

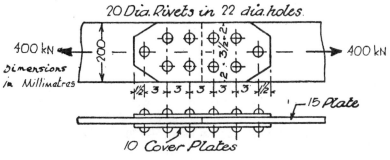

Fig. 4.9 *Joint in tie member.*

Width of tie.
Section 1:
$$(x-d)tf_t = 400 \text{ kN}$$

Allowing 2 mm clearance for rivet hole,
$$(x-22)15 \times 155/1000 = 400$$
$$(x-22) = 172$$
$$x = 194 \text{ mm, say 200 mm}$$

Section 2:
$$(x-2d)tf_t + V = 400$$
$$(x-44)15 \times 155/1000 + 83\cdot 6 = 400$$
$$(x-44) = \frac{316\cdot 4}{2\cdot 325} = 136$$
$$x = 180 \text{ mm}$$

Section 3 will be stronger than section 2,

$$\therefore \text{ necessary width of tie} = 200 \text{ mm}$$

Thickness of covers.
Section 3:
$$(x-2d)2Tf_t = 400$$
$$(200-44)2T \times 155/1000 = 400$$
$$T = 8\cdot 3 \text{ mm, say 10 mm}$$

Practical rule: $T = \tfrac{5}{8}t = \tfrac{5}{8} \times 15 = 9\cdot 4$ mm, say 10 mm.

Efficiency of Joint The efficiency of a joint is the ratio of its actual strength to the strength of the solid plate outside the joint.

$$\text{Percentage efficiency} = \frac{\text{strength of joint}}{\text{strength of solid plate}} \times 100.$$

A low efficiency would mean that the stress in the tie-bar outside the joint was considerably below the economic working value. Taking the joint designed, the rivet strength is $5 \times 83\cdot 6$ kN $= 418$ kN. The tearing strength is governed by section 1 and equals $(200-22) \times 15 \times 155/1000 = 414$ kN.

Cover-plate strength $= (200-44)2 \times 10 \times 155/1000 = 484$ kN.

Actual safe load for joint = 414 kN, i.e. the smallest of its various strengths.

Strength of solid plate outside joint = xtf_t N

$$= (200 \times 15 \times 155/1000) \text{ kN} = 465 \text{ kN}$$

$$\% \text{ efficiency} = \frac{414}{465} \times 100 = 89$$

Connection of beam to stanchion

Fig. 4.1(b) shows a typical connection of a beam to the flange of a stanchion. It will be observed that the seating angle cleat is riveted to the stanchion in the shop, and that site bolts or rivets effect the remaining connection. The top cleat is useful for erection purposes, but the additional strength it provides is not added in when computing the strength of the connection.

A similar type of connection, involving similar calculations, is used to connect a beam to the web of a stanchion. When sufficient rivets cannot be provided in a seating cleat, web cleats are employed, as indicated in fig. 4.10.

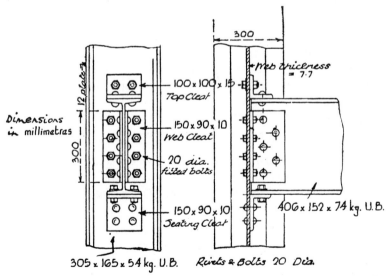

Fig. 4.10 *Connection of beam to web of stanchion.*

Taking the example shown, we have the following strength values.
Web-cleat bolts The bolts are in single shear and bearing in either 7·7 mm plate thickness or 15 mm (angle) thickness.

$$\text{S.S. value of one 20 mm diameter bolt} = \frac{\pi d^2}{4} f_s$$

$$= \left(\frac{\pi \times 20^2}{4} \times \frac{95}{1000}\right) \text{kN} = 29 \cdot 8 \text{ kN}$$

B.V. in 7·7 mm plate $= dtf_b = (20 \times 7 \cdot 7 \times 0 \cdot 8 \times 300/1000)$ kN
$\qquad\qquad\qquad\qquad\qquad = 37$ kN
$\therefore\quad$ Value of one bolt $= 29 \cdot 8$ kN
For 8 bolts the safe load $= 8 \times 29 \cdot 8$ kN
$\qquad\qquad\qquad\qquad\quad = 238 \cdot 4$ kN

Seating-angle-cleat rivets The value of one rivet = 41·8 kN.
Strength for 4 rivets = $4 \times 41 \cdot 8$ kN = 167·2 kN.
Total safe load for the connection from the point of view of the bolts and rivets (in bottom cleat) = $(238 \cdot 4 + 167 \cdot 2)$ kN = 405·6 kN.
The riveting of a web cleat to the web of a beam requires calculations of a special form, illustrated in fig. 4.11.

Connection of beam to beam

The end connections of beams are usually of a standard design, and the detail for any particular standard beam size will be found in the section books issued by steel firms. With the beam size is given a minimum span, so that the greatest load safely carried by the beam will not result in a reaction at the end exceeding the safe strength of the joint.

Fig. 4.12 gives a standard web-cleat end connection for a 305 × 127 × 48 kg UB. Before calculating the maximum safe load for this connection, the effect of rotation on groups of rivets will have to be considered.

Eccentric loading of rivet groups The theory of eccentricity of load application in members is considered in Chapter 11. We may assume here that the eccentric load R kN in fig. 4.11 has two effects: (i) a tendency to push all the rivets in a direction vertically upwards, and (ii) a rotary effect, tending to turn the rivet group round the centroid G of the rivets as centre.

$$\text{The vertical load per rivet} = \frac{R \text{ kN}}{\text{number of rivets}}$$

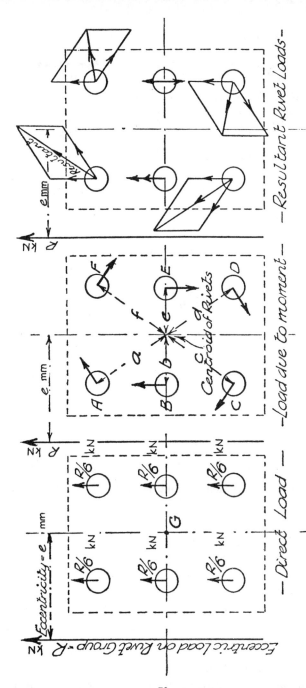

Fig. 4.11 Rivet group subjected to eccentric load.

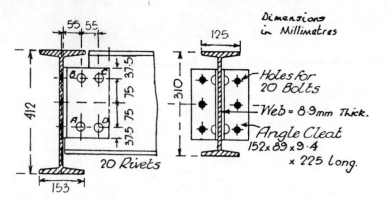

Fig. 4.12 *Web-cleat end connection for beam.*

The vertical load per rivet $= \dfrac{R}{6}$ kN, in the example given.

Let L kN be the load on rivet A due to the rotation effect. Rivet B, not being so far from G, will have a smaller load—in proportion to its distance. A list of loads may therefore be compiled as follows:

$$\text{load on rivet A} = L \text{ kN}$$
$$\text{load on rivet B} = L \times \frac{b}{a} \text{ kN}$$
$$\text{load on rivet C} = L \times \frac{c}{a} \text{ kN, and so on.}$$

The moments of these loads about the centroid G will be respectively $(L \times a)$ kN mm, $(L \times (b/a) \times b)$ kN mm, etc. The sum of all these moments must equal the applied turning moment.

$$\therefore \quad R \times e = La + Lb^2/a + Lc^2/a + \text{ etc.}$$
$$= \frac{L}{a}(a^2 + b^2 + c^2 + \text{etc.})$$
$$= \frac{L}{a}(\Sigma a^2), \text{ where '}\Sigma a^2\text{' ('}sigma\ a^2\text{') means 'the sum of all}$$

such quantities as a^2'.

This equation enables L to be found, from which the load, *due to the turning moment*, can be deduced for any other rivet. The two loads,

Practical Design of Riveted and Bolted Connections

'direct' and that due to turning moment, for any given rivet, are combined by the parallelogram of forces. The resultant load must not exceed the safe load for the rivet, from the point of view of bearing and shear.

Example Find, by means of the foregoing theoretical principles, the safe load for the beam connection shown in fig. 4.12.

The eccentricity of loading is, in this case, the distance from the web face of the main beam to the centroid of the rivets connecting the angle cleat to the web of the secondary beam.

Assume a load of 1 kN to be the reaction load applied to the connection (fig. 4.13). Direct load carried by one rivet = 1 kN/4 = 0·25 kN. The turning moment about the centroid of the rivet group = 1 kN × 82·5 mm = 82·5 kN mm.

Let L kN be the load on rivet A.
In the example, $a = b = c = d = \sqrt{(27\cdot5^2 + 75^2)} = 79\cdot9$ mm.

$$R \times e = \frac{L}{a}(\Sigma a^2)$$

$$\therefore \quad 82\cdot5 = \frac{L}{79\cdot9}(4 \times 79\cdot9^2) = 319\cdot6\, L$$

$$\therefore \quad L = 0\cdot258 \text{ kN}$$

As $a = b = c = d$, the same load will be applied to each rivet.

The maximum resultant load carried by any rivet in the group = 0·42 kN.

The value of one rivet in the connection is the lesser of its double-shear value and bearing value in the thickness of the secondary beam web, i.e. in 8·9 mm plate thickness.

D.S. value of one 20 mm shop rivet = 83·6 kN
B.V.* in 8·9 mm plate = 22 × 8·9 × 315/1000 kN = 61·7 kN
$\therefore \quad V = 61\cdot7$ kN

The reaction load transmitted to the connection can therefore be increased from 1 kN to (61·7/0·42) kN = 146·9 kN, which is the safe maximum load for the connection on the basis of the stated theory.

The connection must now be tested from the point of view of the bolts in the web of the main beam.

*Enclosed bearing.

There is a tendency for some bolts in such a connection to have tensile stress developed in them, in addition to shear. A good deal of experimental work is being carried out at the present time on the end connections of beams. It is the practice of some designers to calculate the strength of the bolts on the basis of a low working stress in shear—about 65 N/mm². Applying this to the given example, the safe load per bolt $= (\pi d^2/4)f_s$ N $= (\pi \times 20^2/4) \times 65/1000 = 20{\cdot}4$ kN.

For 6 bolts the safe reaction load would thus be $6 \times 20{\cdot}4$ kN.

Top-flange cleats are sometimes used, but, as in beam to stanchion connections, no addition to strength is attributed to these. Seating brackets are employed for heavy reaction loads.

Fish-plated beam connections

Two lengths of standard beams, which are in alignment and butt together over a support, are connected by fish-plates. The connections are standardised and will be found in section books.

There are a number of types of joints in steelwork construction, other than those already dealt with. Stanchion lengths are connected, flange and web joints have to be made in plate-girder construction, and so on. Such joints involve theory not yet considered, and their design will be taken up in later chapters.

High strength friction grip bolts

The practice of site riveting has, over the years, decreased to an extent that it is almost non-existent at the present time. This decline has been largely due to the decrease in the number of riveters who, principally owing to the conditions under which they were required to work, have sought other employment. The incidence of the h.s.f.g. bolt has been brought about, principally, by the need to replace site riveting by a method other than the use of close-tolerance turned bolts.

The principle underlying the use of the h.s.f.g. bolt is that, by a process of tightening the bolt to a predetermined amount, the pieces or *plies* to be connected are drawn tightly together so that the load to be carried by the joint is resisted by the friction generated, by the tightening, between the plies.

The dimensions and properties of h.s.f.g. bolts are set out in BS 4395, and the application of the bolts is governed by the provisions of BS 4604.

The strength of a h.s.f.g. bolt is measured by the *proof* load of the bolt. This proof load is the minimum shank tension which must be

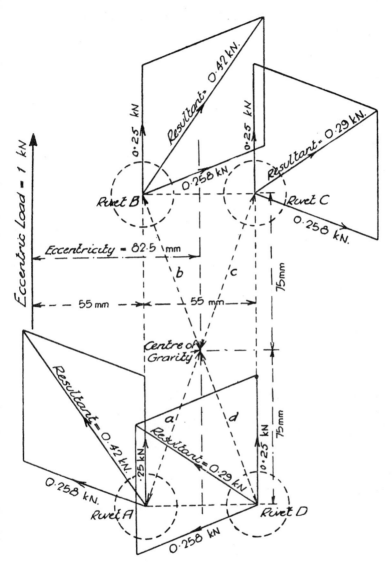

Fig. 4.13 *Rivets in beam connection carrying eccentric reaction.*

applied to the bolt in order to ensure that the parts to be connected are adequately drawn together.

The proof loads of h.s.f.g. bolts are given in the following table, which is taken from BS 4604: part 1.

Normal size and thread diameter (mm)	Proof load (minimum shank tension) (kN)
12	49·4
16	92·1
20	144
22	177
24	207
27	234
30	286
36	418

The working load of the bolt is based on the proof load and is dependent on certain other factors. The working load is given by:

$$\frac{\text{slip factor}}{\text{load factor}} \times \text{number of effective interfaces} \times \text{proof load}$$

Slip factor is the ratio of the load per effective interface required to produce slip in a pure shear joint to the nominal shank tension (proof load) induced in the bolt. In most practical cases the value of the slip factor may be taken as 0·45.

Load factor is the numerical value by which the load which would cause slip in a joint is divided to give the permissible working load on the joint. The load factor may be taken as 1·4 for structures and materials covered by BS 449. Where the effect of wind forces on the structure has to be taken into consideration, this load factor may be reduced to 1·2, provided the connections are adequate when wind forces are neglected. In connections subject only to shear, no additional factor is required to take account of fatigue conditions.

Effective interface is a common contact surface between two load-transmitting plies, excluding packing pieces, through which the bolt passes.

Example Determine the working load of a 16 mm h.s.f.g. bolt which passes through 2 plies (i.e. has only one effective interface).

$$\text{Working load} = \frac{\text{slip factor}}{\text{load factor}} \times \text{number of interfaces} \times \text{proof load}$$

Working load = $\dfrac{0\cdot 45}{1\cdot 4} \times 1 \times 92\cdot 1 = 29\cdot 6$ kN

This value should be checked with the table on page 64. Note that this connection is equivalent to one in single shear.

It must be noted that certain conditions must be satisfied in order that h.s.f.g. bolts may be used. These conditions are set down in BS 4604. They concern such matters as the requirement that all surfaces in contact must be free of paint, oil, dirt, loose scale, etc., and that all holes must be drilled, and all burrs removed. A close study of BS 4604 is required.

The most important point to be observed in the use of h.s.f.g. bolts is to ensure that the tightening of the nut is sufficient to ensure that the proof load (minimum shank tension) has been achieved. This may be done, in general, by three methods:

i) *The part-turn method* This involves a preliminary tightening of each bolt in a joint so that the facing surfaces are brought into close contact. Each bolt is then finally tightened by turning the nut a specified amount. The amount of turn will vary with the grip length, and will be between one-half turn to a full turn.

ii) *The use of a torque-controlled, or -calibrated, wrench* This method is not permitted for bolts with waisted shanks.

iii) *The use of a procedure which provides direct measurement of the shank tension.* There are two common methods of achieving this measurement. One is the use of a load-indicating washer or bolt. This is a washer or bolt which has four protruding lugs which compress on tightening, thus enabling the tension to be measured by a feeler gauge. The other is by the use of a bolt which has a waist outside the grip length. The nut, which is of a special type, is tightened until the bolt snaps off at the waist.

In the foregoing, only connections subject to shear have been dealt with—h.s.f.g. bolts can also be used for connections subject to external tension only in the direction of the bolt axes, as well as for connections subject to external tension in addition to shear. Such connections are considered to be beyond the scope of this book.

Exercises 4

1 Find the strength of one 18 mm diameter black bolt in single shear, and one 22 mm diameter black bolt in double shear, using a working stress in shear of 80 N/mm^2.

2 Calculate the value in kN of one rivet in the following circumstances: rivet diameter = 20 mm; plate thickness = 15 mm; $f_s = 100$ N/mm^2; $f_b = 290$ N/mm^2. Rivet is in D.S.

3 For the connection given in fig. 4.14, calculate (a) safe load for section 1, (b) safe load for section 2, and (c) safe load for rivets.

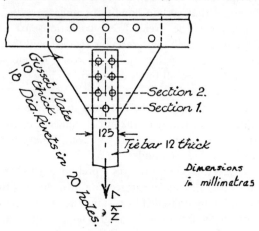

Fig. 4.14 *Connection of tie-bar to frame.*

Hence determine the safe value of L and the percentage efficiency of the joint. ($f_s = 110$ N/mm^2; $f_b = 0.8 \times 315$ N/mm^2; $f_t = 155$ N/mm^2.)

4 Obtain the safe maximum reaction load for the 381 mm × 152 mm UB shown in fig. 4.15. Use the working-stress values appropriate to the case. Angle-bracket thickness = 12·6 mm.

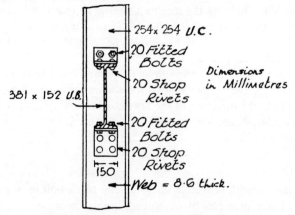

Fig. 4.15 *Beam connection to web of stanchion.*

5 Two 12 mm plates are to be connected by a double-covered butt joint. The load to be transmitted through the connection is 250 kN. The rivets are to be 16 mm diameter in 18 mm diameter holes. Taking the working stresses given in question 3, and adopting a leading rivet in the arrangement of rivets, find the number of rivets required, the necessary width of the plate, and a suitable cover thickness. Evaluate the percentage efficiency of the joint.

6 Fig. 4.16 gives details of an eccentrically loaded connection as found in an existing building. Assuming the 254 × 102 UB to transmit a load of 60 kN to the seating angle cleat, find the maximum load carried by one 22 mm diameter rivet, and test for safety.

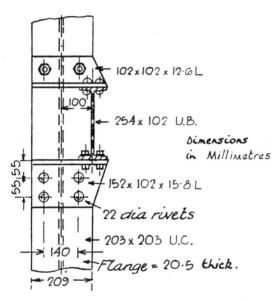

Fig. 4.16 *Eccentric connection of beam and stanchion.*

7 A tension member consisting of two angles is connected to both sides of a gusset plate by h.s.f.g. bolts. The member is required to carry a load of 180 kN. Find the number of 20 mm bolts required.

8 A tension member consisting of two angles is connected to both sides of a gusset plate by h.s.f.g. bolts. The load in the member due to dead and imposed loads (other than wind) is 350 kN and, in addition, there is load due to wind of 45 kN. Find the number of 22 mm bolts required in the connection.

5

Theory of Beam Design

Bending moment and moment of resistance

The subject of beam design may be divided into two parts: (i) the consideration of the effects which the external loads carried have on beams and (ii) the design of beam sections, and details, to resist these effects. The experimental model shown in fig. 5.1 illustrates the nature of the

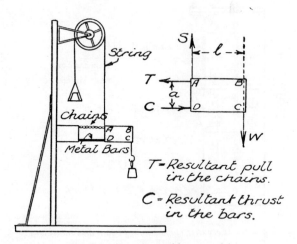

Fig. 5.1   *Experimental beam model.*

forces acting across a vertical section of a loaded beam. The portion ABCD of an originally solid cantilever is assumed to have been cut off from the remainder at a section AD, and to have been removed to the position shown. In order to maintain this portion in equilibrium, it is necessary to introduce at the face AD certain forces, which in the model are supplied by (i) the horizontal pull in the chains, (ii) the horizontal thrust in the metal bars, and (iii) the vertical pull in the string. Removal of any of these forces results in collapse of the cantilever, so that they are independently necessary for equilibrium.

Applying the laws of equilibrium to the detached portion ABCD, we get:

for horizontal equilibrium $T = C$;
for vertical equilibrium $S = W$.

The forces T and C constitute a couple of anticlockwise moment T (or C) × a. Similarly, forces S and W form a couple of clockwise moment W (or S) × l. These two couples must have equal moment for equilibrium, therefore

$$T \text{ (or } C) \times a = W \text{ (or } S) \times l.$$

In a practical beam, T represents the resultant of all the little fibre pulls exerted across the section AD, by the beam to the left of AD. In the same way, C is the resultant force of all the thrusts exerted on AD. The relative positions of T and C are, of course, reversed in a simply supported beam.

W stands for the resultant vertical force (which may be compounded of forces acting vertically upwards or downwards) of all the forces acting to the right of the section AD. S represents the resistance the beam section at AD offers to vertical shear.

The moment, $W \times l$, of the couple tending to bend the beam is termed the *bending moment* at the section AD. The moment, T (or C) × a, which resists the bending of the beam, is termed the *moment of resistance* of the section AD.

When a beam has deflected to its position of equilibrium, the *bending moment* at every beam section will be equalled by the *moment of resistance* of the section. This is a very important result, and will be often used in later chapters.

The force W tending to produce vertical sliding, or shear, at AD is the *shear force* at the section.

We may now express these facts in the form of three important definitions, assuming the usual case of a horizontal beam with vertical loads.

Bending moment (B.M.) The bending moment at any given section of a beam is the resultant moment, about that section, of all the external forces acting to one side of the section.

Shear force (S.F.) The shear force at any given section of a beam is the resultant vertical force of all the external forces acting to one side of the section.

***Moment of resistance* (*M.R.*)** The moment of resistance of a beam section is the moment of the couple which is set up at the section by the longitudinal forces created in the beam by its deflection.

Bending moment and shear force

When computing values of these quantities it should be noted that:
i) in finding B.M. values, it does not matter which side of the given section is taken, but only one side must be considered;
ii) in finding S.F. values, *the actual positions of the loads do not matter,* provided only those loads to one side of the section be taken.

Example Determine the B.M. and S.F. values for the section XX in the beam shown in fig. 5.2.

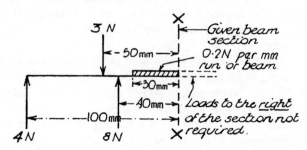

Fig. 5.2 *Beam with general load system.*

Resultant moment about section

$= \{(4 \times 100) + (8 \times 40) - (3 \times 50) - (6 \times 15)\}$ Nmm
$= (400 + 320 - 150 - 90)$ Nmm $= 480$ Nmm

This is the value of the 'bending moment' at the section. It is a clockwise moment tending to bend the beam so that it is concave upwards.

Resultant vertical load at section $= (4 + 8 - 3 - 6)$ N
$= 3$ N

This is the value of the shear force at the section, and the tendency is for the left portion of the beam to move vertically upwards, at section XX, with respect to the right portion.

It will be necessary to distinguish between bending moments which, respectively, produce *concave* and *convex* bending in a beam. Also the two possible types of shearing must be capable of identification. Distinction is made in both cases by the use of positive and negative symbols. The convention of signs adopted in the subsequent calculations is illustrated in fig. 5.3.

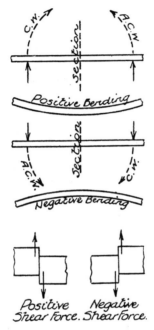

Fig. 5.3   *Convention of signs for bending moments and shear force.*

Certain standard cases of beams and loading will now be considered.

a) Cantilever with single concentrated load W at the free end

To the right of the given typical section XX, fig. 5.4, there is only one load, viz. W, and it is at a distance x from the section. The B.M. at the section is therefore Wx. We may write this as follows:

$$\text{B.M.}_x = Wx$$

x may have any value from 0 to l, and a graph can be drawn showing the change in value of the B.M. as x varies.

86 Structural Steelwork

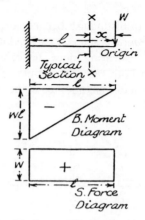

Fig. 5.4 *Cantilever with concentrated end load.*

When $x = 0$, B.M. $= W \times 0 = 0$.
When $x = l$, B.M. $= W \times l = Wl$.

Such a graph, or diagram, exhibiting the value of the B.M. for all points of the beam span, is known as a *bending-moment diagram*. It is not necessary, in most cases, actually to plot a large number of points, as in graph construction. The form of the B.M. expression for the typical section is a clue to the geometrical form of the diagram, and geometrical means may be used to construct it. The expression 'Wx' is of the first degree in x and leads to a *straight-line diagram*.

Shear-force diagrams are constructed on the same general principles.

$$\text{S.F.}_x = \text{resultant vertical load to right of section}$$
$$= W$$

This value is independent of x, so that the graph is a *horizontal straight line*.

It will be noted that, according to the convention of signs adopted, the B.M. is negative and the S.F. positive, in this case.

b) Cantilever with several concentrated loads

The loading on a beam may be disintegrated, as convenient, for the calculation of B.M. (or S.F.) values. The net B.M. (or S.F.) value at any given section will then be the algebraic sum of the various separately calculated values. The principle of addition and subtraction may also be applied to *diagrams* of B.M. (and S.F.).

In fig. 5.5 is shown the building up of the final B.M. and S.F. diagrams

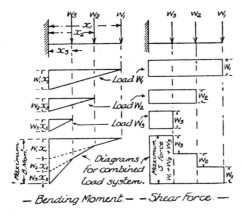

Fig. 5.5 Summation of component diagrams.

from the component diagrams. In practice, the final diagrams are drawn directly.

Example A cantilever carries the load system given in fig. 5.6. Construct the B.M. and S.F. diagrams, and compute the B.M. and S.F. values, respectively, for a section 1 metre from the free end.

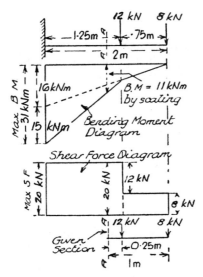

Fig. 5.6 Cantilever with concentrated loads.

88 Structural Steelwork

B.M. at support due to 8 kN load = (8×2) kNm
$= 16$ kNm $(-)$
B.M. at support due to 12 kN load = $(12 \times 1\cdot25)$ kNm
$= 15$ kNm $(-)$
Total B.M. at support $= (16+15)$ kNm $= 31$ kNm
B.M. at given section AA $= [(8 \times 1)+(12 \times 0\cdot25)]$ kNm
$= (8+3)$ kNm $= 11$ kNm $(-)$
S.F. at section AA = total load to right of section
$= (12+8)$ kN $= 20$ kN $(+)$

The diagrams may now be constructed as shown in the figure. The reader is recommended to draw out these diagrams (and those in subsequent worked examples) to suitable scales. Suggested scales: 50 mm = 1 m (for span); 20 mm = 10 kNm (for B.M. values); 20 mm = 10 kN (for S.F. values).

c) Cantilever with uniformly distributed load

The total load on the hatched portion of the cantilever (fig. 5.7) to the right of the given typical section = w per unit run $\times x$ units of length = wx. The centre of gravity of this load is $x/2$ from the section, hence the moment of the load about the section

$$= wx \times \frac{x}{2} = \frac{wx^2}{2}. \qquad \text{B.M.}_x = \frac{wx^2}{2}(-).$$

If this expression were plotted for different values of x, the curve obtained would be a 'parabola', tangential to the beam at the free end. The reader is referred to Appendix 4 for the method of constructing a parabola, and for other useful properties of this important curve.

When $x = 0$,   B.M. $= \dfrac{wx^2}{2} = \dfrac{w \times 0^2}{2} = 0.$

When $x = l$,   B.M. $= \dfrac{wx^2}{2} = \dfrac{w \times l^2}{2} = \dfrac{wl^2}{2}.$

If W = total load on cantilever, $W = wl$.

B.M. at support, i.e. maximum value of the B.M. $= \dfrac{wl^2}{2} = \dfrac{Wl}{2}.$

S.F.$_x$ = total load to right of section
$= wx$

Theory of Beam Design 89

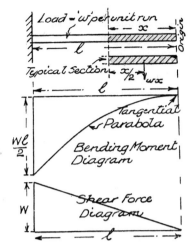

Fig. 5.7 Cantilever with uniformly distributed load.

The S.F. diagram is therefore linear.

When $x = 0$,   S.F. $= w \times 0 = 0$
When $x = l$,   S.F. $= w \times l = wl = W\,(+)$
                   ∴   S.F. maximum $= W$

Example A balcony is carried by universal beams placed at 0·8 m centres (fig. 5.8). The beams project 1·2 m from a wall. The total depth

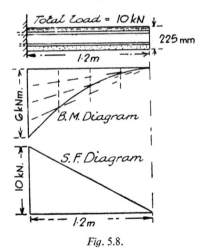

Fig. 5.8.

of the concrete floor, including floor finish, is 225 mm. Construct the B.M. and S.F. diagrams for one beam, assuming a super load of 5 kN/m² on the balcony.

Area of floor carried by one beam = $1 \cdot 2 \times 0 \cdot 8 = 0 \cdot 96$ m²
Volume of concrete = $(0 \cdot 96$ m² $\times 0 \cdot 225$ m$)$ m³ = $0 \cdot 216$ m³
Total weight at 24 kN/m³ = $(0 \cdot 216 \times 24)$ kN = $5 \cdot 184$ kN
Total dead load including steel beam, say 5·2 kN.
Super load per beam = $(0 \cdot 96 \times 5)$ kN = 4·8 kN.
Total uniformly distributed load per beam = 10 kN.
B.M. maximum = $Wl/2 = 10 \times 1 \cdot 2/2$ kNm = 6 kNm.
S.F. maximum = W = 10 kN.

The diagrams may be conveniently drawn to the following scales: 100 mm = 1 m; 12·5 mm = 1 kNm; 5 mm = 1 kN.

d) Simply supported beam with a single concentrated load

In the case of beams, the first step is to calculate the two support reactions. By taking moments about point B (fig. 5.9) we get

$$R_A \times l = W \times b \quad \therefore \quad R_A = Wb/l$$

Similarly, $R_B = Wa/l$. There is only one load to the left of the typical section indicated, viz. the reaction at A ($= Wb/l$).

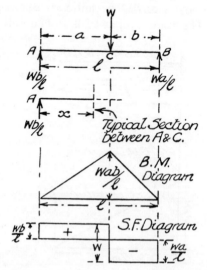

Fig. 5.9 *Simply supported beam with single concentrated load.*

$\text{B.M.}_x = (Wb/l) \times x = Wbx/l$. This value indicates that the B.M. graph from A to C will be a straight line.

When $x = 0$, $\quad$ B.M. $= \dfrac{Wb}{l} \times 0 = 0$.

When $x = a$, $\quad$ B.M. $= \dfrac{Wb}{l} \times a = \dfrac{Wab}{l}$.

We cannot put x greater than a in this expression, but similar reasoning will show that the B.M. gradually increases from zero at the point B to the value Wab/l at the point C.

Maximum B.M. for this case is therefore Wab/l.

$$\text{S.F.}_x = \text{total load to left of given section} = Wb/l.$$

For any beam section to the right of point C, the total load to the *left* of the section would clearly be

$$\frac{Wb}{l} - W = \frac{Wb}{a+b} - W = \frac{Wb - Wa - Wb}{a+b} = -\frac{Wa}{l}$$

i.e. the right-end reaction with a negative sign.

In this type of beam we get, therefore, positive B.M.'s and both positive and negative S.F.'s.

If W is at the centre of the beam, $a = b = l/2$.

$$\text{B.M. maximum becomes } \frac{W \times l/2 \times l/2}{l} = \frac{Wl}{4}.$$

$$\text{S.F. maximum} = \pm \frac{W}{2}.$$

The values for B.M. maximum and S.F. maximum in this case should be memorised.

In the calculation of B.M. and S.F. values for beams simply supported at the ends, the following concise statements will be found useful:

B.M. = reaction moment − load moments, the reactions and loads being taken to one side of the section, and moments taken about the section.

S.F. = left-end reaction − sum of loads up to section. If the left side of section be taken always for S.F., the danger of incorrect sign of S.F. will be avoided.

e) Simply supported beam with several concentrated loads

Fig. 5.10 shows the derivation of the B.M. and S.F. diagrams for this case from the component diagrams for the separate loads. It will be seen that we need only calculate the values of the B.M.'s at C and D

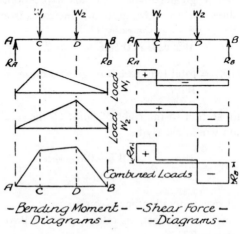

Fig. 5.10 *Composition of B.M. and S.F. diagrams.*

to complete the B.M. diagram. The S.F. diagram is a stepped diagram, the vertical drop at C representing the load W_1, and that at D the load W_2.

Example In fig. 5.11 there is shown a simply supported beam carrying three concentrated loads. Calculate the values of B.M. and S.F. respectively for a section 1·8 m from the left end. Construct the B.M. and S.F. diagrams for the beam.

$$R_A \times 3 = (6 \times 2 \cdot 1) + (4 \times 1 \cdot 5) + (2 \times 0 \cdot 6)$$
$$3R_A = 12 \cdot 6 + 6 \cdot 0 + 1 \cdot 2 = 19 \cdot 8$$
$$R_A = 6 \cdot 6 \text{ kN}$$
$$R_B \times 3 = (2 \times 2 \cdot 4) + (4 \times 1 \cdot 5) + (6 \times 0 \cdot 9)$$
$$3R_B = 4 \cdot 8 + 6 \cdot 0 + 5 \cdot 4 = 16 \cdot 2$$
$$R_B = 5 \cdot 4 \text{ kN}$$

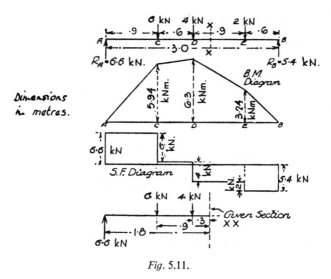

Fig. 5.11.

$R_A + R_B = (6·6 + 5·4)$ kN $= 12$ kN, which agrees with the total load on the beam.

B.M.$_C = R_A \times 0·9 = (6·6 \times 0·9)$ kNm $= 5·94$ kNm
B.M.$_D$ = reaction moment − load moment
$= (6·6 \times 1·5) − (6 \times 0·6)$ kNm $= 6·3$ kNm
B.M.$_E$ (taking loads to right of section for convenience)
$= R_B \times 0·6 = (5·4 \times 0·6)$ kNm $= 3·24$ kNm
B.M. at the given section XX

$= (6·6 \times 1·8) − (6 \times 0·9) − (4 \times 0·3)$ kNm
$= (11·88 − 5·4 − 1·2)$ kNm $= 5·28$ kNm

S.F. at given section = left-end reaction − sum of loads up to section
$= (6·6 − 6 − 4)$ kN $= −3·4$ kN

The reader should check these given section values by taking loads to the *right of the section*, and also by scaling diagrams of B.M. and S.F. for the beam.

Suitable scales: 50 mm = 1 m; 10 mm = 1 kNm; 10 mm = 1 kN.

f) Simply supported beam with uniformly distributed load

The total load on the beam (fig. 5.12) $= wl$, so that each reaction $= wl/2$. Considering the B.M. at the typical section, and taking loads to the left of the section, we get

$$\text{B.M.}_x = \text{reaction moment} - \text{load moment}$$

$$= \frac{wl}{2} \times x - wx \times \frac{x}{2}$$

$$= \frac{wlx}{2} - \frac{wx^2}{2}$$

This expression would give a parabola if plotted.

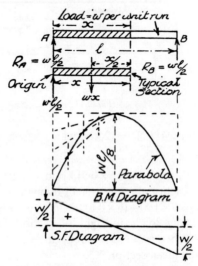

Fig. 5.12 *Simply supported beam with uniformly distributed load.*

When $x = 0$, $\quad$ B.M. $= \dfrac{wl}{2} \times 0 - \dfrac{w}{2} \times (0)^2 = 0$

When $x = \dfrac{l}{2}$, $\quad$ B.M. $= \dfrac{wl}{2} \times \dfrac{l}{2} - \dfrac{w}{2} \times \left(\dfrac{l}{2}\right)^2$

$$= \frac{wl^2}{4} - \frac{wl^2}{8} = \frac{wl^2}{8}$$

Inserting W (the total load) for wl, the B.M. at centre of beam $= Wl/8$. ✓

At B, $x = l$ and B.M. $= (wl/2) \times l - (w/2) \times l^2 = 0$, as we would expect.

The B.M. is clearly a maximum at the beam centre, but readers familiar with the calculus will be able to show that the expression $wlx/2 - wx^2/2$ has a maximum value for $x = l/2$.

$$\text{S.F.}_x = \text{reaction} - \text{total load up to section}$$

$$= \frac{wl}{2} - wx$$

This indicates uniform variation for the shear force, as x is in the first degree.

When $x = 0$, $\quad \text{S.F.} = \dfrac{wl}{2} - w \times 0 = \dfrac{wl}{2} = \dfrac{W}{2}$

When $x = \dfrac{l}{2}$, $\quad \text{S.F.} = \dfrac{wl}{2} - \dfrac{wl}{2} = 0$

When $x = l$, $\quad \text{S.F.} = \dfrac{wl}{2} - wl = -\dfrac{wl}{2} = -\dfrac{W}{2}$

For this case, therefore,

$$\text{B.M. maximum} = \frac{Wl}{8}$$

and

$$\text{S.F. maximum} = \pm \frac{W}{2}.$$

The formulae given for B.M. maximum and S.F. maximum are very important, and are frequently required in the design of beams.

Example Draw the B.M. and S.F. diagrams for one of the steel floor beams given in fig. 5.13. Find the B.M. and S.F. values respectively, for the section of a beam 2·4 m from the left end. Total load (including weight of floor) to be taken as 4·5 kN/m².

Area supported by one joist $= 3 \text{ m} \times 3 \cdot 6 \text{ m} = 10 \cdot 8 \text{ m}^2$.

Uniformly distributed load carried by one beam

$$= (10 \cdot 8 \times 4 \cdot 5) \text{ kN} = 48 \cdot 6 \text{ kN}.$$

$$\text{B.M. maximum} = \frac{Wl}{8} = \frac{48 \cdot 6 \times 3 \cdot 6}{8} \text{ kNm} = 21 \cdot 87 \text{ kNm}$$

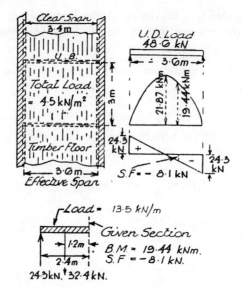

Fig. 5.13.

$$\text{S.F. maximum} = \pm\frac{W}{2} = \pm\frac{48\cdot 6}{2}\text{ kN} = \pm 24\cdot 3\text{ kN}$$

$$\text{Load carried per metre length of beam} = \frac{48\cdot 6}{3\cdot 6}\text{ kN}$$

$$= 13\cdot 5\text{ kN}$$

B.M. at given section = reaction moment − load moment

$$= \{(243\times 2\cdot 4)-(13\cdot 5\times 2\cdot 4\times 1\cdot 2)\}\text{ kNm}$$
$$= (58\cdot 32-38\cdot 88)\text{ kNm}$$
$$= 19\cdot 44\text{ kNm}$$

S.F. at given section = reaction − load up to section

$$= 24\cdot 3\text{ kN} - 32\cdot 4\text{ kN}$$
$$= -8\cdot 1\text{ kN}$$

Suggested scales for diagrams: 50 mm = 1 m; 5 mm = 1 kNm; 5 mm = 1 kN.

g) Beam overhanging its supports and carrying a system of concentrated loads

It will now be appreciated that *a system of concentrated loads will always give a B.M. diagram composed of straight lines,* so that B.M. values at load points only are required to be calculated.

S.F. diagrams for such a system will always be *stepped diagrams, with a step, up or down at a load point, representing the value of the given load.*

Example Construct the B.M. and S.F. diagrams for the case given in fig. 5.14.

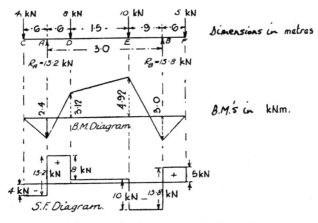

Fig. 5.14 *Overhanging beam with concentrated loads.*

Taking moments about B, to find R_A, we get

$$(R_A \times 3) + (5 \times 0 \cdot 6) = (4 \times 3 \cdot 6) + (8 \times 2 \cdot 4) + (10 \times 0 \cdot 9)$$
$$3R_A + 3 = 14 \cdot 4 + 19 \cdot 2 + 9 \cdot 0 = 42 \cdot 6$$
$$3R_A = 42 \cdot 6 - 3 = 39 \cdot 6$$
$$\therefore \quad R_A = 13 \cdot 2 \text{ kN}$$

Moments about A:

$$(R_B \times 3) + (4 \times 0 \cdot 6) = (5 \times 3 \cdot 6) + (10 \times 2 \cdot 1) + (8 \times 0 \cdot 6)$$
$$3R_B + 2 \cdot 4 = 18 \cdot 0 + 21 \cdot 0 + 4 \cdot 8$$
$$3R_B = 43 \cdot 8 - 2 \cdot 4 = 41 \cdot 4$$
$$\therefore \quad R_B = 13 \cdot 8 \text{ kN}$$

B.M.$_C$ = 0 (there is no load to left of section)
B.M.$_A$ = (4×0.6) kNm = 2.4 kNm (negative)
B.M.$_D$ = reaction moment − load moment
     = $\{(13.2 \times 0.6) - (4 \times 1.2)\}$ kNm
     = $(7.92 - 4.8)$ kNm = 3.12 kNm (positive)
B.M.$_E$ = $\{(13.2 \times 2.1) - (8 \times 1.5) - (4 \times 2.7)\}$ kNm
     = $(27.72 - 12.0 - 10.8)$ kNm
     = 4.92 kNm (positive)

Alternatively, *taking loads to right of section at E*:

B.M.$_E$ = $\{(13.8 \times 0.9) - (5 \times 1.5)\}$ kNm = $(12.42 - 7.5)$ kNm
     = 4.92 kNm
B.M.$_B$ = (5×0.6) kNm = 3.0 kNm (negative)
B.M.$_F$ = 0

Suggested scales for diagrams: 50 mm = 1 m; 10 mm = 1 kNm; 5 mm = 1 kN.

h) Beam overhanging its supports and carrying a uniformly distributed load

The form of the B.M. diagram will be understood from an inspection of fig. 5.15, in which it is built up from component diagrams which are already familiar to the reader.

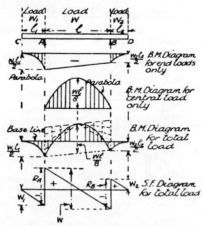

Fig. 5.15 *Overhanging beam with uniformly distributed load.*

The two end portions act as cantilevers. In the practical construction of the B.M. diagram, the usual geometrical construction for a parabola—somewhat modified as shown—can be used for the central span. The S.F. diagram presents no difficulty, if it is borne in mind that a uniformly distributed load causes a uniform slope in the S.F. diagram.

Example A steel and concrete floor is carried by UB's which rest on columns in the manner shown in fig. 5.16. The total load transmitted to one UB is 40 kN per metre run. Construct the B.M. and S.F. diagrams for a beam.

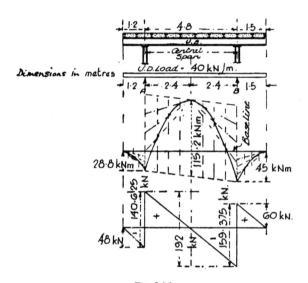

Fig. 5.16.

Portion of beam from left end to A.

$$\text{B.M. at A} = \frac{Wl}{2} = \left(48 \times \frac{1 \cdot 2}{2}\right) \text{kNm} = 28 \cdot 8 \text{ kNm}$$

Portion of beam from B to right end.

$$\text{B.M. at B} = \frac{Wl}{2} = \left(60 \times \frac{1 \cdot 5}{2}\right) \text{kNm} = 45 \text{ kNm}$$

Both these will be negative bending moments.

B.M. maximum for central portion AB, regarded temporarily as a simple beam of length AB,

$$= \frac{Wl}{8} = \frac{192 \times 4 \cdot 8}{8} \text{ kNm} = 115 \cdot 2 \text{ kNm}$$

The B.M. diagram is constructed as shown. For the S.F. diagram we require to calculate the two support reactions.

$R_A \times 4 \cdot 8 = 300 \text{ kN} \times 2 \cdot 25 \text{ m}$ (the c.g. of the whole load is 2·25 m from B).

$$R_A = \frac{300 \times 2 \cdot 25}{4 \cdot 8} = 140 \cdot 625 \text{ kN}$$

Similarly

$$R_B = \frac{300 \times 2 \cdot 55}{4 \cdot 8} = 159 \cdot 38 \text{ kN}$$

(Check $R_A + R_B = 300$ kN.)

Between the left end of the beam and A the diagram slopes uniformly to a negative value of 48 kN at A. There is then a sudden vertical jump of 140·625 kN, owing to the reaction at A. Between A and B the total fall = 192 kN. The diagram finally ends with zero value at the extreme right end of the beam.

Two or more load systems

The self-weight of a beam always involves a distributed load, in addition to any other loads carried. The weight of the beam itself may be small in comparison with the applied loading and may often be neglected in calculations. To obtain the net B.M. (or S.F.) diagrams for two or more simple load systems, the component diagrams should be drawn out to the same scale and arranged one beneath the other. Vertical lines —drawn at suitable intervals in the span—to cut the series of diagrams will give corresponding ordinates in the separate diagrams. These ordinates algebraically added are then plotted on a new base line to give the final net diagram.

The load systems dealt with in this chapter represent those commonly occurring in practical problems. For the treatment of other load types, the reader is referred to books on the theory of structures.

Exercises 5

1 Distinguish between the terms 'bending moment' and 'moment of resistance', and explain the meaning of the term 'shear force'. State the regions of a simply supported beam for which the B.M. and S.F. are, respectively, of relatively greater importance.

2 A cantilever projects 1·5 m horizontally from a wall. It carries a load of 2 kN at its free end, and 4 kN at 0·9 m from the free end. Find the B.M. and S.F. values for the centre of the cantilever, and draw the B.M. and S.F. diagrams for the given loads.

3 A tank is supported outside a building by two UB's which are fixed in the wall and extent 1·2 m horizontally therefrom. The tank when full weighs 20 kN, the load being equally divided between, and uniformly distributed along, the two beams. Draw the B.M. and S.F. diagrams for one beam, and find the B.M. and S.F. respectively to which each beam is subjected at 0·3 m from the wall.

4 A simply supported beam of 10 m span carries the following concentrated load system:

    4 kN at 2 m from left end of beam,
    8 kN at 5 m from left end of beam,
    8 kN at 9 m from left end of beam.

Calculate the B.M. and S.F. values for a section 4 m from the left end, and check the values obtained by the construction of the B.M. and S.F. diagrams for the beam.

5 A room 5·4 m × 4·8 m, has a UB parallel to the longer side and at mid-width of floor, supporting the pitch-pine beams of which the floor is composed. The total load carried by the floor, including the weight of the floor itself, is 7·5 kN/m^2. The floor beams are spaced at 300 mm centres.
(a) Construct the B.M. and S.F. diagrams for one of the timber beams, and (b) determine the maximum B.M. in the steel beam, allowing 2 kN for its self-weight.

6 A lintel of 2 m effective span carries a wall of uniform height and thickness. The thickness is 450 mm and the average density of the wall material is 18 kN/m^3. Calculate the maximum height of the wall so that the B.M., due to the weight of the wall, does not exceed 6·75 kN m. Sketch the B.M. and S.F. diagrams for the lintel, inserting thereon all important values.

7 A beam, resting on two walls, A and B, 4·8 m apart, carries four lifting appliances, which may cause the loads given in fig. 5.17. Assuming the four loads to be simultaneously applied, draw the B.M. and S.F. diagrams for the beam for these loads.

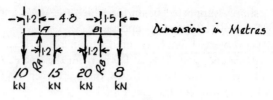

Fig. 5.17.

8 Draw the B.M. and S.F. diagrams for the self-weight of the beam referred to in question 7, assuming it to weigh 300 N per metre run.

6

Design of Simple Beams. Moment of Resistance

Assumptions in the beam theory

The theory which is involved in the derivation of the formula for the *moment of resistance* of a beam section is based upon certain assumptions.

i) The beam section, for the theory to apply strictly, should have a vertical axis of symmetry, in order to ensure the deflection of the beam being in a vertical plane.

ii) *Simple bending* is assumed. This is the type of bending produced by the application of pure couples at the ends of the beam. Such bending does not involve the beam in deflection due to shear strain (which is, however, relatively small in any practical case) and is exemplified by the portion AB of the beam shown in fig. 6·1.

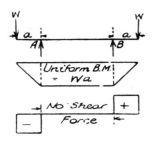

Fig. 6.1 *Example of simple bending.*

iii) Vertical sections of the beam before bending are assumed to remain plane after bending. In fig. 6.3, the plane sections AB and CD remain plane in the positions A'B' and C'D' respectively.

iv) The stress in any given beam fibre is assumed to be proportional to its strain, i.e. Hooke's law is assumed to hold for the beam material. Young's modulus (E) is assumed to have the same value throughout the beam.

It will be instructive, before taking the general theory, to study the applications of these assumptions to the case of a beam having a rectangular cross-section.

Rectangular beams

The beam shown in fig. 6.2 will have its upper fibres in compressive strain (and hence compressive stress) and its lower fibres in tensile

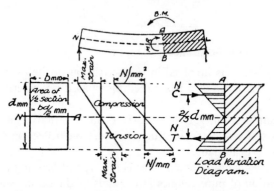

Fig. 6.2 Moment of resistance of a rectangular beam section.

strain (and hence tensile stress). It will be clear that the biggest strains will occur at the top and bottom of the beam respectively. These strains will decrease in value as we proceed inwards from the extreme fibres. At some level in the beam the strains will have completely vanished. Throughout a bent beam there is a layer of material which undergoes no strain and, therefore, is in an unstressed condition. This extremely thin layer is termed the *neutral layer* (N.L.) and the straight line in which this layer cuts any given beam cross-section is known as the *neutral axis* (N.A.) of that section. The N.A. of a rectangular beam section will be at mid-depth.

From assumption (iii) it may be concluded that the strain in the beam fibres will increase uniformly as we proceed, up and down, from the N.A. to the extreme beam fibres. As stress and strain are assumed to vary in proportion, the stress variation will also be of a uniformly changing character. Further, in the case being considered, little horizontal strips of beam cross-section will be of constant width, so that the load distribution (as a result of the stress) will also be of the same uniformly varying type. The diagrams in fig. 6.2 will now be clear to the reader.

The two like parallel systems composed, respectively, of the numerous little thrusts and pulls on section AB, will each have a resultant with a definite line of action. These two resultants (C and T) form the forces in the couple resisting bending, and the distance between their lines of action is the 'arm' of the couple.

Value of C (or T) C will be found by multiplying the average stress in the top half-section by the area of this portion. Taking the symbols and units given in fig. 6.2,

$$C = \frac{f}{2} \text{N/mm}^2 \times \frac{bd}{2} \text{mm}^2$$

$$= \frac{fbd}{4} \text{N}$$

Similarly, $T = \dfrac{fbd}{4}$ N

Value of the 'arm' of the couple C will act through the c.g. of the upper load triangle, and T through that of the lower. The distance between their lines of action will therefore be

$$\left\{ d - \left(\tfrac{1}{3} \text{ of } \tfrac{d}{2}\right) - \left(\tfrac{1}{3} \text{ of } \tfrac{d}{2}\right) \right\} \text{mm} = \tfrac{2}{3} d \text{ mm}$$

Moment of couple

$$\text{Moment} = \text{force} \times \text{arm}$$

$$= \frac{fbd}{4} \text{N} \times \tfrac{2}{3} d \text{ mm}$$

$$= \frac{fbd^2}{6} \text{Nmm}$$

This is the 'moment of resistance' and equals the 'bending moment' at the section. Writing 'M' for 'M.R.' or 'B.M.', the formula becomes (without reference to any special system of units)

$$M = \frac{fbd^2}{6}$$

Example 1   A timber beam is 75 mm wide, 150 mm deep, and has an effective span of 3 m. It carries a total U.D. load of 5250 N. Calculate the maximum stress in the timber.

$$\text{B.M. maximum} = \frac{Wl}{8} = \frac{5250 \times 3 \times 1000}{8} = 1\,968\,750 \text{ Nmm}$$

(Note that the span must always be in mm, as the stress is expressed in mm units.)

$$M = \frac{fbd^2}{6}$$

$$1\,968\,750 = \frac{f \times 75 \times 150 \times 150}{6}$$

$$f = 7 \text{N/mm}^2$$

Example 2 Find the maximum safe central load for a pitch-pine beam 100 mm wide and 300 mm deep, if the effective span (i.e. centre to centre of bearings) is 4·5 m. The working stress may be taken as 8·4 N/mm².

Let W N = safe central load, neglecting weight of beam.

$$\text{B.M. maximum} = \frac{Wl}{4} = \frac{W \times 4\cdot 5 \times 1000}{4} \text{ Nmm}$$

$$= 1125\,W \text{ Nmm}$$

$$M = \frac{fbd^2}{6}$$

$$1125\,W = \frac{8\cdot 4 \times 100 \times 300 \times 300}{6}$$

$$W = 11\,200 \text{ N}$$

Example 3 A floor is composed of timber joists 50 mm wide, 150 mm deep, and 3 m effective span. It is to carry a total load (including the weight of the floor) of 3·33 kN/m². Adopting a working stress of 7 N/mm², determine the maximum spacing for the joists.

Let x mm = spacing.

$$\text{Area of floor carried by one joist} = \left(3 \times \frac{x}{1000}\right) = \frac{3x}{1000} \text{ m}^2$$

Load carried by one joist $= \left(\dfrac{3x}{1000} \times \dfrac{10}{3} \times 1000\right)$ N

$= 10x$ N

B.M. maximum in joist $= \dfrac{Wl}{8} = \dfrac{10x \times 3 \times 1000}{8}$

$= 3750x$ N mm

$$M = \dfrac{fbd^2}{6}$$

$$3750x = \dfrac{7 \times 50 \times 150 \times 150}{6}$$

$$x = 350 \text{ mm}$$

General theory of bending

Fig. 6.3 shows a portion of a beam in the unbent and in the bent conditions. AB and CD are two vertical cross-sections, assumed so close together that the portion of beam between them may be regarded as bending to the arc of a circle.

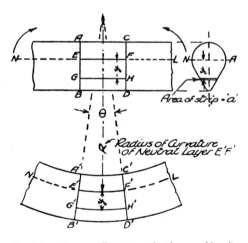

Fig. 6.3   *Diagram illustrating the theory of bending.*

108 Structural Steelwork

EF is the part of the neutral layer intercepted between the sections. GH represents a typical layer of material at a distance y from the neutral axis. R is the radius of curvature of the portion of the neutral layer, in the bent beam. The following are the steps in the development of the theory:

1. determination of the strain in layer G'H' by principles of geometry;
2. evaluation of the stress in this layer by means of Young's modulus;
3. determination of the load carried by the little strip of cross-section at distance y from the N.A.;
4. computation of the moment this load has about the N.A. and, by summation, the total moment of all such strip loads.

Step 1 Extension in layer G'H' = G'H' − GH.

$$\text{Strain in layer G'H'} = \frac{\text{extension}}{\text{original length}}$$

$$= \frac{\text{G'H'} - \text{GH}}{\text{GH}}$$

But GH = EF and EF = E'F' (being on the unstrained layer),

$$\therefore \text{ strain in layer G'H'} = \frac{\text{G'H'} - \text{E'F'}}{\text{E'F'}}$$

Expressing these distances in terms of R and θ (the angle in radians contained by B'A' and D'C'), we have

$$\text{strain in layer G'H'} = \frac{(R+y)\theta - R\theta}{R\theta} = \frac{y}{R}$$

Step 2 $\dfrac{\text{stress in G'H'}}{\text{strain in G'H'}} = E,$

$$\therefore \text{ stress in layer} = E \times \text{strain} = \frac{Ey}{R}$$

If f = the stress, $\qquad f = \dfrac{Ey}{R}$

Step 3 If a = the area of the cross-sectional strip,

$$\text{load carried} = \text{stress} \times \text{area}$$

$$= \frac{Ey}{R} \times a = \frac{E}{R} \times ay$$

Design of Simple Beams. Moment of Resistance

Step 4 Moment of the load on this strip about N.A.

$$= \text{load} \times \text{distance}$$

$$= \left(\frac{E}{R} \times ay\right) \times y = \frac{E}{R} \times ay^2$$

The total 'moment of resistance' of the beam section is made up of all such moments as this.

$$\text{Total moment of resistance} = \Sigma \frac{E}{R} \times ay^2$$

$$= \frac{E}{R} \times \Sigma ay^2$$

Σay^2 ('*sigma ay*^{2}') is a geometrical property of the beam section, with reference to the axis N.A. It is termed the *moment of inertia* of the beam section, and is denoted by the letter I.

Writing M for 'moment of resistance',

$$M = \frac{E}{R} \times I$$

But $f = \frac{Ey}{R}$ or $\frac{E}{R} = \frac{f}{y}$ (step 2),

$$\therefore \quad M = \frac{fI}{y}$$

This is the important formula for finding the moment of resistance of a beam section. Writing the results found as a continued ratio, we get the complete expression for the theory:

$$\frac{E}{R} = \frac{M}{I} = \frac{f}{y}$$

In using this expression, f will normally represent a maximum stress, so that y, in that case, will be the distance from the neutral axis to an extreme fibre, top or bottom of the section, as the case may be.

M will usually be a 'bending moment', such as $Wl/4$, $Wl/8$, etc.

Position of the neutral axis In the theory, we found that the strip load was given by the expression $(E/R) \times ay$.

As long as y is measured downwards, all these strip loads will represent tension. If we put y negative, i.e. measured the distance upwards from N.A., the load would be compression.

110 *Structural Steelwork*

$\Sigma(E/R) \times ay$ or $(E/R)\Sigma ay$ (since E and R are constants) will therefore represent a summation of a large number of positive and negative quantities. But, as the total compressive force = the total tensile force (being forces in a couple), $(E/R)\Sigma ay$ must $= 0$.

This, i.e. $\Sigma ay = 0$, means that the axis, from which y is measured, passes through the centre of gravity of the section. *The neutral axis of a beam section therefore passes through its centre of gravity.*

Moment of inertia

This is one of three very important *properties of section*. It is a geometrical property of the shape and size of the beam section, and has reference to an axis. The material of the beam has nothing to do with its value.

In fig. 6.4(a) we have an extremely small area a at a distance y from an axis XX. The product $(a \times y^2)$ is termed the *moment of inertia* of the area a about the axis XX:

$$I_{XX} = ay^2$$

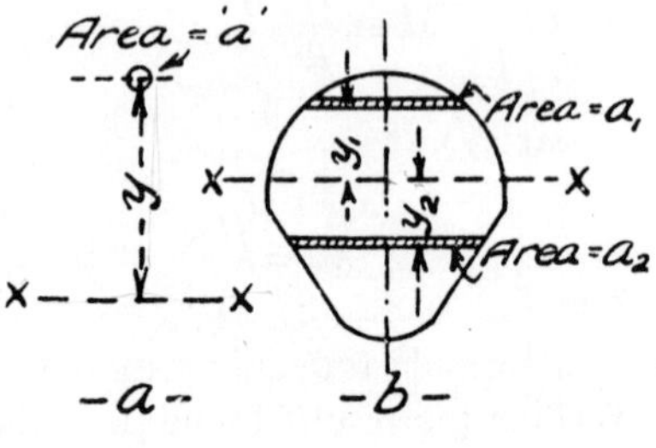

Fig. 6.4.

In fig. 6.4(b) there is shown an area divided up into very small elements a_1, a_2, etc., at distances y_1, y_2, etc., from an axis XX. The sum of all such products as $a_1 y_1^2$, $a_2 y_2^2$, etc., will be the *moment of inertia* of the given area about the axis XX:

$$\begin{aligned} I_{XX} &= a_1 y_1^2 + a_2 y_2^2 + \text{etc.} \\ &= \Sigma ay^2 \end{aligned}$$

In the beam theory, the little strip of beam cross-section had to be extremely small, so that the stress could be regarded as being of the value f all over it. In computing I values, we will necessarily be approximate if we divide the given section into areas of finite size, but we can reduce the error by taking reasonably narrow strips.

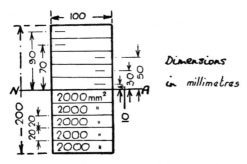

Fig. 6.5 Approximate I_{NA} for a rectangle.

Example Fig. 6.5 shows a rectangle, 100 mm wide and 200 mm deep, divided into 10 strips. Find an approximate value for I_{NA}.

I_{NA} for top strip $= ay^2 = 2000 \text{ mm}^2 \times (90 \text{ mm})^2 = 16{\cdot}2 \times 10^6 \text{ mm}^4$
I_{NA} for 2nd strip $= ay^2 = 2000 \text{ mm}^2 \times (70 \text{ mm})^2 = 9{\cdot}8 \times 10^6 \text{ mm}^4$
I_{NA} for 3rd strip $= ay^2 = 2000 \text{ mm}^2 \times (50 \text{ mm})^2 = 5{\cdot}0 \times 10^6 \text{ mm}^4$
I_{NA} for 4th strip $= ay^2 = 2000 \text{ mm}^2 \times (30 \text{ mm})^2 = 1{\cdot}8 \times 10^6 \text{ mm}^4$
I_{NA} for 5th strip $= ay^2 = 2000 \text{ mm}^2 \times (10 \text{ mm})^2 = 0{\cdot}2 \times 10^6 \text{ mm}^4$
Total I_{NA} for $\frac{1}{2}$ section $= 33 \times 10^6 \text{ mm}^4$
I_{NA} for whole section $= 66 \times 10^6 \text{ mm}^4$

The reader should note carefully the unit in which I values are expressed.

If the strips were taken 10 mm deep, instead of 20 mm, the answer would be slightly higher, and if taken extremely thin the value would be $66{\cdot}666\,667 \times 10^6 \text{ mm}^4$.

I_{NA} for a rectangular section

To obtain a correct value for I_{NA}, the methods of calculus must be employed. Readers unfamiliar with the calculus should omit the work immediately following, and accept the formula given for this case.

$$I_{XX} = \Sigma ay^2 \text{ (fig. 6.6)}$$

Consider top half-section only.

$$I_{XX} = \int_{y=0}^{y=d/2} ay^2 = \int_{y=0}^{y=d/2} b\delta y \times y^2$$

$$= b \int_{y=0}^{y=d/2} y^2 \delta y. \quad = b\left[\frac{y^3}{3}\right]_{y=0}^{y=d/2} = \frac{bd^3}{24}$$

Structural Steelwork

I_{XX} for whole section $= 2 \times \dfrac{bd^3}{24} = \dfrac{bd^3}{12}$

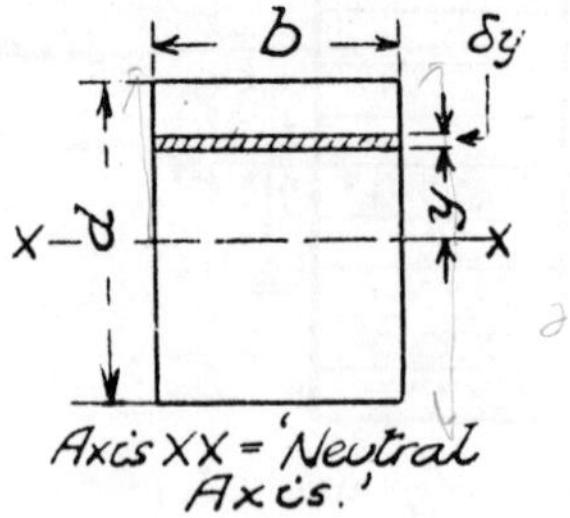

Fig. 6.6.

Applying this formula to the example worked previously,

$$I_{NA} = \frac{bd^3}{12} = \frac{100 \times 200 \times 200 \times 200}{12} \; mm^4 = 66 \cdot 666\,667 \times 10^6 \; mm^4$$

I_{BASE} would be $bd^3/3$ by same method.

$I_{maximum}$ and $I_{minimum}$

In dealing with the strength of beams, the computation of I is made about the appropriate neutral axis of the section. If we calculated the various I values for all the possible axes *passing through the c.g. of the section*, we would find that the biggest value, and the least value, were associated with the axes of symmetry of the section—assuming such axes existed. It is supposed in the beam theory that there is at least a vertical axis of symmetry, and in this case, if the horizontal axis through the c.g. is not an axis of symmetry, it still becomes the axis for either $I_{maximum}$ or $I_{minimum}$. The 'properties of section' for certain standard beams given on pages 122 to 127 (published by kind permission of the British Constructional Steelwork Association and CONSTRADO), illustrate these two I values, $I_{maximum}$ being required for ordinary beam calculations.

Important Note

It will be seen that, in the example worked above, the use of *millimetre* units leads to very large arithmetical values of I. It will also be seen that, in the section tables, the values of I and Z are given in *centimetre* units, in order to make the arithmetical values more manageable. From now on, in this book, all values of I and Z will be calculated in centimetre units.

Example I_{max} for a 381 × 152 × 52 kg UB is given in the table referred to as 16046 cm⁴. Calculate the safe U.D. load for this section for an effective span of 5 m, using a working stress of 165 N/mm².

$$M = \frac{fI}{y}$$

$$\frac{Wl}{8} = \frac{fI}{y} = \frac{165 \times 16046 \times 10^4}{190 \cdot 5} \left(y = \frac{381 \text{ mm}}{2} = 190 \cdot 5 \text{ mm} \right)$$

$$\frac{W \times 5 \times 1000}{8} = \frac{165 \times 16046 \times 10^4}{190 \cdot 5}$$

$$W = 222\,300 \text{ N} = 222 \cdot 3 \text{ kN}.$$

Addition and subtraction of I values

Moments of inertia of component areas about any given axis may be directly added, or subtracted, to obtain the final result for the compound area— *but only provided each component I value has reference to the one common given axis.* In ordinary beam problems, this axis will be the neutral axis of the beam section.

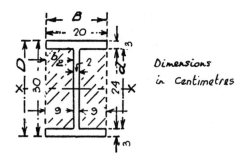

Fig. 6.7.

Example Find the value of I_{xx} for the section given in fig. 6.7.

We first treat the section as a solid rectangle 20 cm wide and 30 cm deep.

$$I_{xx} = \frac{bd^3}{12} = \frac{20 \times 30 \times 30 \times 30}{12} \text{ cm}^4 = 45\,000 \text{ cm}^4$$

To get the net section we have to subtract the two rectangles lying respectively, to either side of the web (shown hatched).

I_{xx} for these two rectangles $= 2\left(\dfrac{bd^3}{12}\right)$

$$= 2\left(\dfrac{9 \times 24^3}{12}\right) \text{cm}^4 = 20736 \text{ cm}^4$$

Net I_{xx} for section $= (45000 - 20736) \text{ cm}^4 = 24264 \; cm^4$

It is useful to remember a formula for such a section, viz.:

$$I_{xx} = \dfrac{BD^3}{12} - \dfrac{bd^3}{12} \quad [b = B - \text{web thickness}]$$

Economy of 'I' type of section

Beams of rectangular section are not economical of material, as the parts of the section situated near the neutral axis are stressed to only low values (fig. 6.8). The small loads carried here are further subject to the

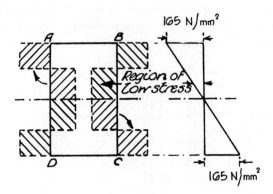

Fig. 6.8 *Derivation of flanged section from rectangular section of same area.*

disadvantage of having a small arm for their resistance moment. In the 'I' form of section, a fair proportion of the steel is placed in such a position as to become highly stressed, and the corresponding loads have a bigger arm of resistance moment.

Example Fig. 6.9 shows three similar plates welded together to form a beam section in two different ways. Compare the safe U.D. loads for these two beams, for an effective span of 4 m.

Design of Simple Beams. Moment of Resistance   115

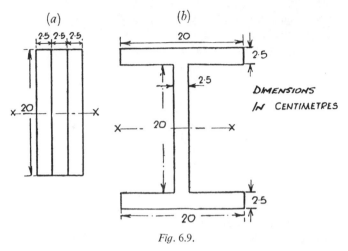

Fig. 6.9.

(a) $M = fbd^2/6$,

$$\frac{W \times 4 \times 1000}{8} = \frac{165 \times 7\cdot5 \times 20 \times 20 \times 1000}{6} \quad (f = 165 \text{ N/mm}^2)$$

$$W = 165\,000 \text{ N}.$$

(b) $M = fI/y$,

$$I_{XX} = \frac{BD^3}{12} - \frac{bd^3}{12} = \left(\frac{20 \times 25^3}{12} - \frac{17\cdot5 \times 20^3}{12}\right) \text{cm}^4$$

$$= 14\,375 \text{ cm}$$

$$\frac{W \times 4 \times 1000}{8} = \frac{165 \times 14\,375 \times 10^4}{12\cdot5 \times 10}$$

$$W = 379\,500 \text{ N}$$

Both beams have the same weight of steel, but the 'I' section can support a far greater load.

Section modulus

In the expression fI/y we have two symbols, I and y, which represent geometrical properties associated with the beam section. For any given geometrical figure, or standard beam section, they can be separately computed. It is convenient to merge these two section properties into one single property, in the form of their ratio I/y. To this property is given the name *section modulus*. The symbol used for this composite property is usually Z. The units for Z will be 'cm^3' being cm^4/cm. Corresponding to $I_{maximum}$ and $I_{minimum}$ there will be $Z_{maximum}$ and $Z_{minimum}$.

$$Z = \frac{I}{y}$$

In most structural sections, as in fig. 6.10(a), the neutral axis is also an axis of symmetry, so that y is half the overall depth of the section. In this case,

$$Z = \frac{\text{moment of inertia (about N.A.)}}{\tfrac{1}{2} \text{ overall depth of section}}$$

In the event of the distances from the neutral axis to the extreme fibres at the top and bottom of the section, respectively, not being equal, as in fig. 6.10(b) and (c), there will be two Z values for this axis.

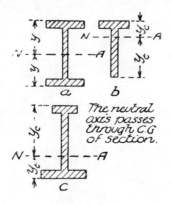

Fig. 6.10.

In a beam composed of material like steel, which is equally reliable in tension and compression, the lower of these two possible values must be used for calculation of strength. Where the working stresses are not equal in tension and compression (as in cast iron), the safe load for the beam must be separately calculated from the two values—Z_t (tension) and Z_c (compression)—using the corresponding working-stress values. The latter case is not common, owing to the extensive modern use of rolled-steel sections.

Values of Z for UB's will be found on pages 123 and 125.

Design of Simple Beams. Moment of Resistance

The formula $M = fI/y$ may now be written

$$M = fZ$$

In the form required for design purposes, the expression for Z becomes $Z = M/f$, or, for a beam of constant section throughout its length, we may write

$$\text{Necessary section modulus for beam section} = \frac{\text{maximum B.M.}}{\text{working stress}}$$

NOTE: This will give the value of the section modulus in cm³, if BM is in kN mm, and the stress is in N/mm².

If L = span in metres for a simply supported beam with a U.D. load = W kN, and if 165 N/mm² be taken as the working stress,

$$\frac{W \times 1000 \times L \times 1000}{8} = 165 \times Z \text{ (in mm}^3\text{)}$$

$$\therefore \quad Z \text{ (in cm}^3\text{)} = \frac{WL}{8} \times \frac{1000}{165} = \text{B.M. in kN m} \times 6 \text{ (approx.)}$$

a rule which may be used by designers for the given case.

$$\text{Also } Z \text{ (in cm}^3\text{)} = \frac{W \times L \times 1000}{1320} = \frac{W \times L}{1\cdot 32}$$

$$= \frac{\text{load in kN} \times \text{span in metres}}{1\cdot 32}$$

See Chapter 16 for the employment of these rules.

Examples

(In the following examples, the self-weights of the steel beams are not taken into account. In practice an allowance may be made, if required—usually the self-weight for an uncased simple beam is small compared with the load carried.)

Example 1 A 254 × 146 × 31 kg UB has a value of Z_{XX} equal to 352 cm³ (see tables). Calculate the safe U.D. load for this beam, for an effective span of 4 m. Use a working stress of 165 N/mm².

(The reader will note that the XX axis is always at right angles to—and the YY axis always parallel to—the direction of the web, in standard sections.)

118 *Structural Steelwork*

$$M = fZ$$

$$\frac{Wl}{8} = fZ$$

$$\frac{W \times 4 \times 1000}{8} = 165 \times 352 \times 1000 \text{ N}$$

$$\therefore W = 116 \cdot 16 \text{ kN}$$

This value may be checked by inspection of the table of safe distributed loads given on page 130.

Example 2 A steel joist of 3 m effective span is required to carry a central load of 59 kN. Select a suitable section from the section tables.

$$f = 165 \text{ N/mm}^2$$

$$M = fZ$$

$$\frac{Wl}{4} = fZ$$

$$\frac{59 \times 3 \times 1000}{4} = 165 \, Z$$

$$\therefore Z = \frac{59 \times 3 \times 1000}{4 \times 165} \text{ cm}^3$$

$$= 268 \cdot 2 \text{ cm}^3$$

The nearest value given in the tables, under the heading *Elastic Modulus* (axis xx), and not less than the required figure, is 278·5 cm³, which corresponds to a 203 × 133 × 30 kg UB. However, a 254 × 102 × 28 kg UB has Z_{xx} of 307·6 cm³, with a saving of 2 kg per metre. The tabular load for this section, for 3 m, is given as 135 kN (U.D.). This is equivalent to 68 kN as a central load.

Example 3 An opening 3·6 m in the clear is to be made in a well-bonded brick wall 337·5 mm thick. The lintel beam for the opening, in addition to carrying the brickwork loading, has to support floor joists which transmit a uniform load of total value 60 kN. Taking the detail given in fig. 6.11, design the lintel. $f = 165 \text{ N/mm}^2$.

Design of Simple Beams. Moment of Resistance

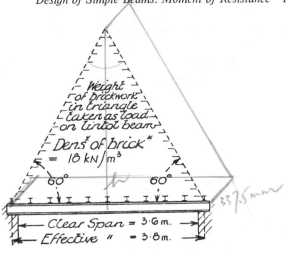

Fig. 6.11 *Beam carrying a brick wall.*

Provided the brickwork is well bonded and extends at least half the span to each side of the opening, it is usual to take the brickwork loading on a lintel as that contained in an equilateral triangle standing on the effective span as base. This load is then treated as a uniformly distributed (U.D.) load. Any loads immediately above the opening, such as floor or roof loads must be treated as additional load.

In the example, allowing 100 mm bearing, at each end, the effective span = 3·8 m.

The weight of brickwork in the equilateral triangle

$$= \left\{ \frac{3 \cdot 8}{2} \times \left(\frac{\sqrt{3}}{2} \times 3 \cdot 8 \right) \times 0 \cdot 3375 \times 18 \right\} \text{ kN}$$

$$= 38 \text{ kN}$$

Total U.D. load = (38+60) kN = 98 kN.

$$M = fZ.$$

In cases such as this, the wall cannot be considered as giving *lateral* support to the beam and the beam must be examined in accordance with the provisions of Clause 19a 2 (see page 26). If two beams are used in this particular case, it might be assumed that the 'needles' do, in fact connect the compression flanges together and therefore that the two beams do not act independently. However, to illustrate the method to be used and the application of Clauses 19a (2) and 26, they will be assumed independent.

Assuming two $178 \times 102 \times 21.54$ kg joists,

$$D/T \text{ for each beam} = 19.8 \text{ (see tables)}$$

$$\frac{l}{r_y} = \frac{0.7 \times 3.8 \times 100}{2.25} = 118$$

The effective length ratio 0·7 is taken in accordance with Clause 26a (3), the beams being built into walls and therefore fully restrained.

From Table 3a (in BS 449) it will be found, by interpolation, that the allowable stress is 150 N/mm².

$$\therefore \frac{98 \times 3.8 \times 1000}{8} = 150 \times Z$$

$$Z = \frac{98 \times 3.8 \times 1000}{8 \times 150} \text{ cm}^3 = 310 \text{ cm}^3$$

Assuming two joists, each should have a Z value of at least 155 cm³.

Two $178 \times 102 \times 21.54$ kg joists would provide (see tables) 2×170.9 cm³ $= 341.8$ cm³, hence these will be suitable.

Example 4 Fig. 6.12 shows a floor supported by a plate girder, with UB's as secondary beams. The floor is constructed of concrete with

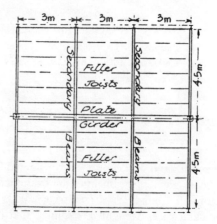

Fig. 6.12 *Steel floor frame.*

fillers. Assuming the inclusive floor load at 8 kN/m² find the necessary section modulus for one of the UB's, and also for the girder (allowing 12 kN for its self-weight). Take $f = 165$ N/mm² for the secondary beams and 155 N/mm² for the plate girder.

Area of floor supported by one UB = $3 \text{ m} \times 4 \cdot 5 \text{ m} = 13 \cdot 5 \text{ m}^2$.
Total carried = $(13 \cdot 5 \times 8)$ kN = 108 kN.

$$\frac{Wl}{8} = fZ$$

$$\frac{108 \times 4 \cdot 5 \times 1000}{8} = 165 \times Z$$

$$\therefore Z = \frac{108 \times 4 \cdot 5 \times 1000}{8 \times 165} \text{ cm}^3 = 368 \text{ cm}^3$$

A $305 \times 102 \times 33$ kg U.B. has a section modulus of 433 cm³, and hence will be suitable. The maximum B.M. for the reaction loads due to the secondary beams in the case of the plate girder will be:

$$(108 \text{ kN} \times 4 \cdot 5 \text{ m}) - (108 \text{ kN} \times 1 \cdot 5 \text{ m})$$

$$= 324 \text{ kNm}$$

Due to self-weight of girder, the B.M. maximum will be

$$\frac{Wl}{8} = \frac{12 \times 9}{8} \text{ kNm} = 13 \cdot 5 \text{ kNm}$$

Total B.M. maximum = 337·5 kNm

$$Z = \frac{M}{f} = \frac{337 \cdot 5 \times 1000}{155} \text{ cm}^3$$

$\therefore$ Necessary section modulus = 2177 cm³.

Safe-load tables for beams

The tables on pages 122 to 142 inclusive are reproduced by permission of the British Constructional Steelwork Association and CONSTRADO.

The tabulated loads on pages 128, 130, 132, 134, 139, and 141 have been calculated on the assumptions that the beams are simply supported, the loads are uniformly distributed *and include the self weights*

UNIVERSAL BEAMS

DIMENSIONS AND PROPERTIES

Serial Size	Mass per metre	Depth of Section D	Width of Section B	Thickness Web t	Thickness Flange T	Root Radius r	Depth between Fillets d	Area of Section
mm	kg	mm	mm	mm	mm	mm	mm	cm²
914 × 419	388	920.5	420.5	21.5	36.6	24.1	791.5	493.9
	343	911.4	418.5	19.4	32.0	24.1	791.5	436.9
914 × 305	289	926.6	307.8	19.6	32.0	19.1	819.2	368.5
	253	918.5	305.5	17.3	27.9	19.1	819.2	322.5
	224	910.3	304.1	15.9	23.9	19.1	819.2	284.9
	201	903.0	303.4	15.2	20.2	19.1	819.2	256.1
838 × 292	226	850.9	293.8	16.1	26.8	17.8	756.4	288.4
	194	840.7	292.4	14.7	21.7	17.8	756.4	246.9
	176	834.9	291.6	14.0	18.8	17.8	756.4	223.8
762 × 267	197	769.6	268.0	15.6	25.4	16.5	681.2	250.5
	173	762.0	266.7	14.3	21.6	16.5	681.2	220.2
	147	753.9	265.3	12.9	17.5	16.5	681.2	187.8
686 × 254	170	692.9	255.8	14.5	23.7	15.2	610.6	216.3
	152	687.6	254.5	13.2	21.0	15.2	610.6	193.6
	140	683.5	253.7	12.4	19.0	15.2	610.6	178.4
	125	677.9	253.0	11.7	16.2	15.2	610.6	159.4
610 × 305	238	633.0	311.5	18.6	31.4	16.5	531.6	303.5
	179	617.5	307.0	14.1	23.6	16.5	531.6	227.7
	149	609.6	304.8	11.9	19.7	16.5	531.6	189.9
610 × 229	140	617.0	230.1	13.1	22.1	12.7	543.1	178.2
	125	611.9	229.0	11.9	19.6	12.7	543.1	159.4
	113	607.3	228.2	11.2	17.3	12.7	543.1	144.3
	101	602.2	227.6	10.6	14.8	12.7	543.1	129.0
610 × 178	91	602.5	178.4	10.6	15.0	12.7	547.1	115.9
	82	598.2	177.8	10.1	12.8	12.7	547.1	104.4
533 × 330	212	545.1	333.6	16.7	27.8	16.5	450.1	269.6
	189	539.5	331.7	14.9	25.0	16.5	450.1	241.2
	167	533.4	330.2	13.4	22.0	16.5	450.1	212.7
533 × 210	122	544.6	211.9	12.8	21.3	12.7	472.7	155.6
	109	539.5	210.7	11.6	18.8	12.7	472.7	138.4
	101	536.7	210.1	10.9	17.4	12.7	472.7	129.1
	92	533.1	209.3	10.2	15.6	12.7	472.7	117.6
	82	528.3	208.7	9.6	13.2	12.7	472.7	104.3
533 × 165	73	528.8	165.6	9.3	13.5	12.7	476.5	93.0
	66	524.8	165.1	8.8	11.5	12.7	476.5	83.6
457 × 191	98	467.4	192.8	11.4	19.6	10.2	404.4	125.2
	89	463.6	192.0	10.6	17.7	10.2	404.4	113.8
	82	460.2	191.3	9.9	16.0	10.2	404.4	104.4
	74	457.2	190.5	9.1	14.5	10.2	404.4	94.9
	67	453.6	189.9	8.5	12.7	10.2	404.4	85.4

Note: These tables are based on Universal Beams with the flange taper shown on page 82. Universal Beams with parallel flanges have properties at least equal to the values given. Both Taper and Parallel Flange Beams comply with the requirements of the British Standard 4: Part 1 : 1971 and are interchangable.

Tables reproduced by permission and courtesy of

UNIVERSAL BEAMS

DIMENSIONS AND PROPERTIES

Serial Size	Moment of Inertia			Radius of Gyration		Elastic Modulus		Ratio $\frac{D}{T}$
	Axis x–x		Axis y–y	Axis x–x	Axis y–y	Axis x–x	Axis y–y	
	Gross	Net						
mm	cm⁴	cm⁴	cm⁴	cm	cm	cm³	cm³	
914 × 419	717325	639177	42481	38.1	9.27	15586	2021	25.2
	623866	555835	36251	37.8	9.11	13691	1733	28.5
914 × 305	503781	469903	14793	37.0	6.34	10874	961.3	29.0
	435796	406504	12512	36.8	6.23	9490	819.2	32.9
	375111	350209	10425	36.3	6.05	8241	685.6	38.1
	324715	303783	8632	35.6	5.81	7192	569.1	44.7
838 × 292	339130	315153	10661	34.3	6.08	7971	725.9	31.8
	278833	259625	8384	33.6	5.83	6633	573.6	38.7
	245412	228867	7111	33.1	5.64	5879	487.6	44.4
762 × 267	239464	221138	7699	30.9	5.54	6223	574.6	30.3
	204747	189341	6376	30.5	5.38	5374	478.1	35.3
	168535	156213	5002	30.0	5.16	4471	377.1	43.1
686 × 254	169843	156106	6225	28.0	5.36	4902	486.8	29.2
	150015	137965	5391	27.8	5.28	4364	423.7	32.7
	135972	125156	4789	27.6	5.18	3979	377.5	36.0
	117700	108580	3992	27.2	5.00	3472	315.5	41.8
610 × 305	207252	192203	14973	26.1	7.02	6549	961.3	20.2
	151312	140269	10571	25.8	6.81	4901	688.6	26.2
	124341	115233	8471	25.6	6.68	4079	555.9	30.9
610 × 229	111673	101699	4253	25.0	4.88	3620	369.6	27.9
	98408	89675	3676	24.8	4.80	3217	321.1	31.2
	87260	79645	3184	24.6	4.70	2874	279.1	35.1
	75549	69132	2658	24.2	4.54	2509	233.6	40.7
610 × 178	63970	57238	1427	23.5	3.51	2124	160.0	40.2
	55779	50076	1203	23.1	3.39	1865	135.3	46.7
533 × 330	141682	121777	16064	22.9	7.72	5199	963.2	19.6
	125618	107882	14093	22.8	7.64	4657	849.6	21.6
	109109	93647	12057	22.6	7.53	4091	730.3	24.2
533 × 210	76078	68719	3208	22.1	4.54	2794	302.8	25.6
	66610	60218	2755	21.9	4.46	2469	261.5	28.7
	61530	55671	2512	21.8	4.41	2293*	239.2	30.8
	55225	50040	2212	21.7	4.34	2072	211.3	34.2
	47363	43062	1826	21.3	4.18	1793	175.0	40.0
533 × 165	40414	35752	1027	20.8	3.32	1528	124.1	39.2
	35083	31144	863	20.5	3.21	1337	104.5	45.6
457 × 191	45653	40469	2216	19.1	4.21	1954	229.9	23.8
	40956	36313	1960	19.0	4.15	1767	204.2	26.2
	37039	32869	1746	18.8	4.09	1610	182.6	28.8
	33324	29570	1547	18.7	4.04	1458	162.4	31.5
	29337	26072	1328	18.5	3.95	1293	139.9	35.7

Note: One hole is deducted from each flange under 300mm wide (serial size) and two holes from each flange 300mm and over (serial size), in calculating the Net Moment of Inertia about x–x.

the British Constructional Steelwork Association.

UNIVERSAL BEAMS

DIMENSIONS AND PROPERTIES

Serial Size	Mass per metre	Depth of Section D	Width of Section B	Thickness		Root Radius r	Depth between Fillets d	Area of Section
				Web t	Flange T			
mm	kg	mm	mm	mm	mm	mm	mm	cm²
457 × 152	82	465.1	153.5	10.7	18.9	10.2	404.4	104.4
	74	461.3	152.7	9.9	17.0	10.2	404.4	94.9
	67	457.2	151.9	9.1	15.0	10.2	404.4	85.3
	60	454.7	152.9	8.0	13.3	10.2	407.7	75.9
	52	449.8	152.4	7.6	10.9	10.2	407.7	66.5
406 × 178	74	412.8	179.7	9.7	16.0	10.2	357.4	94.9
	67	409.4	178.8	8.8	14.3	10.2	357.4	85.4
	60	406.4	177.8	7.8	12.8	10.2	357.4	76.1
	54	402.6	177.6	7.6	10.9	10.2	357.4	68.3
406 × 152	74	416.3	153.7	10.1	18.1	10.2	357.4	94.8
	67	412.2	152.9	9.3	16.0	10.2	357.4	85.3
	60	407.9	152.2	8.6	13.9	10.2	357.4	75.8
406 × 140	46	402.3	142.4	6.9	11.2	10.2	357.4	58.9
	39	397.3	141.8	6.3	3.6	10.2	357.4	49.3
381 × 152	67	388.6	154.3	9.7	16.3	10.2	333.2	85.4
	60	384.8	153.4	8.7	14.4	10.2	333.2	75.9
	52	381.0	152.4	7.8	12.4	10.2	333.2	66.4
356 × 171	67	364.0	173.2	9.1	15.7	10.2	309.1	85.3
	57	358.6	172.1	8.0	13.0	10.2	309.1	72.1
	51	355.6	171.5	7.3	11.5	10.2	309.1	64.5
	45	352.0	171.0	6.9	9.7	10.2	309.1	56.9
356 × 127	39	352.8	126.0	6.5	10.7	10.2	309.1	49.3
	33	348.5	125.4	5.9	8.5	10.2	309.1	41.7
305 × 165	54	310.9	166.8	7.7	13.7	8.9	262.6	68.3
	46	307.1	165.7	6.7	11.8	8.9	262.6	58.8
	40	303.8	165.1	6.1	10.2	8.9	262.6	51.4
305 × 127	48	310.4	125.2	8.9	14.0	8.9	262.6	60.8
	42	306.6	124.3	8.0	12.1	8.9	262.6	53.1
	37	303.8	123.5	7.2	10.7	8.9	262.6	47.4
305 × 102	33	312.7	102.4	6.6	10.8	7.6	275.3	41.8
	28	308.9	101.9	6.1	8.9	7.6	275.3	36.3
	25	304.8	101.6	5.8	6.8	7.6	275.3	31.4
254 × 146	43	259.6	147.3	7.3	12.7	7.6	216.2	55.0
	37	256.0	146.4	6.4	10.9	7.6	216.2	47.4
	31	251.5	146.1	6.1	8.6	7.6	216.2	39.9
254 × 102	28	260.4	102.1	6.4	10.0	7.6	224.5	36.2
	25	257.0	101.9	6.1	8.4	7.6	224.5	32.1
	22	254.0	101.6	5.8	6.8	7.6	224.5	28.4
203 × 133	30	206.8	133.8	6.3	9.6	7.6	169.9	38.0
	25	203.2	133.4	5.8	7.8	7.6	169.9	32.3

Note: These tables are based on Universal Beams with the flange taper shown on page 82. Universal Beams with parallel flanges have properties at least equal to the values given. Both Taper and Parallel Flange Beams comply with the requirements of the British Standard 4: Part 1 : 1971 and are interchangable.

Tables reproduced by permission and courtesy of

UNIVERSAL BEAMS

DIMENSIONS AND PROPERTIES

Serial Size	Moment of Inertia			Radius of Gyration		Elastic Modulus		Ratio $\dfrac{D}{T}$
	Axis x–x		Axis y–y	Axis x–x	Axis y–y	Axis x–x	Axis y–y	
	Gross	Net						
mm	cm⁴	cm⁴	cm⁴	cm	cm	cm³	cm³	
457 × 152	36160	32058	1093	18.6	3.24	1555	142.5	24.6
	32380	28731	963	18.5	3.18	1404	126.1	27.1
	28522	25342	829	18.3	3.12	1248	109.1	30.5
	25464	22613	794	18.3	3.23	1120	104.0	34.2
	21345	19034	645	17.9	3.11	949.0	84.61	41.3
406 × 178	27279	23981	1448	17.0	3.91	1322	161.2	25.8
	24279	21357	1269	16.9	3.85	1186	141.9	28.6
	21520	18928	1108	16.8	3.82	1059	124.7	31.8
	18576	16389	922	16.5	3.67	922.8	103.8	36.9
406 × 152	26938	23811	1047	16.9	3.32	1294	136.2	23.0
	23798	21069	908	16.7	3.26	1155	118.8	25.8
	20619	18283	768	16.5	3.18	1011	100.9	29.3
406 × 140	15603	13699	500	16.3	2.92	775.6	70.26	35.9
	12408	10963	373	15.9	2.75	624.7	52.61	46.2
381 × 152	21276	18817	947	15.8	3.33	1095	122.7	23.8
	18632	16489	814	15.7	3.27	968.4	106.2	26.7
	16046	14226	685	15.5	3.21	842.3	89.96	30.7
356 × 171	19483	17002	1278	15.1	3.87	1071	147.6	23.2
	16038	14018	1026	14.9	3.77	894.3	119.2	27.6
	14118	12349	885	14.8	3.71	794.0	103.3	30.9
	12052	10578	730	14.6	3.58	684.7	85.39	36.3
356 × 127	10054	8688	333	14.3	2.60	570.0	52.87	33.0
	8167	7099	257	14.0	2.48	468.7	40.99	41.0
305 × 165	11686	10119	988	13.1	3.80	751.8	118.5	22.7
	9924	8596	825	13.0	3.74	646.4	99.54	26.0
	8500	7368	691	12.9	3.67	559.6	83.71	29.8
305 × 127	9485	8137	438	12.5	2.68	611.1	69.94	22.2
	8124	6978	367	12.4	2.63	530.0	58.99	25.3
	7143	6142	316	12.3	2.58	470.3	51.11	28.4
305 × 102	6482	5792	189	12.5	2.13	414.6	37.00	29.0
	5415	4855	153	12.2	2.05	350.7	30.01	34.7
	4381	3959	116	11.8	1.92	287.5	22.85	44.8
254 × 146	6546	5683	633	10.9	3.39	504.3	85.97	20.4
	5544	4814	528	10.8	3.34	433.1	72.11	23.5
	4427	3859	406	10.5	3.19	352.1	55.53	29.2
254 × 102	4004	3565	174	10.5	2.19	307.6	34.13	26.0
	3404	3041	144	10.3	2.11	264.9	28.23	30.6
	2863	2572	116	10.0	2.02	225.4	22.84	37.4
203 × 133	2880	2469	354	8.71	3.05	278.5	52.85	21.5
	2348	2020	280	8.53	2.94	231.1	41.92	26.1

Note: One hole is deducted from each flange under 300mm wide (serial size) and two holes from each flange 300mm and over (serial size), in calculating the Net Moment of Inertia about x–x.

the British Constructional Steelwork Association.

UNIVERSAL COLUMNS
Parallel Flanges

DIMENSIONS AND PROPERTIES

Serial Size	Mass per metre	Depth of Section D	Width of Section B	Thickness		Root Radius r	Depth between Fillets d	Area of Section
				Web t	Flange T			
mm	kg	mm	mm	mm	mm	mm	mm	cm²
356 × 406	634	474.7	424.1	47.6	77.0	15.2	290.1	808.1
	551	455.7	418.5	42.0	67.5	15.2	290.1	701.8
	467	436.6	412.4	35.9	58.0	15.2	290.1	595.5
	393	419.1	407.0	30.6	49.2	15.2	290.1	500.9
	340	406.4	403.0	26.5	42.9	15.2	290.1	432.7
	287	393.7	399.0	22.6	36.5	15.2	290.1	366.0
	235	381.0	395.0	18.5	30.2	15.2	290.1	299.8
Column Core	477	427.0	424.4	48.0	53.2	15.2	290.1	607.2
356 × 368	202	374.7	374.4	16.8	27.0	15.2	290.1	257.9
	177	368.3	372.1	14.5	23.8	15.2	290.1	225.7
	153	362.0	370.2	12.6	20.7	15.2	290.1	195.2
	129	355.6	368.3	10.7	17.5	15.2	290.1	164.9
305 × 305	283	365.3	321.8	26.9	44.1	15.2	246.6	360.4
	240	352.6	317.9	23.0	37.7	15.2	246.6	305.6
	198	339.9	314.1	19.2	31.4	15.2	246.6	252.3
	158	327.2	310.6	15.7	25.0	15.2	246.6	201.2
	137	320.5	308.7	13.8	21.7	15.2	246.6	174.6
	118	314.5	306.8	11.9	18.7	15.2	246.6	149.8
	97	307.8	304.8	9.9	15.4	15.2	246.6	123.3
254 × 254	167	289.1	264.5	19.2	31.7	12.7	200.2	212.4
	132	276.4	261.0	15.6	25.1	12.7	200.2	167.7
	107	266.7	258.3	13.0	20.5	12.7	200.2	136.6
	89	260.4	255.9	10.5	17.3	12.7	200.2	114.0
	73	254.0	254.0	8.6	14.2	12.7	200.2	92.9
203 × 203	86	222.3	208.8	13.0	20.5	10.2	160.8	110.1
	71	215.9	206.2	10.3	17.3	10.2	160.8	91.1
	60	209.6	205.2	9.3	14.2	10.2	160.8	75.8
	52	206.2	203.9	8.0	12.5	10.2	160.8	66.4
	46	203.2	203.2	7.3	11.0	10.2	160.8	58.8
152 × 152	37	161.8	154.4	8.1	11.5	7.6	123.4	47.4
	30	157.5	152.9	6.6	9.4	7.6	123.4	38.2
	23	152.4	152.4	6.1	6.8	7.6	123.4	29.8

Tables reproduced by permission and courtesy of

UNIVERSAL COLUMNS
Parallel Flanges
DIMENSIONS AND PROPERTIES

Serial Size	Moment of Inertia			Radius of Gyration		Elastic Modulus		Ratio $\dfrac{D}{T}$
	Axis x–x		Axis y–y	Axis x–x	Axis y–y	Axis x–x	Axis y–y	
	Gross	Net						
mm	cm⁴	cm⁴	cm⁴	cm	cm	cm³	cm³	
356 × 406	275140	243076	98211	18.5	11.0	11592	4632	6.2
	227023	200312	82665	18.0	10.9	9964	3951	6.8
	183118	161331	67905	17.5	10.7	8388	3293	7.5
	146765	129159	55410	17.1	10.5	7004	2723	8.5
	122474	107667	46816	16.8	10.4	6027	2324	9.5
	99994	87843	38714	16.5	10.3	5080	1940	10.8
	79110	69424	31008	16.2	10.2	4153	1570	12.6
Column Core	172391	152936	68057	16.8	10.6	8075	3207	8.0
356 × 368	66307	57806	23632	16.0	9.57	3540	1262	13.9
	57153	49798	20470	15.9	9.52	3104	1100	15.5
	48525	42250	17470	15.8	9.46	2681	943.8	17.5
	40246	35040	14555	15.6	9.39	2264	790.4	20.3
305 × 305	78777	72827	24545	14.8	8.25	4314	1525	8.3
	64177	59295	20239	14.5	8.14	3641	1273	9.4
	50832	46935	16230	14.2	8.02	2991	1034	10.8
	38740	35766	12524	13.9	7.89	2368	806.3	13.1
	32838	30314	10672	13.7	7.82	2049	691.4	14.8
	27601	25472	9006	13.6	7.75	1755	587.0	16.8
	22202	20488	7268	13.4	7.68	1442	476.9	20.0
254 × 254	29914	27171	9796	11.9	6.79	2070	740.6	9.1
	22416	20350	7444	11.6	6.66	1622	570.4	11.0
	17510	15890	5901	11.3	6.57	1313	456.9	13.0
	14307	12976	4849	11.2	6.52	1099	378.9	15.1
	11360	10297	3873	11.1	6.46	894.5	305.0	17.9
203 × 203	9462	8374	3119	9.27	5.32	851.5	298.7	10.8
	7647	6758	2536	9.16	5.28	708.4	246.0	12.5
	6088	5383	2041	8.96	5.19	581.1	199.0	14.8
	5263	4653	1770	8.90	5.16	510.4	173.6	16.5
	4564	4035	1539	8.81	5.11	449.2	151.5	18.5
152 × 152	2218	1932	709	6.84	3.87	274.2	91.78	14.1
	1742	1515	558	6.75	3.82	221.2	73.06	16.8
	1263	1104	403	6.51	3.68	165.7	52.95	22.4

Note: One hole is deducted from each flange under 300mm wide (serial size) and two holes from each flange 300mm and over (serial size), in calculating the Net Moment of Inertia about x–x.

the British Constructional Steelwork Association.

UNIVERSAL BEAMS

SAFE LOADS FOR GRADE 43 STEEL

BASED ON BS 449 1969

Serial Size mm	Mass per metre in kg	SAFE DISTRIBUTED LOADS IN KILONEWTONS FOR SPANS IN METRES AND DEFLECTION COEFFICIENTS															
		4.00	5.00	6.00	7.00	8.00	9.00	10.00	11.00	12.00	13.00	14.00	15.00	16.00	18.00	20.00
		28.00	17.92	12.44	9.143	7.000	5.531	4.480	3.702	3.111	2.651	2.286	1.991	1.750	1.383	1.120
914 × 419	388		†3958	3429	2939	2572	2286	2057	1870	1714	1583	1470	1372	1286	1143	1029
	343		†3536	3012	2582	2259	2008	1807	1643	1506	1390	1291	1205	1130	1004	904
914 × 305	289	3588	2871	2392	2050	1794	1595	1435	1305	1196	1104	1025	957	897	797	718
	253	3132	2505	2088	1789	1566	1392	1253	1139	1044	964	895	835	783	696	626
	224	2720	2176	1813	1554	1360	1209	1088	989	907	837	777	725	680	604	544
	201	2373	1899	1582	1356	1187	1055	949	863	791	730	678	633	593	527	475
838 × 292	226	2630	2104	1754	1503	1315	1169	1052	957	877	809	752	701	658	585	526
	194	2189	1751	1459	1251	1094	973	876	796	730	674	625	584	547	486	438
	176	1940	1552	1293	1109	970	862	776	705	647	597	554	517	485	431	388
762 × 267	197	2054	1643	1369	1173	1027	913	821	747	685	632	587	548	513	456	411
	173	1773	1419	1182	1013	887	788	709	645	591	546	507	473	443	394	355
	147	1475	1180	984	843	738	656	590	537	492	454	422	393	369	328	295
686 × 254	170	1618	1294	1079	924	809	719	647	588	539	498	462	431	404	360	324
	152	1440	1152	960	823	720	640	576	524	480	443	411	384	360	320	288
	140	1313	1050	875	750	656	584	525	477	438	404	375	350	328	292	263
	125	1146	917	764	655	573	509	458	417	382	353	327	306	286	255	229
610 × 305	238	2161	1729	1441	1235	1081	960	864	786	720	665	617	576	540	480	432
	179	1617	1294	1078	924	809	719	647	588	539	498	462	431	404	359	323
	149	1346	1077	897	769	673	598	538	490	449	414	385	359	337	299	269
610 × 229	140	1195	956	796	683	597	531	478	434	398	368	341	319	299	265	239
	125	1061	849	708	607	531	472	425	386	354	327	303	283	265	236	212
	113	948	759	632	542	474	421	379	345	316	292	271	253	237	211	190
	101	828	662	552	473	414	368	331	301	276	255	237	221	207	184	166
610 × 178	91	701	561	467	400	350	311	280	255	234	216	200	187	175	156	140
	82	615	492	410	352	308	274	246	224	205	189	176	164	154	137	123
533 × 330	212	1716	1372	1144	980	858	762	686	624	572	528	490	457	429	381	
	189	1537	1229	1025	878	768	683	615	559	512	473	439	410	384	342	
	167	1350	1080	900	771	675	600	540	491	450	415	386	360	338	300	
533 × 210	122	922	738	615	527	461	410	369	335	307	284	263	246	231	205	
	109	815	652	543	466	407	362	326	296	272	251	233	217	204	181	
	101	757	605	504	432	378	336	303	275	252	233	216	202	189	168	
	92	684	547	456	391	342	304	273	249	228	210	195	182	171	152	
	82	592	473	394	338	296	263	237	215	197	182	169	158	148		
533 × 165	73	504	404	336	288	252	224	202	183	168	155	144	135	126		
	66	441	353	294	252	221	196	176	160	147	136	126	118	110		
457 × 191	98	645	516	430	368	322	287	258	234	215	198	184	172			
	89	583	467	389	333	292	259	233	212	194	179	167	156			
	82	531	425	354	304	266	236	212	193	177	163	152	142			
	74	481	385	321	275	241	214	192	175	160	148	137	128			
	67	427	341	285	244	213	190	171	155	142	131	122	114			

Generally, tabular loads are based on a flexural stress of $165 N/mm^2$, assuming adequate lateral support. Beams without adequate lateral support must not exceed the critical span Lc. unless the allowable compressive stress is reduced in accordance with clause 19.a.(ii) of BS 449 : 1969.
Tabular loads printed in bold face type exceed the buckling capacity of the unstiffened web without allowance for actual length of bearing: the load bearing capacity should be checked, see page 109.
Tabular loads marked thus † are based on the maximum shear value of the web and are less than the permissible flexural load.

Tables reproduced by permission and courtesy of

BASED ON
BS 449
1969

UNIVERSAL BEAMS
DIMENSIONS AND PROPERTIES
GRADE 43 STEEL

Actual Size D×B mm	Critical Span Lc metres	Area of section cm²	Moment of Inertia		Radius of Gyration	Elastic Modulus		Ratio D/T
			Axis x—x cm⁴	Axis y—y cm⁴	Axis y—y cm	Axis x—x cm³	Axis y—y cm³	
920.5 × 420.5	8.610	493.9	717325	42481	9.27	15586	2021	25.2
911.4 × 418.5	8.456	436.9	623866	36251	9.11	13691	1733	28.5
926.6 × 307.8	5.882	368.5	503781	14793	6.34	10874	961.3	29.0
918.5 × 305.5	5.783	322.5	435796	12512	6.23	9490	819.2	32.9
910.3 × 304.1	5.615	284.9	375111	10425	6.05	8241	685.6	38.1
903.0 × 303.4	5.390	256.1	324715	8632	5.81	7192	569.1	44.7
850.9 × 293.8	5.644	288.4	339130	10661	6.08	7971	725.9	31.8
840.7 × 292.4	5.410	246.9	278833	8384	5.83	6633	573.6	38.7
834.9 × 291.6	5.232	223.8	245412	7111	5.64	5879	487.6	44.4
769.6 × 268.0	5.147	250.5	239464	7699	5.54	6223	574.6	30.3
762.0 × 266.7	4.995	220.2	204747	6376	5.38	5374	478.1	35.3
753.9 × 265.3	4.791	187.8	168535	5002	5.16	4471	377.1	43.1
692.9 × 255.8	4.980	216.3	169843	6225	5.36	4902	486.8	29.2
687.6 × 254.5	4.898	193.6	150015	5391	5.28	4364	423.7	32.7
683.5 × 253.7	4.810	178.4	135972	4789	5.18	3979	377.5	36.0
677.9 × 253.0	4.645	159.4	117700	3992	5.00	3472	315.5	41.8
633.0 × 311.5	6.982	303.5	207252	14973	7.02	6549	961.3	20.2
617.5 × 307.0	6.326	227.7	151312	10571	6.81	4901	688.6	26.2
609.6 × 304.8	6.201	189.9	124341	8471	6.68	4079	555.9	30.9
617.0 × 230.1	4.535	178.2	111673	4253	4.88	3620	369.6	27.9
611.9 × 229.0	4.458	159.4	98408	3676	4.80	3217	321.1	31.2
607.3 × 228.2	4.361	144.3	87260	3184	4.70	2874	279.1	35.1
602.2 × 227.6	4.213	129.0	75549	2658	4.54	2509	233.6	40.7
602.5 × 178.4	3.258	115.9	63970	1427	3.51	2124	160.0	40.2
598.2 × 177.8	3.151	104.4	55779	1203	3.39	1865	135.3	46.7
545.1 × 333.6	7.757	269.6	141682	16064	7.72	5199	963.2	19.6
539.5 × 331.7	7.414	241.2	125618	14093	7.64	4657	849.6	21.6
533.4 × 330.2	7.047	212.7	109109	12057	7.53	4091	730.3	24.2
544.6 × 211.9	4.215	155.6	76078	3208	4.54	2794	302.8	25.6
539.5 × 210.7	4.141	138.4	66610	2755	4.46	2469	261.5	28.7
536.7 × 210.1	4.095	129.1	61530	2512	4.41	2293	239.2	30.8
533.1 × 209.3	4.026	117.6	55225	2212	4.34	2072	211.3	34.2
528.3 × 208.7	3.884	104.3	47363	1826	4.13	1793	175.0	40.0
528.8 × 165.6	3.085	93.0	40414	1027	3.32	1528	124.1	39.2
524.8 × 165.1	2.983	83.6	35083	863	3.21	1337	104.5	45.6
467.4 × 192.8	3.957	125.2	45653	2216	4.21	1954	229.9	23.8
463.6 × 192.0	3.853	113.8	40956	1960	4.15	1767	204.2	26.2
460.2 × 191.3	3.796	104.4	37039	1746	4.09	1610	182.6	28.8
457.2 × 190.5	3.749	94.9	33324	1547	4.04	1458	162.4	31.5
453.6 × 189.9	3.663	85.4	29337	1328	3.95	1293	139.9	35.7

Tabular loads printed in italic type are within the web buckling capacity of the unstiffened web and produce a total deflection not exceeding 1/360th of the span.
Tabular loads printed in ordinary type should be checked for deflection. see page 110.
For explanation of tables, see notes commencing page 102.
1 kilonewton may be taken as 0.102 metric tonne (megagramme) force. but see page 102.

the British Constructional Steelwork Association.

UNIVERSAL BEAMS

SAFE LOADS FOR GRADE 43 STEEL

BASED ON BS 449 1969

Serial Size mm	Mass per metre in kg	SAFE DISTRIBUTED LOADS IN KILONEWTONS FOR SPANS IN METRES AND DEFLECTION COEFFICIENTS														
		2.00	2.50	3.00	3.50	4.00	4.50	5.00	5.50	6.00	6.50	7.00	7.50	8.00	9.00	10.00
		112.0	71.68	49.78	36.57	28.00	22.12	17.92	14.81	12.44	10.60	9.143	7.964	7.000	5.531	4.480
457 × 152	82	†995	821	684	586	513	456	411	373	342	316	293	274	257	228	205
	74	†913	741	618	529	463	412	371	337	309	285	265	247	232	206	185
	67	823	659	549	471	412	366	329	299	274	253	235	220	206	183	165
	60	†728	591	493	422	370	329	296	269	246	227	211	197	185	164	148
	52	626	501	418	358	313	278	251	228	209	193	179	167	157	139	125
406 × 178	74	†801	698	582	499	436	388	349	317	291	268	249	233	218	194	174
	67	†721	626	522	447	391	348	313	285	261	241	224	209	196	174	157
	60	†634	559	466	399	349	311	280	254	233	215	200	186	175	155	140
	54	609	487	406	348	305	271	244	221	203	187	174	162	152	135	122
406 × 152	74	†841	683	569	488	427	380	342	311	285	263	244	228	214	190	171
	67	762	610	508	435	381	339	305	277	254	234	218	203	191	169	152
	60	667	534	445	381	334	297	267	243	222	205	191	178	167	148	133
406 × 140	46	512	410	341	293	256	228	205	186	171	158	146	137	128	114	102
	39	412	330	275	236	206	183	165	150	137	127	118	110	103	92	82
381 × 152	67	723	578	482	413	361	321	289	263	241	222	206	193	181	161	145
	60	639	511	426	365	320	284	256	232	213	197	183	170	160	142	128
	52	556	445	371	318	278	247	222	202	185	171	159	148	139	124	111
356 × 171	67	†662	565	471	404	353	314	283	257	236	217	202	188	177	157	141
	57	†574	472	394	337	295	262	236	215	197	182	169	157	148	131	118
	51	†519	419	349	299	262	233	210	191	175	161	150	140	131	116	105
	45	452	362	301	258	226	201	181	164	151	139	129	121	113	100	90
356 × 127	39	376	301	251	215	188	167	150	137	125	116	107	100	94	84	75
	33	309	247	206	177	155	137	124	112	103	95	88	82	77	69	62
305 × 165	54	†479	397	331	284	248	221	198	180	165	153	142	132	124	110	99
	46	†412	341	284	244	213	190	171	155	142	131	122	114	107	95	85
	40	369	295	246	211	185	164	148	134	123	114	106	98	92	82	74
305 × 127	48	403	323	269	230	202	179	161	147	134	124	115	108	101	90	81
	42	350	280	233	200	175	155	140	127	117	108	100	93	87	78	70
	37	310	248	207	177	155	138	124	113	103	96	89	83	78	69	62
305 × 102	33	274	219	182	156	137	122	109	100	91	84	78	73	68	61	55
	28	231	185	154	132	116	103	93	84	77	71	66	62	58	51	46
	25	190	152	126	108	95	84	76	69	63	58	54	51	47	42	38
254 × 146	43	333	266	222	190	166	148	133	121	111	102	95	89	83		
	37	286	229	191	163	143	127	114	104	95	88	82	76	71		
	31	232	186	155	133	116	103	93	84	77	71	66	62	58		
254 × 102	28	203	162	135	116	102	90	81	74	68	62	58	54	51		
	25	175	140	117	100	87	78	70	64	58	54	50	47	44		
	22	149	119	99	85	74	66	60	54	50	46	43	40	37		
203 × 133	30	184	147	123	105	92	82	74	67	61	57	53				
	25	153	122	102	87	76	68	61	55	51	47					

Generally, tabular loads are based on a flexural stress of 165N/mm², assuming adequate lateral support. Beams without adequate lateral support must not exceed the critical span Lc. unless the allowable compressive stress is reduced in accordance with clause 19.a.(ii) of BS 449 : 1969.
Tabular loads printed in bold face type exceed the buckling capacity of the unstiffened web without allowance for actual length of bearing: the load bearing capacity should be checked, see page 109.
Tabular loads marked thus † are based on the maximum shear value of the web and are less than the permissible flexural load.

Tables reproduced by permission and courtesy of

BASED ON
BS 449
1969

UNIVERSAL BEAMS

DIMENSIONS AND PROPERTIES
GRADE 43 STEEL

Actual Size D×B mm	Critical Span Lc metres	Area of section cm²	Moment of Inertia		Radius of Gyration	Elastic Modulus		Ratio D — T
			Axis x—x cm⁴	Axis y—y cm⁴	Axis y—y cm	Axis x—x cm³	Axis y—y cm³	
465.1 × 153.5	3.017	104.4	36160	1093	3.24	1555	142.5	24.6
461.3 × 152.7	2.957	94.9	32380	963	3.18	1404	126.1	27.1
457.2 × 151.9	2.894	85.3	28522	829	3.12	1248	109.1	30.5
454.7 × 152.9	3.003	75.9	25464	794	3.23	1120	104.0	34.2
449.8 × 152.4	2.891	66.5	21345	645	3.11	949.0	84.61	41.3
412.8 × 179.7	3.627	94.9	27279	1448	3.91	1322	161.2	25.8
409.4 × 178.8	3.578	85.4	24279	1269	3.85	1186	141.9	28.6
406.4 × 177.8	3.543	76.1	21520	1108	3.82	1059	124.7	31.8
402.6 × 177.6	3.410	68.3	18576	922	3.67	922.8	103.8	36.9
416.3 × 153.7	3.157	94.8	26938	1047	3.32	1294	136.2	23.0
412.2 × 152.9	3.029	85.3	23798	908	3.26	1155	118.8	25.8
407.9 × 152.2	2.954	75.8	20619	768	3.18	1011	100.9	29.3
402.3 × 142.4	2.706	58.9	15603	500	2.92	775.6	70.26	35.9
397.3 × 141.8	2.553	49.3	12408	373	2.75	624.7	52.61	46.2
388.6 × 154.3	3.131	85.4	21276	947	3.33	1095	122.7	23.8
384.8 × 153.4	3.040	75.9	18632	814	3.27	968.4	106.2	26.7
381.0 × 152.4	2.983	66.4	16046	685	3.21	842.3	89.96	30.7
364.0 × 173.2	3.670	85.3	19483	1278	3.87	1071	147.6	23.2
358.6 × 172.1	3.502	72.1	16038	1026	3.77	894.3	119.2	27.6
355.6 × 171.5	3.440	64.5	14118	885	3.71	794.0	103.3	30.9
352.0 × 171.0	3.326	56.9	12052	730	3.58	684.7	85.39	36.3
352.8 × 126.0	2.413	49.3	10054	333	2.60	570.0	52.87	33.0
348.5 × 125.4	2.303	41.7	8167	257	2.48	468.7	40.99	41.0
310.9 × 166.8	3.630	68.3	11686	988	3.80	751.8	118.5	22.7
307.1 × 165.7	3.476	58.8	9924	825	3.74	646.4	99.54	26.0
303.8 × 165.1	3.403	51.4	8500	691	3.67	559.6	83.71	29.8
310.4 × 125.2	2.581	60.8	9485	438	2.68	611.1	69.94	22.2
306.6 × 124.3	2.439	53.1	8124	367	2.63	530.0	58.99	25.3
303.8 × 123.5	2.396	47.4	7143	316	2.58	470.3	51.11	28.4
312.7 × 102.4	1.977	41.8	6482	189	2.13	414.6	37.00	29.0
308.9 × 101.9	1.905	36.3	5415	153	2.05	350.7	30.01	34.7
304.8 × 101.6	1.786	31.4	4381	116	1.92	287.5	22.85	44.8
259.6 × 147.3	3.354	55.0	6546	633	3.39	504.3	85.97	20.4
256.0 × 146.4	3.151	47.4	5544	528	3.34	433.1	72.11	23.5
251.5 × 146.1	2.958	39.9	4427	406	3.19	352.1	55.53	29.2
260.4 × 102.1	2.037	36.2	4004	174	2.19	307.6	34.13	26.0
257.0 × 101.9	1.963	32.1	3404	144	2.11	264.9	28.23	30.6
254.0 × 101.6	1.876	28.4	2863	116	2.02	225.4	22.84	37.4
206.8 × 133.8	2.963	38.0	2880	354	3.05	278.5	52.85	21.5
203.2 × 133.4	2.732	32.3	2348	280	2.94	231.1	41.92	26.1

Tabular loads printed in italic type are within the web buckling capacity of the unstiffened web and produce a total deflection not exceeding 1/360th of the span.
Tabular loads printed in ordinary type should be checked for deflection, see page 110.
For explanation of tables, see notes commencing page 102.
1 kilonewton may be taken as 0.102 metric tonne (megagramme) force, but see page 102.

the British Constructional Steelwork Association.

UNIVERSAL BEAMS

SAFE LOADS FOR GRADE 50 STEEL

BASED ON BS 449 1969

Serial Size mm	Mass per metre in kg	SAFE DISTRIBUTED LOADS IN KILONEWTONS FOR SPANS IN METRES AND DEFLECTION COEFFICIENTS															
		4.00	5.00	6.00	7.00	8.00	9.00	10.00	11.00	12.00	13.00	14.00	15.00	16.00	18.00	20.00
		28.00	17.92	12.44	9.143	7.000	5.531	4.480	3.702	3.111	2.651	2.286	1.991	1.750	1.383	1.120
914 × 419	388		†5541	4780	4097	3585	*3186*	2868	2607	2390	2206	2048	1912	1792	1593	1434
	343		†4951	4199	3599	3149	2799	*2519*	*2290*	2099	1938	1799	1679	1574	1400	1260
914 × 305	289	5002	4002	3335	2858	*2501*	*2223*	*2001*	1819	1667	1539	1429	1334	1250	1112	1000
	253	4365	3492	2910	2494	2183	*1940*	*1746*	*1587*	1455	1343	1247	1164	1091	970	873
	224	3791	3033	2527	2166	1895	1685	*1516*	*1379*	1264	1166	1083	1011	948	842	758
	201	3308	2647	2206	1891	1654	1470	*1323*	1203	1103	1018	945	882	827	735	662
838 × 292	226	3667	2933	2444	2095	1833	*1630*	*1467*	1333	1222	1128	1048	978	917	815	733
	194	3051	2441	2034	1744	1526	*1356*	*1220*	1110	1017	939	872	814	763	678	610
	176	2704	2163	1803	1545	1352	1202	*1082*	983	901	832	773	721	676	601	541
762 × 267	197	2863	2290	1908	1636	*1431*	*1272*	1145	1041	954	881	818	763	716	636	
	173	2472	1978	1648	1413	*1236*	*1099*	989	899	824	761	706	659	618	549	
	147	2057	1645	1371	1175	*1028*	914	823	748	686	633	588	548	514	457	
686 × 254	170	2255	1804	1503	*1289*	*1128*	1002	902	820	752	694	644	601	564		
	152	2007	1606	1338	*1147*	*1004*	892	803	730	669	618	573	535	502		
	140	1830	1464	1220	1046	*915*	813	732	666	610	563	523	488	458		
	125	1597	1278	1065	913	*799*	710	639	581	532	491	456	426	399		
610 × 305	238	3012	2410	*2008*	*1721*	1506	1339	1205	1095	1004	927	861	803			
	179	2254	1804	*1503*	*1288*	1127	1002	902	820	751	694	644	601			
	149	1877	1501	1251	1072	938	834	751	682	626	577	536				
610 × 229	140	1665	1332	*1110*	*952*	833	740	666	606	555	512	476	444			
	125	1480	1184	986	*845*	740	658	592	538	493	455	423				
	113	1322	1057	881	755	661	587	529	481	441	407	378				
	101	1154	923	769	659	577	513	462	420	385	355	330				
610 × 178	91	977	781	*651*	558	488	434	391	355	326	301	279				
	82	858	686	*572*	490	429	381	343	312	286	264	245				
533 × 330	212	2391	1913	*1594*	1366	1196	1063	957	870	797	736					
	189	2142	1714	*1428*	1224	1071	952	857	779	714	659					
	167	1882	1506	1255	1075	941	836	753	684	627						
533 × 210	122	1285	*1028*	857	734	643	571	514	467	428	395					
	109	1136	909	757	649	568	505	454	413	379	350					
	101	1055	844	703	603	527	469	422	384	352	325					
	92	953	762	635	545	476	424	381	347	318						
	82	825	660	550	471	412	367	330	300	275						
533 × 165	73	703	562	*469*	402	352	312	281	256	234						
	66	615	492	*410*	351	308	273	246	224	205						
457 × 191	98	899	*719*	599	514	449	399	359	327							
	89	813	*650*	542	464	406	361	325	296							
	82	740	*592*	494	423	370	329	296	269							
	74	671	*536*	447	383	335	298	268	244							
	67	595	*476*	397	340	297	264	238	216							

Generally, tabular loads are based on a flexural stress of 230N/mm², assuming adequate lateral support. Beams without adequate lateral support must not exceed the critical span Lc, unless the allowable compressive stress is reduced in accordance with clause 19.a.(ii) of BS 449 : 1969.
Tabular loads printed in bold face type exceed the buckling capacity of the unstiffened web without allowance for actual length of bearing: the load bearing capacity should be checked, see page 109.
Tabular loads marked thus + are based on the maximum shear value of the web and are less than the permissible flexural load.

Tables reproduced by permission and courtesy of

UNIVERSAL BEAMS

BASED ON BS 449 1969

DIMENSIONS AND PROPERTIES
GRADE 50 STEEL

Actual Size D×B mm	Critical Span Lc metres	Area of section cm²	Moment of Inertia Axis x–x cm⁴	Moment of Inertia Axis y–y cm⁴	Radius of Gyration Axis y–y cm	Elastic Modulus Axis x–x cm³	Elastic Modulus Axis y–y cm³	Ratio D/T
920.5 × 420.5	7.724	493.9	717325	42481	9.27	15586	2021	25.2
911.4 × 418.5	7.586	436.9	623866	36251	9.11	13691	1733	28.5
926.6 × 307.8	5.277	368.5	503781	14793	6.34	10874	961.3	29.0
918.5 × 305.5	5.188	322.5	435796	12512	6.23	9490	819.2	32.9
910.3 × 304.1	5.038	284.9	375111	10425	6.05	8241	685.6	38.1
903.0 × 303.4	4.835	256.1	324715	8632	5.81	7192	569.1	44.7
850.9 × 293.8	5.063	288.4	339130	10661	6.08	7971	725.9	31.8
840.7 × 292.4	4.854	246.9	278833	8384	5.83	6633	573.6	38.7
834.9 × 291.6	4.694	223.8	245412	7111	5.64	5879	487.6	44.4
769.6 × 268.0	4.617	250.5	239464	7699	5.54	6223	574.6	30.3
762.0 × 266.7	4.481	220.2	204747	6376	5.38	5374	478.1	35.3
753.9 × 265.3	4.298	187.8	168535	5002	5.16	4471	377.1	43.1
692.9 × 255.8	4.468	216.3	169843	6225	5.36	4902	486.8	29.2
687.6 × 254.5	4.394	193.6	150015	5391	5.28	4364	423.7	32.7
683.5 × 253.7	4.315	178.4	135972	4789	5.18	3979	377.5	36.0
677.9 × 253.0	4.168	159.4	117700	3992	5.00	3472	315.5	41.8
633.0 × 311.5	6.206	303.5	207252	14973	7.02	6549	961.3	20.2
617.5 × 307.0	5.675	227.7	151312	10571	6.81	4901	688.6	26.2
609.6 × 304.8	5.563	189.9	124341	8471	6.68	4079	555.9	30.9
617.0 × 230.1	4.068	178.2	111673	4253	4.88	3620	369.6	27.9
611.9 × 229.0	4.000	159.4	98408	3676	4.80	3217	321.1	31.2
607.3 × 228.2	3.912	144.3	87260	3184	4.70	2874	279.1	35.1
602.2 × 227.6	3.780	129.0	75549	2658	4.54	2509	233.6	40.7
602.5 × 178.4	2.923	115.6	63970	1427	3.51	2124	160.0	40.2
598.2 × 177.8	2.827	104.4	55779	1203	3.39	1865	135.3	46.7
545.1 × 333.6	6.886	269.6	141682	16064	7.72	5199	963.2	19.6
539.5 × 331.7	6.612	241.2	125618	14093	7.64	4657	849.6	21.6
533.4 × 330.2	6.315	212.7	109109	12057	7.53	4091	730.3	24.2
544.6 × 211.9	3.781	155.6	76078	3208	4.54	2794	302.8	25.6
539.5 × 210.7	3.715	138.4	66610	2755	4.46	2469	261.5	28.7
536.7 × 210.1	3.673	129.1	61530	2512	4.41	2293	239.2	30.8
533.1 × 209.3	3.612	117.6	55225	2212	4.34	2072	211.3	34.2
528.3 × 208.7	3.484	104.3	47363	1826	4.18	1793	175.0	40.0
528.8 × 165.6	2.767	93.0	40414	1027	3.32	1528	124.1	39.2
524.8 × 165.1	2.676	83.6	35083	863	3.21	1337	104.5	45.6
467.4 × 192.8	3.543	125.2	45653	2216	4.21	1954	229.9	23.8
463.6 × 192.0	3.456	113.8	40956	1960	4.15	1767	204.2	26.2
460.2 × 191.3	3.406	104.4	37039	1746	4.09	1610	182.6	28.8
457.2 × 190.5	3.363	94.9	33324	1547	4.04	1458	162.4	31.5
453.6 × 189.9	3.286	85.4	29337	1328	3.95	1293	139.9	35.7

Tabular loads printed in italic type are within the web buckling capacity of the unstiffened web and produce a total deflection not exceeding 1/360th of the span.
Tabular loads printed in ordinary type should be checked for deflection, see page 110.
For explanation of tables, see notes commencing page 102.
1 kilonewton may be taken as 0.102 metric tonne (megagramme) force, but see page 102.

the British Constructional Steelwork Association.

UNIVERSAL BEAMS

BASED ON
BS 449
1969

SAFE LOADS FOR GRADE 50 STEEL

Serial Size mm	Mass per metre in kg	SAFE DISTRIBUTED LOADS IN KILONEWTONS FOR SPANS IN METRES AND DEFLECTION COEFFICIENTS												
		2.00	2.50	3.00	3.50	4.00	4.50	5.00	5.50	6.00	6.50	7.00	7.50	8.00	9.00	10.00
		112.0	71.68	49.78	36.57	28.00	22.12	17.92	14.81	12.44	10.60	9.143	7.964	7.000	5.531	4.480
457 × 152	82	†1393	1144	954	817	715	636	572	520	477	440	409	381	358	318	286
	74	†1279	1033	861	738	646	574	517	470	431	397	369	344	323	287	258
	67	1148	918	765	656	574	510	459	417	383	353	328	306	287	255	230
	60	†1019	824	687	589	515	458	412	375	344	317	294	275	258	229	206
	52	873	698	582	499	437	388	349	317	291	269	249	233	218	194	175
406 × 178	74	†1121	973	811	695	608	540	486	442	405	374	347	324	304	270	243
	67	†1009	873	727	623	546	485	436	397	364	336	312	291	273	242	
	60	† 888	779	650	557	487	433	390	354	325	300	278	260	244	217	
	54	849	679	566	485	425	377	340	309	283	261	243	226	212	189	
406 × 152	74	†1177	952	794	680	595	529	476	433	397	366	340	317	298	265	238
	67	1062	850	708	607	531	472	425	386	354	327	303	283	266	236	212
	60	930	744	620	531	465	413	372	338	310	286	266	248	233	207	
406 × 140	46	714	571	476	408	357	317	285	259	238	220	204	190	178	159	
	39	575	460	383	328	287	255	230	209	192	177	164	153	144	128	
381 × 152	67	1007	806	672	576	504	448	403	366	336	310	288	269	252	224	
	60	891	713	594	509	445	396	356	324	297	274	255	238	223	198	
	52	775	620	517	443	387	344	310	282	258	238	221	207	194	172	
356 × 171	67	† 927	788	657	563	492	438	394	358	328	303	281	263	246		
	57	† 803	658	549	470	411	366	329	299	274	253	235	219	206		
	51	† 727	584	487	417	365	325	292	266	243	225	209	195	183		
	45	630	504	420	360	315	280	252	229	210	194	180	168	157		
356 × 127	39	524	419	350	300	262	233	210	191	175	161	150	140	131		
	33	431	345	287	246	216	192	172	157	144	133	123	115	108		
305 × 165	54	† 670	553	461	395	346	307	277	252	231	213	198	184			
	46	† 576	476	396	340	297	264	238	216	198	183	170				
	40	515	412	343	294	257	229	206	187	172	158	147				
305 × 127	48	562	450	375	321	281	250	225	204	187	173	161	150			
	42	488	390	325	279	244	217	195	177	163	150	139				
	37	433	346	288	247	216	192	173	157	144	133	124				
305 × 102	33	381	305	254	218	191	170	153	139	127	117	109	102			
	28	323	258	215	184	161	143	129	117	108	99	92	86			
	25	264	212	176	151	132	118	106	96	88	81	76				
254 × 146	43	464	371	309	265	232	206	186	169	155						
	37	398	319	266	228	199	177	159	145	133						
	31	324	259	216	185	162	144	130	118	108						
254 × 102	28	283	226	189	162	142	126	113	103	94						
	25	244	195	162	139	122	108	97	89	81						
	22	207	166	138	119	104	92	83	75	69						
203 × 133	30	256	205	171	146	128	114	103								
	25	213	170	142	121	106	94									

Generally, tabular loads are based on a flexural stress of 230N/mm², assuming adequate lateral support. Beams without adequate lateral support must not exceed the critical span Lc. unless the allowable compressive stress is reduced in accordance with clause 19.a.(ii) of BS 449 : 1969.
Tabular loads printed in bold face type exceed the buckling capacity of the unstiffened web without allowance for actual length of bearing; the load bearing capacity should be checked, see page 109.
Tabular loads marked thus † are based on the maximum shear value of the web and are less than the permissible flexural load.

Tables reproduced by permission and courtesy of

BASED ON
BS 449
1969

UNIVERSAL BEAMS

DIMENSIONS AND PROPERTIES
GRADE 50 STEEL

Actual Size $D \times B$ mm	Critical Span Le metres	Area of section cm^2	Moment of Inertia		Radius of Gyration		Elastic Modulus		Ratio $\dfrac{D}{T}$
			Axis x—x cm^4	Axis y—y cm^4	Axis y—y cm	Axis x—x cm^3	Axis y—y cm^3		
465.1 × 153.5	2.705	104.4	36160	1093	3.24	1555	142.5	24.6
461.3 × 152.7	2.652	94.9	32380	963	3.18	1404	126.1	27.1
457.2 × 151.9	2.596	85.3	28522	829	3.12	1248	109.1	30.5
454.7 × 152.9	2.694	75.9	25464	794	3.23	1120	104.0	34.2
449.8 × 152.4	2.593	66.5	21345	645	3.11	949.0	84.61	41.3
412.8 × 179.7	3.254	94.9	27279	1448	3.91	1322	161.2	25.8
409.4 × 178.8	3.210	85.4	24279	1269	3.85	1186	141.9	28.6
406.4 × 177.8	3.179	76.1	21520	1108	3.82	1059	124.7	31.8
402.6 × 177.6	3.059	68.3	18576	922	3.67	922.8	103.8	36.9
416.3 × 153.7	2.823	94.8	26938	1047	3.32	1294	136.2	23.0
412.2 × 152.9	2.717	85.3	23798	908	3.26	1155	118.8	25.8
407.9 × 152.2	2.650	75.8	20619	768	3.18	1011	100.9	29.3
402.3 × 142.4	2.428	58.9	15603	500	2.92	775.6	70.26	35.9
397.3 × 141.8	2.290	49.3	12408	373	2.75	624.7	52.61	46.2
388.6 × 154.3	2.804	85.4	21276	947	3.33	1095	122.7	23.8
384.8 × 153.4	2.727	75.9	18632	814	3.27	968.4	106.2	26.7
381.0 × 152.4	2.676	66.4	16046	685	3.21	842.3	89.96	30.7
364.0 × 173.2	3.283	85.3	19483	1278	3.87	1071	147.6	23.2
358.6 × 172.1	3.142	72.1	16038	1026	3.77	894.3	119.2	27.6
355.6 × 171.5	3.086	64.5	14118	885	3.71	794.0	103.3	30.9
352.0 × 171.0	2.984	56.9	12052	730	3.58	684.7	85.39	36.3
352.8 × 126.0	2.164	49.3	10054	333	2.60	570.0	52.87	33.0
348.5 × 125.4	2.066	41.7	8167	257	2.48	468.7	40.99	41.0
310.9 × 166.8	3.244	68.3	11686	988	3.80	751.8	118.5	22.7
307.1 × 165.7	3.118	58.8	9924	825	3.74	646.4	99.54	26.0
303.8 × 165.1	3.053	51.4	8500	691	3.67	559.6	83.71	29.8
310.4 × 125.2	2.304	60.8	9485	438	2.68	611.1	69.94	22.2
306.6 × 124.3	2.188	53.1	8124	367	2.63	530.0	58.99	25.3
303.8 × 123.5	2.149	47.4	7143	316	2.58	470.3	51.11	28.4
312.7 × 102.4	1.774	41.8	6482	189	2.13	414.6	37.00	29.0
308.9 × 101.9	1.709	36.3	5415	153	2.05	350.7	30.01	34.7
304.8 × 101.6	1.602	31.4	4381	116	1.92	287.5	22.85	44.8
259.6 × 147.3	2.983	55.0	6546	633	3.39	504.3	85.97	20.4
256.0 × 146.4	2.820	47.4	5544	528	3.34	433.1	72.11	23.5
251.5 × 146.1	2.654	39.9	4427	406	3.19	352.1	55.53	29.2
260.4 × 102.1	1.828	36.2	4004	174	2.19	307.6	34.13	26.0
257.0 × 101.9	1.761	32.1	3404	144	2.11	264.9	28.23	30.6
254.0 × 101.6	1683	28.4	2863	116	2.02	225.4	22.84	37.4
206.8 × 133.8	2.642	38.0	2880	354	3.05	278.5	52.85	21.5
203.2 × 133.4	2.451	32.3	2348	280	2.94	231.1	41.92	26.1

Tabular loads printed in italic type are within the web buckling capacity of the unstiffened web and produce a total deflection not exceeding 1/360th of the span.
Tabular loads printed in ordinary type should be checked for deflection, see page 110.
For explanation of tables, see notes commencing page 102.
1 kilonewton may be taken as 0.102 metric tonne (megagramme) force, but see page 102.

the British Constructional Steelwork Association.

JOISTS

DIMENSIONS AND PROPERTIES

Nominal Size	Mass per metre	Depth of Section D	Width of Section B	Thickness Web t	Thickness Flange T	Radius Root r_1	Radius Toe r_2	Depth between Fillets d	Area of Section
mm	kg	mm	mm	mm	mm	mm	mm	mm	cm²
203 × 102	25.33	203.2	101.6	5.8	10.4	9.4	3.2	161.0	32.3
178 × 102	21.54	177.8	101.6	5.3	9.0	9.4	3.2	138.2	27.4
152 × 89	17.09	152.4	88.9	4.9	8.3	7.9	2.4	117.9	21.8
127 × 76	13.36	127.0	76.2	4.5	7.6	7.9	2.4	94.2	17.0
102 × 64	9.65	101.6	63.5	4.1	6.6	6.9	2.4	73.2	12.3
76 × 51	6.67	76.2	50.8	3.8	5.6	6.9	2.4	50.3	8.49

JOISTS

DIMENSIONS AND PROPERTIES

Nominal Size	Moment of Inertia Axis x–x Gross	Moment of Inertia Axis x–x Net	Moment of Inertia Axis y–y	Radius of Gyration Axis x–x	Radius of Gyration Axis y–y	Elastic Modulus Axis x–x	Elastic Modulus Axis y–y	Ratio $\dfrac{D}{T}$
mm	cm⁴	cm⁴	cm⁴	cm	cm	cm³	cm³	
203 × 102	2294	2023	162.6	8.43	2.25	225.8	32.02	19.5
178 × 102	1519	1340	139.2	7.44	2.25	170.9	27.41	19.8
152 × 89	881.1	762.1	85.98	6.36	1.99	115.6	19.34	18.4
127 × 76	475.9	399.8	50.18	5.29	1.72	74.94	13.17	16.7
102 × 64	217.6	181.9	25.30	4.21	1.43	42.84	7.97	15.4
76 × 51	82.58	68.85	11.11	3.12	1.14	21.67	4.37	13.6

Note: One hole is deducted from each flange in calculating the Net Moment of Inertia about x–x.

Tables reproduced by permission and courtesy of

CHANNELS

DIMENSIONS AND PROPERTIES

Nominal Size	Mass per metre	Depth of Section	Width of Section	Thickness		Radius		Depth between Fillets	Ratio	Area of Section
		D	B	Web t	Flange T	Root r_1	Toe r_2	d	D — T	
mm	in kg	mm	mm	mm	mm	mm	mm	mm		cm²
432 × 102	65.54	431.8	101.6	12.2	16.8	15.2	4.8	362.5	25.7	83.49
381 × 102	55.10	381.0	101.6	10.4	16.3	15.2	4.8	312.4	23.4	70.19
305 × 102	46.18	304.8	101.6	10.2	14.8	15.2	4.8	239.3	20.6	58.83
305 × 89	41.69	304.8	88.9	10.2	13.7	13.7	3.2	245.4	22.2	53.11
254 × 89	35.74	254.0	88.9	9.1	13.6	13.7	3.2	194.8	18.7	45.52
254 × 76	28.29	254.0	76.2	8.1	10.9	12.2	3.2	203.7	23.3	36.03
229 × 89	32.76	228.6	88.9	8.6	13.3	13.7	3.2	169.9	17.2	41.73
229 × 76	26.06	228.6	76.2	7.6	11.2	12.2	3.2	178.1	20.4	33.20
203 × 89	29.78	203.2	88.9	8.1	12.9	13.7	3.2	145.3	15.8	37.94
203 × 76	23.82	203.2	76.2	7.1	11.2	12.2	3.2	152.4	18.1	30.34
178 × 89	26.81	177.8	88.9	7.6	12.3	13.7	3.2	120.9	14.5	34.15
178 × 76	20.84	177.8	76.2	6.6	10.3	12.2	3.2	128.8	17.3	26.54
152 × 89	23.84	152.4	88.9	7.1	11.6	13.7	3.2	97.0	13.1	30.36
152 × 76	17.88	152.4	76.2	6.4	9.0	12.2	2.4	105.9	16.9	22.77
127 × 64	14.90	127.0	63.5	6.4	9.2	10.7	2.4	84.1	13.8	18.98
102 × 51	10.42	101.6	50.8	6.1	7.6	9.1	2.4	65.8	13.4	13.28
76 × 38	6.70	76.2	38.1	5.1	6.8	7.6	2.4	45.7	11.2	8.53

the British Constructional Steelwork Association.

CHANNELS

DIMENSIONS AND PROPERTIES

Nominal Size	Dimension	Moment of Inertia			Radius of Gyration		Elastic Modulus	
	p	Axis x–x		Axis y–y	Axis x–x	Axis y–y	Axis x–x	Axis y–y
		Gross	Net					
mm	cm	cm⁴	cm⁴	cm⁴	cm	cm	cm³	cm³
432 × 102	2.32	21399	17602	628.6	16.0	2.74	991.1	80.15
381 × 102	2.52	14894	12060	579.8	14.6	2.87	781.8	75.87
305 × 102	2.66	8214	6587	499.5	11.8	2.91	539.0	66.60
305 × 89	2.18	7061	5824	325.4	11.5	2.48	463.3	48.49
254 × 89	2.42	4448	3612	302.4	9.88	2.58	350.2	46.71
254 × 76	1.86	3367	2673	162.6	9.67	2.12	265.1	28.22
229 × 89	2.53	3387	2733	285.0	9.01	2.61	296.4	44.82
229 × 76	2.00	2610	2040	158.7	8.87	2.19	228.3	28.22
203 × 89	2.65	2491	1996	264.4	8.10	2.64	245.2	42.34
203 × 76	2.13	1950	1506	151.4	8.02	2.23	192.0	27.59
178 × 89	2.76	1753	1397	241.0	7.16	2.66	197.2	39.29
178 × 76	2.20	1337	1028	134.0	7.10	2.25	150.4	24.73
152 × 89	2.86	1166	923.7	215.1	6.20	2.66	153.0	35.70
152 × 76	2.21	851.6	654.3	113.8	6.12	2.24	111.8	21.05
127 × 64	1.94	482.6	367.5	67.24	5.04	1.88	75.99	15.25
102 × 51	1.51	207.7	167.9	29.10	3.96	1.48	40.89	8.16
76 × 38	1.19	74.14	54.52	10.66	2.95	1.12	19.46	4.07

One hole is deducted from each flange in calculating the Net Moment of Inertia about x–x.

Tables reproduced by permission and courtesy of

JOISTS

SAFE LOADS FOR GRADE 43 STEEL

BASED ON BS 449 1969

Nominal Size mm	Mass per metre in kg	SAFE DISTRIBUTED LOADS IN KILONEWTONS FOR SPANS IN METRES AND DEFLECTION COEFFICIENTS												
		1.00	1.25	1.50	1.75	2.00	2.25	2.50	2.75	3.00	3.25	3.50	3.75	4.00	4.25
		448.0	286.7	199.1	146.3	112.0	88.49	71.68	59.24	49.78	42.41	36.57	31.86	28.00	24.80
203 × 102	25.33		†236	195	170	149	132	119	108	99	92	85	79	75	70
178 × 102	21.54	†188	180	150	129	113	100	90	82	75	69	64	60	56	53
152 × 89	17.09	†149	122	102	87	76	68	61	56	51	47	44	41	38	36
127 × 76	13.36	99	79	66	57	49	44	40	36	33	30	28	26	25	23
102 × 64	9.65	57	45	38	32	28	25	23	21	19	17				
76 × 51	6.67	29	23	19	16	14	13	11							

CHANNELS

SAFE LOADS FOR GRADE 43 STEEL

BASED ON BS 449 1969

Nominal Size mm	Mass per metre in kg	SAFE DISTRIBUTED LOADS IN KILONEWTONS FOR SPANS IN METRES AND DEFLECTION COEFFICIENTS												
		1.00	1.50	2.00	2.50	3.00	3.50	4.00	4.50	5.00	5.50	6.00	7.00	8.00	9.00
		448.0	199.1	112.0	71.68	49.78	36.57	28.00	22.12	17.92	14.81	12.44	9.143	7.000	5.531
432 × 102	65.54	†1054	872	654	523	436	374	327	291	262	238	218	187	164	145
381 × 102	55.10	†792	688	516	413	344	295	258	229	206	188	172	147	129	115
305 × 102	46.18	†622	474	356	285	237	203	178	158	142	129	119	102	89	79
305 × 89	41.69	612	408	306	245	204	175	153	136	122	111	102	87	76	68
254 × 89	35.74	462	308	231	185	154	132	116	103	92	84	77	66	58	
254 × 76	28.29	350	233	175	140	117	100	87	78	70	64	58	50	44	
229 × 89	32.76	391	261	196	156	130	112	98	87	78	71	65	56		
229 × 76	26.06	301	201	151	121	100	86	75	67	60	55	50	43		
203 × 89	29.78	324	216	162	129	108	92	81	72	65	59	54			
203 × 76	23.82	253	169	127	101	84	72	63	56	51	46	42			
178 × 89	26.81	260	174	130	104	87	74	65	58	52	47	43			
178 × 76	20.84	199	132	99	79	66	57	50	44	40	36	33			
152 × 89	23.84	202	135	101	81	67	58	50	45	40					
152 × 76	17.88	148	98	74	59	49	42	37	33	30					
127 × 64	14.90	100	67	50	40	33	29	25							
102 × 51	10.42	54	36	27	22	18									
76 × 38	6.70	26	17	13	10										

Generally, tabular loads are based on a flexural stress of 165N/mm², assuming adequate lateral support. Beams without adequate lateral support must not exceed the critical span Lc, unless the allowable compressive stress is reduced in accordance with clause 19.a.(ii) of BS 449 : 1969.
Tabular loads printed in bold face type exceed the buckling capacity of the unstiffened web without allowance for actual length of bearing; the load bearing capacity should be checked, see page 109.

the British Constructional Steelwork Association.

BASED ON
BS 449
1969

JOISTS

DIMENSIONS AND PROPERTIES
GRADE 43 STEEL

Actual Size D×B mm	Critical Span Lc metres	Area of section cm²	Moment of Inertia		Radius of Gyration	Elastic Modulus		Ratio D — T
			Axis x—x cm⁴	Axis y—y cm⁴	Axis y—y cm	Axis x—x cm³	Axis y—y cm³	
203.2 × 101.6	2.259	32.26	2294	162.6	2.25	225.8	32.02	19.5
177.8 × 101.6	2.257	27.44	1519	139.2	2.25	170.9	27.41	19.8
152.4 × 88.9	2.053	21.77	881.1	85.98	1.99	115.6	19.34	18.4
127.0 × 76.2	1.856	17.02	475.9	50.18	1.72	74.94	13.17	16.7
101.6 × 63.5	1.621	12.29	217.6	25.30	1.43	42.84	7.97	15.4
76.2 × 50.8	1.396	8.49	82.58	11.11	1.14	21.67	4.37	13.6

BASED ON
BS 449
1969

CHANNELS

DIMENSIONS AND PROPERTIES
GRADE 43 STEEL

Actual Size D×B mm	Critical Span Lc metres	Area of section cm²	Moment of Inertia		Radius of Gyration	Elastic Modulus		Ratio D T	Dimension p cm
			Axis x—x cm⁴	Axis y—y cm⁴	Axis y—y cm	Axis x—x cm³	Axis y—y cm³		
431.8 × 101.6	2.547	83.49	21399	628.6	2.74	991.1	80.15	25.7	2.32
381.0 × 101.6	2.718	70.19	14894	579.8	2.87	781.8	75.87	23.4	2.52
304.8 × 101.6	2.874	58.83	8214	499.5	2.91	539.0	66.60	20.6	2.66
304.8 × 88.9	2.377	53.11	7061	325.4	2.48	463.3	48.49	22.2	2.18
254.0 × 88.9	2.643	45.52	4448	302.4	2.58	350.2	46.71	18.7	2.42
254.0 × 76.2	2.011	36.03	3367	162.6	2.12	265.1	28.22	23.3	1.86
228.6 × 88.9	2.785	41.73	3387	285.0	2.61	296.4	44.82	17.2	2.53
228.6 × 76.2	2.163	33.20	2610	158.7	2.19	228.3	28.22	20.4	2.00
203.2 × 88.9	2.945	37.94	2491	264.4	2.64	245.2	42.34	15.8	2.65
203.2 × 76.2	2.320	30.34	1950	151.4	2.23	·192.0	27.59	18.1	2.13
177.8 × 88.9	3.117	34.15	1753	241.0	2.66	197.2	39.29	14.5	2.76
177.8 × 76.2	2.389	26.54	1337	134.0	2.25	150.4	24.73	17.3	2.20
152.4 × 88.9	3.326	30.36	1166	215.1	2.66	153.0	35.70	13.1	2.86
152.4 × 76.2	2.400	22.77	851.6	113.8	2.24	111.8	21.05	16.9	2.21
127.0 × 63.5	2.275	18.98	482.6	67.24	1.88	75.99	15.25	13.8	1.94
101.6 × 50.8	1.828	13.28	207.7	29.10	1.48	40.89	8.16	13.4	1.51
76.2 × 38.1	1.573	8.53	74.14	10.66	1.12	19.46	4.07	11.2	1.19

Tabular loads marked thus † are based on the maximum shear value of the web and are less than the permissible flexural load.
Tabular loads printed in italic type are within the web buckling capacity of the unstiffened web and produce a total deflection not exceeding 1/360th of the span.
Tabular loads printed in ordinary type should be checked for deflection. see page 110.
For explanation of tables see notes commencing page 102.
1 kilonewton may be taken as 0.102 metric tonne (megagramme) force, but see page 102.

Tables reproduced by permission and courtesy of

JOISTS

SAFE LOADS FOR GRADE 50 STEEL

BASED ON BS 449 1969

Nominal Size mm		Mass per metre in kg	SAFE DISTRIBUTED LOADS IN KILONEWTONS FOR SPANS IN METRES AND DEFLECTION COEFFICIENTS												
			1.00	1.25	1.50	1.75	2.00	2.25	2.50	2.75	3.00	3.25	3.50	3.75	4.00	4.25
			448.0	286.7	199.1	146.3	112.0	88.49	71.68	59.24	49.78	42.41	36.57	31.86	28.00	24.80
203 ×	102	25.33		†330	277	237	208	185	166	151	138	128	119	111	104	98
178 ×	102	21.54	†264	252	210	180	157	140	126	114	105	97	90	84	79	74
152 ×	89	17.09	†209	170	142	122	106	95	85	77	71	65	61			
127 ×	76	13.36	138	110	92	79	69	61	55	50	46					
102 ×	64	9.65	79	63	53	45	39	35								
76 ×	51	6.67	40	32	27	23										

CHANNELS

SAFE LOADS FOR GRADE 50 STEEL

BASED ON BS 449 1969

Nominal Size mm		Mass per metre in kg	SAFE DISTRIBUTED LOADS IN KILONEWTONS FOR SPANS IN METRES AND DEFLECTION COEFFICIENTS												
			1.00	1.50	2.00	2.50	3.00	3.50	4.00	4.50	5.00	5.50	6.00	7.00	8.00	9.00
			448.0	199.1	112.0	71.68	49.78	36.57	28.00	22.12	17.92	14.81	12.44	9.143	7.000	5.531
432 ×	102	65.54	†1475	1216	912	729	608	521	456	405	365	332	304	261	228	203
381 ×	102	55.10	†1109	959	719	575	480	411	360	320	288	262	240	206	180	160
305 ×	102	46.18	†871	661	496	397	331	283	248	220	198	180	165	142		
305 ×	89	41.69	852	568	426	341	284	244	213	189	170	155	142	122		
254 ×	89	35.74	644	430	322	258	215	184	161	143	129	117	107			
254 ×	76	28.29	488	325	244	195	163	139	122	108	98	89	81			
229 ×	89	32.76	545	364	273	218	182	156	136	121	109	99				
229 ×	76	26.06	420	280	210	168	140	120	105	93	84	76				
203 ×	89	29.78	451	301	226	180	150	129	113	100						
203 ×	76	23.82	353	235	177	141	118	101	88	78						
178 ×	89	26.81	363	242	181	145	121	104	91							
178 ×	76	20.84	277	185	138	111	92	79	69							
152 ×	89	23.84	281	188	141	113	94	80								
152 ×	76	17.88	206	137	103	82	69	59								
127 ×	64	14.90	140	93	70	56	47									
102 ×	51	10.42	75	50	38											
76 ×	38	6.70	36	24												

Generally, tabular loads are based on a flexural stress of 230N/mm², assuming adequate lateral support. Beams without adequate lateral support must not exceed the critical span Lc. unless the allowable compressive stress is reduced in accordance with clause 19.a.(ii) of BS 449 : 1969.
Tabular loads printed in bold face type exceed the buckling capacity of the unstiffened web without allowance for actual length of bearing; the load bearing capacity should be checked, see page 109.

the British Constructional Steelwork Association.

BASED ON BS 449 1969

JOISTS

DIMENSIONS AND PROPERTIES
GRADE 50 STEEL

Actual Size D×B mm	Critical Span Lc metres	Area of section cm²	Moment of Inertia		Radius of Gyration	Elastic Modulus		Ratio D/T
			Axis x–x cm⁴	Axis y–y cm⁴	Axis y–y cm	Axis x–x cm³	Axis y–y cm³	
203.2 × 101.6	2.005	32.26	2294	162.6	2.25	225.8	32.02	19.5
177.8 × 101.6	2.004	27.44	1519	139.2	2.25	170.9	27.41	19.8
152.4 × 88.9	1.816	21.77	881.1	85.98	1.99	115.6	19.34	18.4
127.0 × 76.2	1.632	17.02	475.9	50.18	1.72	74.94	13.17	16.7
101.6 × 63.5	1.419	12.29	217.6	25.30	1.43	42.84	7.97	15.4
76.2 × 50.8	1.212	8.49	82.58	11.11	1.14	21.67	4.37	13.6

BASED ON BS 449 1969

CHANNELS

DIMENSIONS AND PROPERTIES
GRADE 50 STEEL

Actual Size D×B mm	Critical Span Lc metres	Area of section cm²	Moment of Inertia		Radius of Gyration	Elastic Modulus		Ratio D/T	Dimension p cm
			Axis x–x cm⁴	Axis y–y cm⁴	Axis y–y cm	Axis x–x cm³	Axis y–y cm³		
431.8 × 101.6	2.285	83.49	21399	628.6	2.74	991.1	80.15	25.7	2.32
381.0 × 101.6	2.432	70.19	14894	579.8	2.87	781.8	75.87	23.4	2.52
304.8 × 101.6	2.557	58.83	8214	499.5	2.91	539.0	66.60	20.6	2.66
304.8 × 88.9	2.123	53.11	7061	325.4	2.48	463.3	48.49	22.2	2.18
254.0 × 88.9	2.340	45.52	4448	302.4	2.58	350.2	46.71	18.7	2.42
254.0 × 76.2	1.799	36.03	3367	162.6	2.12	265.1	28.22	23.3	1.86
228.6 × 88.9	2.454	41.73	3387	285.0	2.61	296.4	44.82	17.2	2.53
228.6 × 76.2	1.924	33.20	2610	158.7	2.19	228.3	28.22	20.4	2.00
203.2 × 88.9	2.581	37.94	2491	264.4	2.64	245.2	42.34	15.8	2.65
203.2 × 76.2	2.050	30.34	1950	151.4	2.23	192.0	27.59	18.1	2.13
177.8 × 88.9	2.717	34.15	1753	241.0	2.66	197.2	39.29	14.5	2.76
177.8 × 76.2	2.105	26.54	1337	134.0	2.25	150.4	24.73	17.3	2.20
152.4 × 88.9	2.881	30.36	1166	215.1	2.66	153.0	35.70	13.1	2.86
152.4 × 76.2	2.113	22.77	851.6	113.8	2.24	111.8	21.05	16.9	2.21
127.0 × 63.5	1.977	18.98	482.6	67.24	1.88	75.99	15.25	13.8	1.94
101.6 × 50.8	1.585	13.28	207.7	29.10	1.48	40.89	8.16	13.4	1.51
76.2 × 38.1	1.350	8.53	74.14	10.66	1.12	19.46	4.07	11.2	1.19

Tabular loads marked thus + are based on the maximum shear value of the web and are less than the permissible flexural load.
Tabular loads printed in italic type are within the web buckling capacity of the unstiffened web and produce a total deflection not exceeding 1/360th of the span.
Tabular loads printed in ordinary type should be checked for deflection, see page 110.
For explanation of tables, see notes commencing page 102.
1 kilonewton may be taken as 0.102 metric tonne (megagramme) force, but see page 102.

Table reproduced by permission and courtesy of the British Constructional Steelwork Association.

of the beams, and that the maximum permissible flexural stress appropriate to the grade of steel can be used. It will be seen that there are three different types used for the tabular loads. The significance of these different types will be discussed in Chapter 8.

Meanwhile, the following example will illustrate how the tabular loads are obtained.

A 461 × 152·7 × 74 kg UB, spans 6 m.

$$\text{B.M.} = \frac{WL}{8} = \frac{W \times 6 \times 1000}{8} \text{ kNmm}$$

$$fZ = 165 \times 1404$$

$$\therefore W = \frac{165 \times 1404 \times 8}{6 \times 1000} = 309 \text{ kN}$$

The questions of concrete casing, web buckling, and shear are dealt with later in the book.

Readers should consult with care the notices at the foot of all the safe-load tables given in the various section books.

Exercises 6

(Where necessary the section tables on pages 122 to 142 must be consulted. These tables may also be used for checking numerical answers. The self-weight of beams may be neglected. Tabular values correspond to $f = 165 \text{ N/mm}^2$.)

1 Calculate the safe U.D. load for a 381 × 152 × 67 kg UB for an effective span of 6 m. Working stress = 165 N/mm².

2 Select a suitable standard section for a simply supported beam which has to carry a central load of 120 kN, the effective span being 5·5 m. Maximum stress not to exceed 165 N/mm².

3 A 203 × 133 × 30 kg UB ($Z = 279 \text{ cm}^3$) carries a U.D. load of 75 kN. If the span is 3·6 m, calculate the maximum steel fibre stress.

4 A beam of 3·6 m span carries two concentrated loads: 20 kN at 0·9 m from the left support, and 42 kN at 1·2 m from the right support. Assuming a maximum stress of 120 N/mm², find the necessary section modulus for the beam.

5 Select suitable steel sections for the floor given in fig. 5.13. The timber beams are of 3 m span, and spaced at 350 mm centres. Calculate suitable dimensions for their breadth and depth. Assume a working stress of 7 N/mm² in the timber.

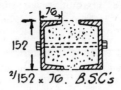

Fig. 6.13.

6 Fig. 6.13 shows a lintel composed of two standard channels, with concrete filling. The effective span is 2·8 m. Calculate the safe total U.D. load for the lintel, from the point of view of the channels alone. I_{xx} for the channel = 851·6 cm^4; maximum stress = 165 N/mm^2.

7

Properties of Compound Beam Sections

Parallel translation of areas

The compounding of several smaller areas into one composite section, or the disintegration of a complex figure into a number of simpler diagrams, are methods sometimes employed in dealing with the computation of section properties. The principle employed is known as the *parallel translation of areas*.

If an element of area a, situated at a distance y from a given axis, be moved parallel to the axis, its moment of inertia about that axis will retain the value ay^2 for all positions of the area. Areas of finite size may be regarded as being composed of a very large number of such elemental areas, so that we may move any given section—in whole or part—parallel to an axis, without altering its I value about the axis. Fig. 7.1

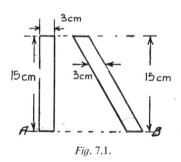

Fig. 7.1.

illustrates the equality of I value for any horizontal axis, such as AB, of a rectangle and a parallelogram. In this case, rectangles, 3 cm wide, and of very small depth, have been moved horizontally to form the parallelogram.

$$I_{AB} \text{ in each case} = \frac{3 \times 15^3}{3} \text{ cm}^4 = 3375 \text{ cm}^4 \text{ (see p. 111)}.$$

In fig. 7.2 the principle is applied to a number of type sections. The diagrams indicate methods of dealing with a half-trough section.

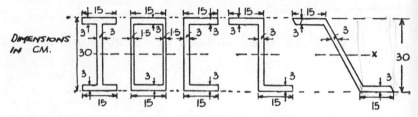

Fig. 7.2 *Principle of parallel translation of areas.*

I_{XX} for each of the sections given

$$= \frac{BD^3}{12} - \frac{bd^3}{12}$$

$$= \left(\frac{15 \times 30^3}{12} - \frac{12 \times 24^3}{12}\right) \text{cm}^4 = 19926 \text{ cm}^4$$

Principle of parallel axes

In fig. 7.3, the axis AB is assumed to be the axis about which the moment of inertia of the given section is required. CG is an axis

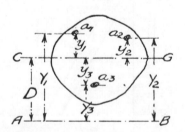

Fig. 7.3 *Principle of parallel axes.*

parallel to AB, passing through the centre of gravity of the section. The distance between the two axes $= D$, and $A =$ the total area of the section.

Taking a_1, a_2, and a_3 as typical small component areas of the section, we get from the definition of I_{AB},

$$I_{AB} = a_1Y_1{}^2 + a_2Y_2{}^2 + a_3Y_3{}^2 + \text{etc.}$$
$$= a_1(y_1+D)^2 + a_2(y_2+D)^2 + a_3(D-y_3)^2 + \text{etc.}$$
$$= a_1y_1{}^2 + 2a_1y_1D + a_1D^2 + a_2y_2{}^2 + 2a_2y_2D + a_2D^2$$
$$\quad + a_3D^2 - 2a_3y_3D + a_3y_3{}^2 + \text{etc.}$$
$$= (a_1y_1{}^2 + a_2y_2{}^2 + a_3y_3{}^2 + \text{etc.})$$
$$\quad + (a_1D^2 + a_2D^2 + a_3D^2 + \text{etc.})$$
$$\quad + (2a_1y_1D + 2a_2y_2D - 2a_3y_3D + \text{etc.})$$

The expression in the first bracket gives I_{CG}. The second bracket may be written $D^2(a_1+a_2+a_3+\text{etc.})$, i.e. $= D^2 \times A$. The third bracket $= 2D(a_1y_1+a_2y_2-a_3y_3+\text{etc.})$. As each term in the latter expression represents an area-moment about an axis passing through the c.g. of the section, the algebraic sum of all the terms will be zero. We may write therefore

$$I_{AB} = I_{CG} + AD^2$$

In a practical example, I_{CG} will usually be obtained by means of a standard formula.

Example 1 Find I_{AB} for the rectangle given in fig. 7.4.

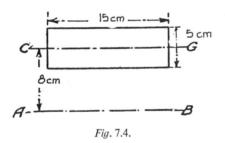

Fig. 7.4.

$$I_{CG} = \frac{bd^3}{12} = \frac{15 \times 5^3}{12} \text{ cm}^4 = 156 \cdot 25 \text{ cm}^4$$

$$A = 15 \text{ cm} \times 5 \text{ cm} = 75 \text{ cm}^2$$
$$I_{AB} = I_{CG} + AD^2$$
$$= (156 \cdot 25 + 75 \times 8^2) \text{ cm}^4 = 4956 \cdot 25 \text{ cm}^4$$

Example 2 Given the formula $bd^3/12$ for I for an axis at mid-depth of a rectangular section, find the formula for I about the base.

In this case, $I_{CG} = \dfrac{bd^3}{12}$

$$A = b \times d$$

$$D = \dfrac{d}{2}$$

$$I_{base} = I_{CG} + AD^2 = \dfrac{bd^3}{12} + (b \times d) \times \left(\dfrac{d}{2}\right)^2$$

$$= \dfrac{bd^3}{12} + \dfrac{bd^3}{4} = \dfrac{bd^3}{3}$$

This formula is a very useful one, and worth remembering.

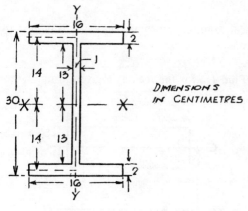

Fig. 7.5 R.S.J. section.

Example 3 Find the value of I_{XX} for the section given in fig. 7.5.

I_{XX} for top flange $= I_{CG} + AD^2$

$$= \dfrac{16 \times 2^3}{12} + (16 \times 2) \times 14^2$$

$$= (10\tfrac{2}{3} + 6272) \text{ cm}^4$$

$$= 6282 \cdot 67 \text{ cm}^4.$$

I_{XX} for bottom flange $= 6282 \cdot 67$ cm^4.

I_{XX} for web (taken as a rectangle with axis at mid-depth)

$$= \frac{bd^3}{12} = \frac{1 \times 26^3}{12} \text{ cm}^4 = 1464 \cdot 67 \text{ cm}^4$$

Total $I_{XX} = (2 \times 6282 \cdot 67) + 1464 \cdot 67 = 14030 \text{ cm}^4$

This value may be checked by the formula already given for this case:

$$I_{XX} = \frac{BD^3}{12} - \frac{bd^3}{12} = \left(\frac{16 \times 30^3}{12} - \frac{15 \times 26^3}{12} \right) \text{ cm}^4$$

$$= 14030 \text{ cm}^4$$

To find I_{YY}, in this case, regard the section as being made up of three rectangles.

$$I_{YY} = 2 \left(\frac{2 \times 16^3}{12} \right) + \frac{26 \times 1^3}{12} \text{ cm}^4$$

$$= 684 \cdot 8 \text{ cm}^3$$

$$Z_{XX} \text{ for the section} = \frac{I_{XX}}{y} = \frac{14030}{15} = 935 \cdot 3 \text{ cm}^3$$

$$Z_{YY} \text{ for the section} = \frac{I_{YY}}{y} = \frac{684 \cdot 8}{8} = 85 \cdot 6 \text{ cm}^3$$

Example 4 Obtain I_{XX} for the channel section given in fig. 7.6.

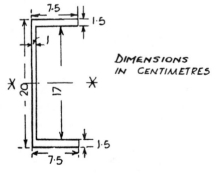

Fig. 7.6 Channel section.

Method (*a*): by subtraction of *I* values.

$$I_{xx} = \left(\frac{7 \cdot 5 \times 20^3}{12} - \frac{6 \cdot 5 \times 17^3}{12}\right) \text{cm}^4$$

$$= (5000 - 2661 \cdot 2) \text{ cm}^4$$

$$= 2338 \cdot 8 \text{ cm}^4$$

Method (*b*): by employing the principle of parallel axes.
Top and bottom flanges:

$$I_{xx} = 2\left(\frac{7 \cdot 5 \times 1 \cdot 5^3}{12} + 7 \cdot 5 \times 1 \cdot 5 \times 9 \cdot 25^2\right) \text{cm}^4$$

$$= 1929 \cdot 4 \text{ cm}^4$$

Web: $\quad I_{xx} = \dfrac{bd^3}{12} = \dfrac{1 \times 17^3}{12} \text{ cm}^4 = 409 \cdot 4 \text{ cm}^4$

Total $I_{xx} = 2338 \cdot 8 \text{ cm}^4$

Example 5 Calculate the safe total uniformly distributed load, for an effective span of 2·4 m, for the T-section shown in fig. 7.7. Working stress = 165 N/mm².

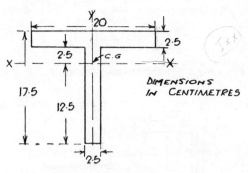

Fig. 7.7 T-section.

We must first find the position of the neutral axis of the section.
Taking moments about the base of the section,

$$\{(20 \times 2\tfrac{1}{2}) + (15 \times 2\tfrac{1}{2})\} \times \bar{y} = (50 \times 16 \cdot 25) + (37 \cdot 5 \times 7 \cdot 5)$$
$$87 \cdot 5\bar{y} = 1093 \cdot 75$$
$$\bar{y} = 12 \cdot 5 \text{ cm.}$$

Properties of Compound Beam Sections 151

I_{xx} for flange $= I_{CG} + AD^2$

$$= \left(\frac{20 \times 2 \cdot 5^3}{12} + (20 \times 2 \cdot 5) \times 3 \cdot 75^2\right) \text{cm}^4 = 729 \cdot 2 \text{ cm}^4$$

I_{xx} for web—treated as being composed of two rectangles, with respective axes at the base:

$$\frac{bd^3}{3} = \left(\frac{2 \cdot 5 \times 2 \cdot 5^3}{3} + \frac{2 \cdot 5 \times 12 \cdot 5^3}{3}\right) = 1640 \cdot 3 \text{ cm}^4$$

Total $I_{xx} = 2369 \cdot 5$ cm^4

$$Z_{xx} = \frac{I}{y} = \frac{2369 \cdot 5}{12 \cdot 5} \text{ cm}^3 = 189 \cdot 56 \text{ cm}^3$$

($y = 12 \cdot 5$ cm, as this is the greater value.)

$$\frac{W \times l}{8} = fZ$$

$$\frac{W \times 2 \cdot 4 \times 1000}{8} = 165 \times 189 \cdot 56$$

$$W = 104 \cdot 3 \text{ kN}$$

Example 6 Find I_{XX} and I_{YY} for the angle section given in fig. 7.8.

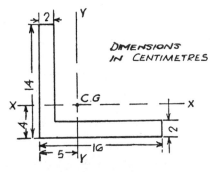

Fig. 7.8 *Angle section.*

Determination of c.g. position:

$$\{(12 \times 2)+(16 \times 2)\}\bar{x} = (24 \times 1)+(32 \times 8)$$
$$56\bar{x} = 280$$
$$\bar{x} = 5 \text{ cm}$$
$$56\bar{y} = (24 \times 8)+(32 \times 1)$$
$$\bar{y} = 4 \text{ cm}$$

I_{xx} for portion beneath the XX axis, treated as the difference of two rectangles

$$= \left(\frac{16 \times 4^3}{3} - \frac{14 \times 2^3}{3}\right) \text{cm}^4 = 304 \text{ cm}^4$$

I_{xx} for portion above XX axis

$$= \frac{bd^3}{3} = \frac{2 \times 10^3}{3} \text{ cm}^4 = 666 \cdot 7 \text{ cm}^4$$

Total $I_{xx} = 970 \cdot 6 \text{ cm}^4$

I_{YY} for portion to left of YY axis

$$= \left(\frac{14 \times 5^3}{3} - \frac{12 \times 3^3}{3}\right) \text{cm}^4 = 508 \cdot 7 \text{ cm}^4$$

I_{YY} for portion to right of YY axis

$$= \frac{2 \times 11^3}{3} \text{ cm}^4 = 887 \cdot 3 \text{ cm}^4$$

Total $I_{YY} = 1396 \text{ cm}^4$.

[The XX axis is parallel to the shorter leg in BS sections.]

Tabular method for unsymmetrical sections

Beam sections for which the neutral axis is not an axis of symmetry may be conveniently dealt with by means of a table. The value of I is first determined for an *axis at the base*. The table is arranged in a form suitable for slide-rule use.

Example Obtain I_{xx} for the girder section given in fig. 7.9.

The values of D, shown in the table in fig. 7.10, are the respective distances from the centres of gravity of the component areas to the axis AB, as indicated in fig. 7.9.

Properties of Compound Beam Sections

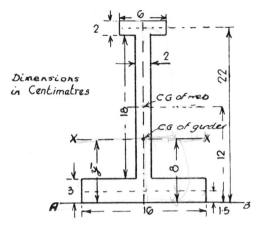

Fig. 7.9 Unequal flanged section.

Taking the top flange, we have

$$I_{CG} = \frac{bd^3}{12} = \frac{6 \times 2^3}{12} \text{ cm}^4 = 4 \text{ cm}^4$$

A = area = 6 cm × 2 cm = 12 cm²
D = distance from its c.g. to axis AB = 22 cm

The values for the web and bottom flange are similarly obtained.

$$I_{AB} = 12247 \text{ cm}^4$$

$$\bar{y} = \frac{\Sigma AD}{\Sigma A} = \frac{768}{96} \text{ cm} = 8 \text{ cm}$$

$$I_{AB} = I_{XX} + AD^2 \quad \text{or} \quad I_{XX} = I_{AB} - AD^2$$
$$\therefore \quad I_{XX} = I_{AB} - A\bar{y}^2$$
$$= (12247 - 96 \times 8^2) \text{ cm}^4$$
$$= 6103 \text{ cm}^4$$

The tabular method is useful when a beam section is rendered unsymmetrical by the deduction of connection holes in one flange only. The section is then treated as gross in the first instance, the holes being subsequently regarded as negative areas and entered as such in the table. In most practical cases the effect of the hole allowance on the N.A. position may be neglected.

154 Structural Steelwork

Part of section	I_{CG}	A	D	D^2	AD	AD^2	$I_{AB} = I_{CG} + AD^2$
Top flange	4	12	22	484	264	5808	5812
Web	972	36	12	144	432	5184	6156
Btm flange	36	48	1·5	2·25	72	243	279
		$\Sigma A = 96$			$\Sigma AD = 768$		$\Sigma I_{AB} = 12247$

Fig. 7.10 *Tabular method for moment of inertia.*

Compound girders

It is usual to allow for rivet or bolt holes in both flanges, in dealing with compound girder sections. The number of holes allowed will depend upon the style of connection. The properties of the UB's in the following examples are obtained from the tables on pages 122 to 142.

Example 1 Find the maximum moment of inertia, and the maximum section modulus, for the plated beam section given in fig. 7.11.

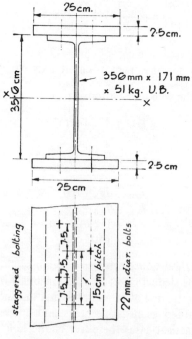

Fig. 7.11 *Plated joist section.*

I_{XX} for the gross section must first be determined.

Flange plates These may be dealt with in several ways. Section books often provide tables giving the I values for plates at stated distances apart. The principle of parallel axes may be employed, or the plates may be computed by regarding them as the difference between an outer and an inner rectangle.

a) Difference method:

$$I_{XX} = \left(\frac{25 \times 40 \cdot 6^3}{12} - \frac{25 \times 35 \cdot 6^3}{12}\right) \text{cm}^4 = 45428 \text{ cm}^4$$

b) Parallel-axis method:

$$I_{XX} = 2\left(\frac{25 \times 2 \cdot 5^3}{12} + (25 \times 2 \cdot 5) \times 19 \cdot 05^2\right) \text{cm}^4 = 45428 \text{ cm}^4$$

UB section

$$I_{XX} = 14118 \text{ cm}^4 \text{ (from tables)}$$

Total I_{XX} (gross) = $(45428 + 14118)$ cm^4
= 59546 cm^4

Hole allowance I_{XX} for a bolt-hole $= I_{CG} + AD^2$.

I_{CG} is extremely small, and may be neglected. The bolt-hole allowance is therefore AD^2 per bolt. A 22 mm diameter bolt will require a 24 mm diameter hole. The flange thickness of a $356 \times 171 \cdot 5 \times 51$ kg UB is 11·5 mm (see section tables), so that the area of one bolt hole = (3·65 cm × 2·4 cm) and its c.g. distance from axis XX = 18·475 cm.
One bolt hole must be allowed for in each flange in this case.

I_{XX} for bolt holes $= 2AD^2$
$= (2 \times 3 \cdot 65 \times 2 \cdot 4 \times 18 \cdot 475^2)$ cm^4
$= 5980$ cm^4

Net I_{XX} for girder section $= (59546 - 5980)$ cm^4
$= 53566$ cm^4

$$Z_{XX} = \frac{I_{XX}}{y} = \frac{53566}{20 \cdot 3} \text{ cm}^3 = 2639 \text{ cm}^3$$

Example 2 Calculate the safe total U.D. load, for an effective span of 6 m, for a compound girder of section given in fig. 7.12. Working stress = 165 N/mm².

156 *Structural Steelwork*

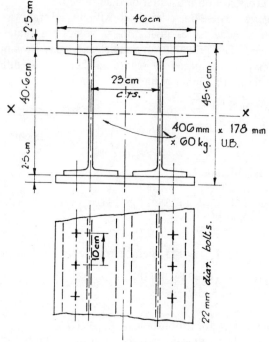

Fig. 7.12 *Compound girder.*

I_{XX} (gross section):   cm⁴
Plates

$$I_{XX} = 2\left(\frac{46 \times 2.5^3}{12} + (46 \times 2.5) \times 21.55^2\right) = 106\,920$$

UB sections

$$I_{XX} = 2 \times 21\,520 = \underline{43\,040}$$
$$\text{Total } I_{XX} \text{ (gross)} = \underline{149\,960}$$

Hole allowance

$$ 149\,960$$
$$4 \times 2.4 \times 3.78 \times 20.91^2 = \underline{15\,860}$$
$$I_{XX} \text{ (net section)} = \underline{134\,100}$$

$$Z \text{ maximum} = \frac{134\,100}{22.8} = 5880 \text{ cm}^3$$

$$M = fZ$$

$$\frac{W \times 6 \times 1000}{8} = 165 \times 5880$$

$$W = 1294 \text{ kN}$$

Plate-girder sections

Plate-girder sections may be designed in one of two ways: either by using the moment of inertia of the *gross* section, in which case the maximum flexural stress is assessed on the ratio of the gross area to the effective area of the flange section, or by using the *net* moment of inertia. In the latter case it is usual to deduct rivet holes from *both* flanges, in order to simplify the calculations by taking the neutral axis at the half depth of the section.

Example Determine I_{xx} for the plate-girder section given in fig. 7.13, allowing 22 mm diameter holes in the flanges.

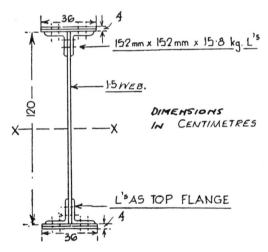

Fig. 7.13 *Plate-girder section.*

I_{xx} (gross section):

Plates

$$I_{xx} = 2\left(\frac{36 \times 4^3}{12} + (36 \times 4) \times 62^2\right) \text{cm}^4 = 1\,107\,456 \text{ cm}^4$$

Angles From angle section tables, it is found that a $150 \times 150 \times 15$ angle has an area of $43 \cdot 0 \text{ cm}^2$, and that the position of the c.g. of the section is 4·25 cm from the back of the angle. In the given example, therefore, the c.g. of each angle is $(60-4 \cdot 25) \text{ cm} = 55 \cdot 75 \text{ cm}$ from axis XX. I_{CG} for one angle is given as 898 cm^4 (gross).

I_{XX} for the four angles

$$= 4\{898 + (43 \cdot 0 \times 55 \cdot 75^2)\} \text{ cm}^4$$
$$= 538\,180 \text{ cm}^4.$$

Web

$$I_{XX} = \frac{bd^3}{12} = \frac{1 \cdot 5 \times 120^3}{12} \text{ cm}^4 = 216\,000 \text{ cm}^4$$

Total I_{XX} (gross section) $= 1\,896\,252 \text{ cm}^4$.

Hole allowance Allowance must be made in each flange for two 22 mm diameter holes (through flange plates and angles) and one 22 mm diameter hole (through angles and web plate). For the latter, the hole farther from axis XX is taken. The hole positions in standard sections are themselves standardised (see Appendix 3), and in the case of a $150 \times 150 \times 15$ angle, the first hole centre is 5·5 cm from the back of the angle. In the example, this will mean a distance of 54·5 cm from axis XX.

I_{XX} for flange holes

$$= 4 \times 2 \cdot 2 \times 5 \cdot 5 \times 61 \cdot 25^2 = 181\,576 \text{ cm}^4$$

I_{XX} for web holes

$$= 2 \times 2 \cdot 2 \times 4 \cdot 66 \times 54 \cdot 5^2 = 60\,900 \text{ cm}^4$$

Total I_{XX} for the holes $= 242\,476 \text{ cm}^4$

Net I_{XX} for girder section
$$= (1\,896\,252 - 242\,476) \text{ cm}^4$$
$$= 1\,653\,776 \text{ cm}^4$$

Exercises 7

(Consult pages 122 to 125 for section properties of UB's.)
1 A steel beam has the following dimensions: overall depth = 300 mm flange width = 150 mm, flange and web thickness, 20 mm and 10 mm respectively. Calculate I_{XX}, (a) using the principle of parallel axes, (b) by any standard formula. Calculate also I_{YY} by any method.

Properties of Compound Beam Sections 159

2 Two rectangular sections, each 150 mm deep × 20 mm wide, are placed side by side. Calculate the distance between their central vertical axes if $I_{XX} = I_{YY}$, for the two sections combined.

3 Check the value of I_{XX} for the T-section, given in fig. 7.7, by means of the tabular method.

4 Find I_{XX} for the portion of trough section given in fig. 7.14.

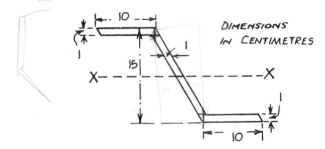

Fig. 7.14.

5 A 500 mm × 240 mm compound girder is composed of one 457 × 191 × 82 kg UB, with plates on each flange to form 240 mm × 20 mm. Calculate I (maximum) and Z (maximum), allowing one 24 mm diameter hole in each flange.

6 A 350 mm × 300 mm compound girder is built up of two 305 × 127 × 48 kg UB's with one 300 mm × 20 mm flange plate, top and bottom. The bolts used are 20 mm diameter. Calculate the safe total uniformly distributed load for this girder, for an effective span of 6 m, working stress = 165 N/mm². (Allow two bolt holes in each flange.)

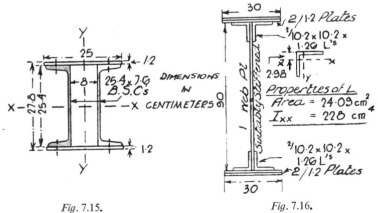

Fig. 7.15.   *Fig.* 7.16.

7 Find the value of I (maximum) and Z (maximum) for the girder section shown in fig. 7.15. The value of I_{XX} for one $254 \times 76 \times 28 \cdot 29$ kg BSC = 3367 cm^4. Allow two 20 mm holes in each flange. Flange thickness of BSC = 10·9 mm.

8 A plate girder, of the section given in fig. 7.16, has an effective span of 12 m. Allowing two 22 mm diameter holes in each flange, calculate the total safe uniformly distributed load for the girder (maximum stress allowable in each flange = 155 N/mm^2, $A = 24 \cdot 09$ cm^2, $I_{XX} = 228$ cm^4).

8

Deflection of Beams. Theory and Practice

Introduction

Not only have beams to be designed to support the applied loads without unduly stressing the fibres of the beam material, but *they must be made stiff enough to prevent excessive deflection*. The result of a large deflection in a beam carrying a plastered ceiling will be apparent, but the importance of 'stiffness' in beams lies deeper than this. The steel frame is composed of a large number of separate units which are, however, often connected together in a manner which prevents them acting entirely independently. A beam which is connected to a stanchion by a rigid form of connection tends to transmit to the stanchion the end slope it would have if it were freely supported.

Permissible deflection values

For many years, the maximum permissible deflection of a steel beam (with minor exceptions) was limited to 1/325 of the span. This limitation was in respect of the *full* load on the beam and took no consideration of the proportions of dead and applied load which made up the total load.

Clause 15 of BS 449 states:

> The maximum deflection of a beam shall be such as will not impair the strength and efficiency of the structure, lead to damage of the finishings or be unsightly.
> Measures may be necessary to nullify the effects of the deflection due to dead load by cambering or by adjustment in the casing.
> The maximum deflection due to loads other than the weight of the structural floors and roof, steelwork and casing, if any, shall not exceed 1/360 of the span.

This modification to previous regulations underlies the fact that the deflection due to dead load is always present and that, if the provisions of the first two paragraphs of the amended Clause are met, calculations are required only in the case of the applied load.

The span–depth ratio for beams is calculated on page 176.

The deflection allowance in any given case is governed by the particular application of the beam. For beams carrying brick walls—if the span exceeds 3·5 m—the maximum deflection should not exceed 2 mm per metre of span. Timber beams of large span are especially liable to excessive deflection if designed for strength alone. Usually the deflection consideration governs the design in such cases. Beams supporting plastered ceilings should not deflect more than 3 mm per metre of span.

As already referred to in a previous chapter, the suitability of a UB section, from the deflection point of view, is indicated in tables of safe loads by the type in which the loads are printed.

Bold type   Safe loads printed in **bold** type are greater than the web-buckling capacity provided by the beam, joist, or channel alone, and therefore web stiffeners may be necessary if sufficient additional capacity is not provided by the stiff bearing and bearing plate (if any).

Ordinary type   Safe loads printed in ordinary type are such that a beam supporting the full amount of the tabular load will have a total maximum deflection exceeding 1/360th of the span. For such cases, the deflection capacity should be inspected.

Italic type   Safe loads printed in *italic* type are within the web-buckling capacity of the unstiffened web and produce a total deflection not exceeding 1/360th of the span.

Deflection coefficients

The safe-load tables give these coefficients for each span.

The coefficient, $C = \dfrac{384E}{5 \times 360 \times 100L^2} = \dfrac{448}{L^2}$

where E = modulus of elasticity, taken as 210000 N/mm^2, and L = span in m.

It will be seen that the deflection coefficient is merely a function of the span. The detailed application of the deflection coefficient will be discussed later in this chapter.

The complete deflection problem is to determine the deflection anywhere in a beam, and not merely the maximum value. The various methods employed in such calculations will now be considered.

Circular bending

In the theory of bending relationship $E/R = M/I$, $R =$ the radius of curvature of the beam (i.e. of its neutral layer). For the beam to deflect to the arc of a circle, R must be constant, i.e. M/EI must be constant. Assuming E constant, this means that M/I must remain constant throughout the beam span.

i) *If I is constant,* i.e. if the beam has a constant section, *M must be constant.* This is not the usual case, as the bending moment generally varies from point to point in the span. Beams of constant section normally, therefore, do not bend in a circular manner. Fig. 8.1 illustrates a special case of a beam undergoing circular deflection.

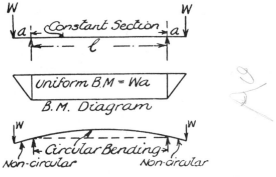

Fig. 8.1   Circular bending.

ii) *If M varies, I must vary in proportion* so as to make the ratio M/I constant. I does vary as M in the economical design of a plate girder, flange plates being dispensed with, as the bending moment falls away. Plate girders, so designed, may be regarded as having circular deflection.

By the geometry of the circle (see fig. 8.2),

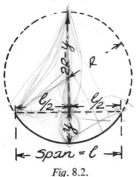

Fig. 8.2.

$$y(2R-y) = \frac{l}{2} \times \frac{l}{2}$$

$$2Ry - y^2 = \frac{l^2}{4}$$

y^2 is extremely small compared with the other quantities and may be neglected.

$$\therefore \quad 2Ry = \frac{l^2}{4}$$

$$y = \frac{l^2}{8R}$$

But $1/R = M/EI$ $\quad \therefore \quad y = \dfrac{Ml^2}{8EI}$

i.e. Maximum deflection in circular bending $= \dfrac{Ml^2}{8EI}$

(See page 338 for use of this formula in a plate-girder design.)

Example A plate girder, fulfilling the requirements for circular deflection, has an effective span of 10 m, and carries a U.D. load of 600 kN. The value of I for the girder section at the centre being 320 000 cm^4, and taking E as 210 kN/mm^2, calculate the maximum deflection.

$$y_{\text{maximum}} = \frac{Ml^2}{8EI}$$

$$M = \frac{Wl}{8} = \frac{600 \times 10 \times 1000}{8} \text{ kNmm} = 750\,000 \text{ kNmm}$$

$$\therefore \quad y_{\text{maximum}} = \frac{750\,000 \times 10 \times 1000 \times 10 \times 1000}{8 \times 210 \times 320\,000 \times 10\,000} = 13 \cdot 95 \text{ mm}$$

E is sometimes taken as 205 kN/mm^2 in beam-deflection problems.

General case of deflection

It will be useful to consider the question of the *slope* of a beam at the same time as that of its deflection. The 'slope' is expressed in radians, the angle being measured to the horizontal.

Mohr's theorem for the deflection and slope of a cantilever

In fig. 8.3, C and D are assumed to be so close together that the radius of curvature is R throughout the length δx of the cantilever.

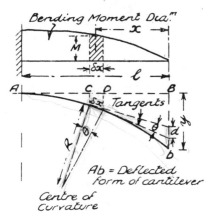

Fig. 8.3 *Deflection of a cantilever.*

The circular measure of the angle between the two radii shown

$$= \frac{\text{arc}}{\text{radius}} = \frac{\delta x}{R} \quad \therefore \quad \theta = \frac{\delta x}{R}$$

The tangents to the bent beam, at points C and D respectively, will also contain the angle θ, and its circular measure may be expressed very closely as d/x.

$$\therefore \quad \theta = \frac{\delta x}{R} = \frac{d}{x} \quad \therefore \quad d = \frac{x \delta x}{R}$$

But $1/R = M/EI$, $\quad \therefore \quad d = \dfrac{xM\delta x}{EI}$

The total deflection is the sum of all such little intercepts as d,

$$\therefore \quad y_{\text{maximum}} = \Sigma d = \frac{\Sigma x M \delta x}{EI}$$

$M\delta x$ represents the area of the little element of B.M. diagram standing on δx as base, and x is its distance from the end B of the cantilever.

∴ $\Sigma x M \delta x = A\bar{x}$, where A is the total area of the B.M. diagram and $\bar{x}$ is its centre-of-gravity distance from B.

$$\therefore y_{\text{maximum}} = \frac{A\bar{x}}{EI}$$

The diagrams in fig. 8.4 show how the result can apply to any point in the cantilever.

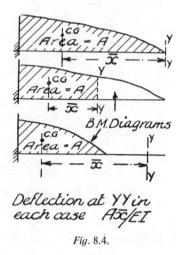

Fig. 8.4.

The total slope of the cantilever at B (fig. 8.3) will be

$$\Sigma \theta = \Sigma \frac{\delta x}{R} = \Sigma \frac{M \delta x}{EI}$$

$$= \frac{A}{EI}$$

The theorem will now be applied to a few standard cases of beams.

a) Cantilever with concentrated load W at the free end

In fig. 8.5
$$A = \tfrac{1}{2} \text{ base} \times \text{height}$$

$$= \tfrac{1}{2} l \times Wl = \frac{Wl^2}{2}$$

$$\bar{x} = \tfrac{2}{3} l$$

Deflection of Beams. Theory and Practice 167

$$y_{\text{maximum}} = \frac{A\bar{x}}{EI} = \frac{Wl^2/2 \times \tfrac{2}{3}l}{EI}$$

$$= \tfrac{1}{3}\frac{Wl^3}{EI}$$

Slope at B $= \dfrac{A}{EI} = \tfrac{1}{2}\dfrac{Wl^2}{EI}$

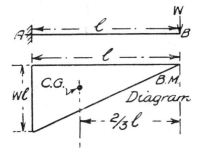

Fig. 8.5 *Cantilever with single end load.*

b) Cantilever with uniformly distributed load of total value W

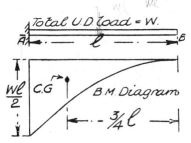

Fig. 8.6 *Cantilever with U.D. load.*

In fig. 8.6, $\quad A$ = area of parabola
  $= \tfrac{1}{3}$ base $\times$ height (see Appendix 4)
  $= \tfrac{1}{3}l \times \dfrac{Wl}{2} = \dfrac{Wl^2}{6}$

  $\bar{x} = \tfrac{3}{4}l$

$$y_{\text{maximum}} = \frac{A\bar{x}}{EI} = \frac{Wl^2/6 \times \tfrac{3}{4}l}{EI}$$

$$= \tfrac{1}{8}\frac{Wl^3}{EI}$$

$$\text{Slope at B} = \frac{A}{EI} = \tfrac{1}{6}\frac{Wl^2}{EI}$$

Example A cantilever projecting 1·5 m from its support (fig. 8.7) carries a load of 20 kN at its free end. The moment of inertia of the section = 3200 cm^4 and E = 210 kN/mm^2. Calculate the maximum deflection (a) by the standard formula, (b) by Mohr's theorem.

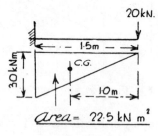

Fig. 8.7.

(a) $$y_{\text{maximum}} = \tfrac{1}{3}\frac{Wl^3}{EI} = \frac{\tfrac{1}{3} \times 20 \times (1·5 \times 1000)^3}{210 \times 3200 \times 10000} = 3·35 \text{ mm}$$

b) B.M. maximum = 20 kN × 1·5 m = 30 kN m.

$$A = \tfrac{1}{2} \text{ base} \times \text{height} = (\tfrac{1}{2} \times 1·5 \times 30) \text{ kN m}^2$$
$$= 22·5 \text{ kN m}^2$$
$$\bar{x} = \tfrac{2}{3} \times 1·5 \text{ m} = 1·0 \text{ m}$$

$$y_{\text{maximum}} = \frac{A\bar{x}}{EI} = \frac{22·5 \times 1·0 \times 1000^3}{210 \times 3200 \times 10000}$$

(The 1000^3 in the numerator is required to reduce the 'metres cubed' to 'millimetre' dimensions.)

$$y_{\text{maximum}} = 3·35 \text{ mm, as before.}$$

Simply supported beams may—by regarding them as pairs of inverted cantilevers—be made adaptable to Mohr's theorem, in the form given. A more direct mode of solution may, however, be employed in such cases.

Secondary B.M. method for deflection

It can be shown that *deflection* bears the same type of relationship to *bending moment* as *bending moment* does to *loading*. If, therefore, we treat the B.M. diagram for a beam as its load system, and recalculate the B.M. value for a given beam section on this basis, the value obtained will be a measure of the deflection at the section. The product EI has to be introduced to obtain the exact value.

The principle is illustrated in fig. 8.8. The deflection at C is given by

$$y_C = \frac{M'}{EI}$$

where M' is the secondary B.M. at C.

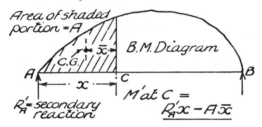

Fig. 8.8 *Deflection by secondary bending moment.*

c) Simply supported beam with single concentrated central load W

In fig. 8.9. $R_A' = $ secondary reaction $= \frac{1}{2}$ total area of B.M. diagram.

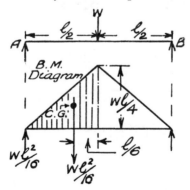

Fig. 8.9 *Simply supported beam with single concentrated central load.*

i.e.
$$R_A' = \left(\tfrac{1}{2} \times l \times \frac{Wl}{4}\right) \div 2$$
$$= \frac{Wl^2}{16}$$

$$M' \text{ (at centre of span)} = \left\{\frac{Wl^2}{16} \times \frac{l}{2}\right\} - \left\{\frac{Wl^2}{16} \times \left(\tfrac{1}{3} \times \frac{l}{2}\right)\right\}$$
$$= \frac{Wl^3}{32} - \frac{Wl^3}{96} = \frac{Wl^3}{48}$$
$$y_{\text{maximum}} = \frac{M'}{EI} = \frac{1}{48}\frac{Wl^3}{EI}$$

d) Simply supported beam with U.D. load of total value W

In fig. 8.10, $R_A' = \tfrac{1}{2}(\tfrac{2}{3} \times \text{base} \times \text{height})$
$$= \tfrac{1}{2}\left(\tfrac{2}{3} \times l \times \frac{Wl}{8}\right) = \frac{Wl^2}{24}$$

$$M' \text{ (at centre of span)} = \left\{\frac{Wl^2}{24} \times \frac{l}{2}\right\} - \left\{\frac{Wl^2}{24} \times \left(\tfrac{3}{8} \times \frac{l}{2}\right)\right\}$$
$$= \frac{Wl^3}{48} - \frac{3Wl^3}{384} = \frac{5}{384}Wl^3$$
$$y_{\text{maximum}} = \frac{M'}{EI} = \frac{5}{384}\frac{Wl^3}{EI}$$

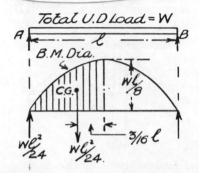

Fig. 8.10 *Simply supported beam with U.D. load.*

The reader should refer to Appendix 4 for the properties of a parabola.

Example 1 Calculate the maximum deflection of a 203 × 133 × 25 kg UB when it is carrying 80 kN U.D. load, for an effective span of 3·6 m (fig. 8.11).

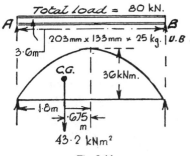

Fig. 8.11.

$I_{maximum}$ for section = 2348 cm^4, E = 210 kN/mm^2

$$y_{maximum} = \frac{5}{384}\frac{Wl^3}{EI} = \frac{5}{384} \times \frac{80 \times (3 \cdot 6 \times 1000)^3}{210 \times 2348 \times 10000} \text{ mm}$$

$$= 0.99 \text{ mm}$$

The result may be obtained by using the secondary B.M. method as follows:

$$\frac{Wl}{8} = \frac{80 \times 3 \cdot 6}{8} \text{ kNm} = 36 \text{ kNm}$$

Area of parabola = $\frac{2}{3}$ base × height
$= (\frac{2}{3} \times 36 \times 3 \cdot 6) \text{ kNm}^2$
$= 86 \cdot 4 \text{ kNm}^2$

Secondary reaction at A = $R_A' = \frac{86 \cdot 4}{2}$ kNm^2

$= 43 \cdot 2 \text{ kNm}^2$

Taking moments about the centre of the beam for the secondary bending moment, we get

$$M' = \{(43 \cdot 2 \times 1 \cdot 8) - (43 \cdot 2 \times 0 \cdot 675)\} \text{ kNm}^3$$

172 Structural Steelwork

Care must be taken with the dimensions of the results in these calculations, in order that correct reduction from 'metre' to 'millimetre' units may be effected, when desired.

$$M' \text{ at centre} = 48·6 \text{ kN m}^3$$
$$= (48·6 \times 1000^3) \text{ kN mm}^3$$

$$y_{maximum} = \frac{M'}{EI} = \frac{48·6 \times 1000^3}{210 \times 2348 \times 10000}$$

$$= 0·99 \text{ mm.}$$

Example 2 A $178 \times 102 \times 21·54$ kg joist carries a single concentrated load of 30 kN, as shown in fig. 8.12. I_{max} for this section $= 1519$ cm^4, and $E = 210$ kN/mm^2. Calculate the deflection under the load, and also determine the position and value of the maximum deflection.

$$\text{B.M.}_c = \frac{30 \times 1·2 \times 2·4}{3·6} = 24 \text{ kNm}$$

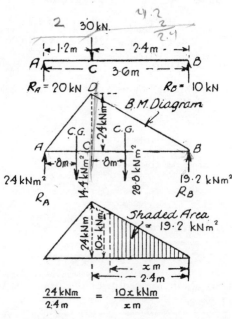

Fig. 8.12.

In the case of a single load, as in this example, it will be found that the maximum deflection will always occur at a point in the larger portion of the span. It will be convenient therefore to calculate R_B'.

Area of triangle ADC = $\frac{1}{2} \times 1\cdot 2 \times 24$ kNm² = 14·4 kNm²
Area of triangle DCB = $\frac{1}{2} \times 2\cdot 4 \times 24$ kNm² = 28·8 kNm²

Moments about A

$$R_B' \times 3\cdot 6 = \{28\cdot 8 \times (1\cdot 2 + 0\cdot 8)\} + \{14\cdot 4 \times (\tfrac{2}{3} \times 1\cdot 2)\}$$
$$3\cdot 6 R_B' = (28\cdot 8 \times 2) + (14\cdot 4 \times 0\cdot 8)$$
$$R_B' = 19\cdot 2 \text{ kNm}^2$$
$$M' \text{ at C} = (19\cdot 2 \times 2\cdot 4) - (28\cdot 8 \times 0\cdot 8)$$
$$= 23\cdot 04 \text{ kNm}^3$$

$$y_C = \frac{M_C'}{EI} = \frac{23\cdot 04 \times 1000^3}{210 \times 1519 \times 10000} = 7\cdot 2 \text{ mm}.$$

To find the position of maximum deflection, we make use of a rule (explained in Chapter 9) which states that the maximum B.M. in a beam occurs at the point where the shear force is zero. As deflection = M'/EI, maximum deflection corresponds to maximum secondary B.M. value.

To find the position of maximum deflection a point will have to be found at such distance from B, that the area of the B.M. diagram up to this point = R_B'.

Let x m = the distance.

B.M. at this section = $10x$ kNm (see fig. 8.12). This is because at 2·4 m the B.M. = 24 kNm, i.e. a numerical ratio of 10:1.

$$\therefore \quad \frac{x \times 10x}{2} = 19\cdot 2 \quad \text{i.e. } x^2 = 3\cdot 84$$

$$\therefore \quad x = 1\cdot 96 \text{ m}$$

M' at this section = $\{(19\cdot 2 \times 1\cdot 96) - (19\cdot 2 \times 1\cdot 96/3)\}$
$$= 25\cdot 09 \text{ kNm}^3$$

$$y_{\text{maximum}} = \frac{M'}{EI} = \frac{25\cdot 09 \times 1000^3}{210 \times 1519 \times 10000} \text{ mm} = 7\cdot 85 \text{ mm}$$

Graphical method for deflection

The reader will recall that, if a link polygon be drawn for the load system on a beam, it forms, with the closing line, the B.M. diagram for the beam. By treating the B.M. diagram (as in the previous examples) as the load diagram, and drawing a link polygon for this new 'load system', we will have a diagram which, to a certain scale, will give deflection for all points on the beam (see fig. 8.13). The graphical method

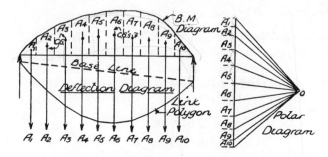

Fig. 8.13 *Graphical method for deflection.*

is by far the easiest method of dealing with deflection, if the loading is at all complicated. It has the advantage of exhibiting the deflection for all points in the beam span. Great care must be exercised in arriving at the correct scale for reading off the deflections from a diagram thus obtained.

Example Check the value of the deflection at the load point, and also the value of maximum deflection, for the example given in fig. 8.12.

The B.M. diagram is divided into a convenient number of strips, strips 0·3 m wide being taken in the given example (fig. 8.14). The areas of these strips are computed, and set down in the polar diagram to scale —just as loads are treated in the usual construction. The polar distance is made a round figure, e.g. 75 mm, to yield a convenient deflection scale. The latter is obtained by multiplying together the span scale, 'load' scale, and the polar distance, and then dividing the result by EI. Thus the diagram in fig. 8.14, as originally set out, had a scale for deflections:

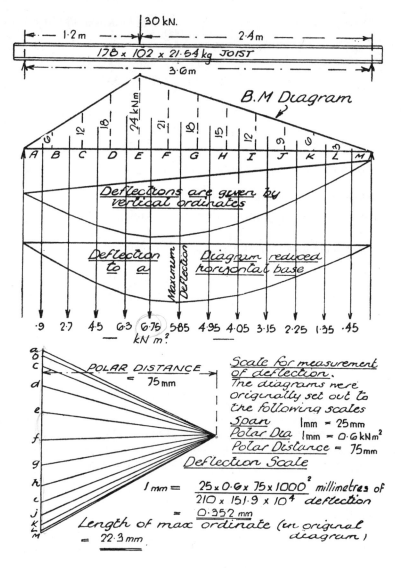

Fig. 8.14 Example of graphical method.

$$1 \text{ mm} = \frac{25 \times 0.6 \times 75 \times 1000^2}{210 \times 1522 \times 10000} \text{ mm of deflection}$$
$$= 0.352 \text{ mm}$$

Length of maximum ordinate was 22·3 mm, therefore
$$\text{maximum deflection} = (22.3 \times 0.352) \text{ mm}$$
$$= 7.85 \text{ mm}$$
$$\text{ordinate at load point} = 20.5 \text{ mm (actual mm)}$$
$$\therefore \text{deflection at load point} = (20.5 \times 0.352) \text{ mm}$$
$$= 7.22 \text{ mm}$$

With average care, the graphical method gives very accurate results.

Relationship between span and deflection

Taking the case of uniformly distributed loading, we have

$$\text{maximum deflection} = y_{\text{maximum}} = \frac{5}{384} \frac{Wl^3}{EI}$$

$$= \frac{5}{48} \times \frac{Wl}{8} \times \frac{l^2}{EI}$$

$$= \frac{5}{48} \times \frac{M}{EI} \times l^2 \qquad \left(M = \frac{Wl}{8}\right)$$

But $M/I = f/y$,

$$\therefore y_{\text{maximum}} = \frac{5}{48} \times \frac{f}{y} \times \frac{l^2}{E}$$

Assuming a maximum stress of 165 N/mm², putting $y = D/2$ ($\frac{1}{2}$ depth of beam) and inserting 210 kN/mm² for E, we get—for the usual case of beam sections, symmetrical about the neutral axis—

$$y_{\text{maximum}} = \frac{5}{48} \times \frac{330}{D} \times \frac{l^2}{210000}$$

$$\frac{y_{\text{maximum}}}{l} = \frac{l}{6109D}$$

Taking the BS 449 Clause 15 value for maximum deflection, i.e. 1/360 of the span,

$$\frac{1}{360} = \frac{l}{6109D}$$

Deflection of Beams. Theory and Practice 177

$$\text{or } l = \frac{6109D}{360} = 16 \cdot 97D$$

The corresponding figure for grade 50 steel, maximum permissible stress 230 N/mm², is 12·17, and that for grade 55 steel, maximum permissible stress 280 N/mm², is 10.

It must be remembered that these ratios concern the load that *causes the deflection*, since the 1/360 limit imposed by Clause 15 refers to what may be termed the *applied* load, over and above the weight of the construction itself, which is carried by a beam.

If this applied load is designated W_D, i.e. the load causing deflection, then $W_D = CI/1000$, where C is the deflection coefficient (see page 162) and I is the moment of inertia of the beam, in cm⁴.

The load W_D will be less than the tabular load if the span exceeds 16·97 times the beam depth for grade 43 steel, or lower ratios for grades 50 and 55 steel. For such cases, it may be necessary to confirm not only that the total load is within the capacity of the beam but also that the load to be considered for deflection purposes does not exceed W_D.

The following table gives limiting values of the span-to-depth ratio for simply supported beams carrying uniformly distributed loading.

Material	Limiting values of span-to-depth ratio for W_D/W_T					
	1·0	0·9	0·8	0·7	0·6	0·5
Grade 43 steel	16·97	18·86	21·21	24·24	28·28	33·94
Grade 50 steel	12·17	13·53	15·22	17·39	20·29	24·35
Grade 55 steel	10·00	11·11	12·50	14·29	16·67	20·00

W_D = load considered for deflection purposes,
W_T = total load on beam.

If the appropriate span-to-depth ratio is exceeded, then the deflection caused by W_D will exceed 1/360 of the span unless the bending stress is reduced.

The following examples illustrate the method to be adopted, and also illustrate the use of the deflection coefficient.

Example 1 A 406 × 178 × 60 kg UB of grade 43 steel carries a total U.D. load of 150 kN over a simply supported span of 9·0 m. The weight of the structural floor, steelwork, and casing is 30 kN. Check if the beam complies with Clause 15.

178 *Structural Steelwork*

The tabular load (see page 130) = 155 kN,

∴ the beam satisfies the condition of total load
$$W_D = 150 - 30 = 120 \text{ kN}$$
Deflection coefficient = 5·531

∴ I required $= \dfrac{120 \times 1000}{5 \cdot 531} = 21\,700 \text{ cm}^4$

The I of the beam is 21 520 cm⁴, and consequently the deflection will exceed 1/360 of the span.

To check by the table given above,

$$\frac{W_D}{W_T} = \frac{120}{150} = 0 \cdot 8$$

∴ limiting span–depth ratio = 21·21

Actual span–depth ratio $= \dfrac{9 \cdot 0 \times 1000}{406 \cdot 4} = 22 \cdot 14$

which also shows that the allowable deflection will be exceeded, so that the beam does not comply with Clause 15.

In the above example, the load considered for deflection purposes is an abnormal high proportion of the total load, and the span–depth ratio is also somewhat higher than those that will be normally encountered, and yet it will be seen that the beam only just fails to comply with Clause 15.

In most normal cases, *if actual deflection is the only criterion*, it will be unnecessary to make the check for deflection, because the ratio of W_D/W_T will normally be of the order of 0·5 to 0·7.

For the student it will be advisable to check each case, but the experienced designer will quickly select the abnormal cases which require a check.

Example 2 A mild-steel beam carries a total U.D. load of 100 kN over a span of 6·0 m. The part of the load to be considered for deflection purposes is 60 kN. If the maximum permissible stress is 165 N/mm², select the lightest U.B. which will carry this load in accordance with the requirements of Clause 15.

Minimum Z required $= \dfrac{\text{B.M.}}{f} = \dfrac{100 \times 6 \times 1000}{8 \times 165} = 454 \cdot 5 \text{ cm}^3$

Deflection coefficient for 6 m span = 12·44

$$W_D = 60 \text{ kN}$$

$$\therefore \text{ minimum } I \text{ required} = \frac{60 \times 1000}{12 \cdot 44} = 4824 \text{ cm}^4$$

Reference to the table on page 125 will show that the Z is the criterion, and the lightest section which will satisfy both requirements is a 356 × 127 × 33 kg UB.

In comparison with the previous example, it will be seen that the ratio $W_D/W_T = 0.6$, and yet it is still the Z which is the criterion and not the I. This confirms the statement made above that only abnormal cases will require to be checked.

Mathematical treatment of deflection

Readers not familiar with the calculus should omit the remainder of this chapter.

It is intended to exemplify this method of solution in only a few simple cases. A full treatment of this part of the subject will be found in the many excellent books on the theory of structures.

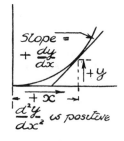

Fig. 8.15.

Convention of signs In fig. 8.15, x is positive to the right and y is positive upwards. The slope dy/dx is positive, y increasing as x increases. Slopes upwards, as we proceed to the right, will therefore be positive. The curvature shown is such that dy/dx increases positively as x increases, therefore d^2y/dx^2 will be positive in this case. Also this type of bending will have been brought about by bending moments which are positive, according to the convention adopted. We must therefore associate positive B.M.'s with $+d^2y/dx^2$. The application of the signs is illustrated in the selected examples given below.

Curvature is given in the calculus by the expression d^2y/dx^2, and in elasticity by the expression M/EI. Equating these values, we get the relationship which forms the basis of the integrations leading to deflection values:

$$\frac{d^2y}{dx^2} = \frac{M}{EI}$$

Case (i)–cantilever with single end load

In fig. 8.16,   B.M. at $x = -W(l-x)$

$$-\frac{d^2y}{dx^2} = \frac{W(l-x)}{EI}$$

$$-EI\frac{d^2y}{dx^2} = W(l-x)$$

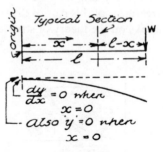

Fig. 8.16.

Integrating,

$$-EI\frac{dy}{dx} = Wlx - \frac{Wx^2}{2} + C_1 \text{ (a possible constant)}$$

To find the constant (if any), known values must be inserted. dy/dx (i.e. the slope) = 0, when $x = 0$; therefore $C_1 = 0$.

$$\therefore \quad -EI\frac{dy}{dx} = Wlx - \frac{Wx^2}{2}$$

Integrating again,

$$-EIy = \frac{Wlx^2}{2} - \frac{Wx^3}{6} + C_2 \text{ (another possible constant)}$$

$y = 0$, when $x = 0$. $\therefore$ C_2, in this case, also $= 0$.

$$\therefore \quad -EIy = \frac{Wlx^2}{2} - \frac{Wx^3}{6}$$

This expression will give the deflection at any point in the cantilever, by inserting the proper value of x.

For maximum deflection, put $x = l$.

$$-EIy = \frac{Wl}{2} \times l^2 - \frac{W}{6} \times l^3$$

$$= \frac{Wl^3}{3}$$

$$y_{\text{maximum}} = -\tfrac{1}{3}\frac{Wl^3}{EI}$$

The minus sign indicates that the deflection is downwards.

Case (ii)-simply supported beam with U.D. load

In fig. 8.17, B.M. at $x = \dfrac{wlx}{2} - \dfrac{wx^2}{2}$ (positive)

$$EI\frac{d^2y}{dx^2} = \frac{wlx}{2} - \frac{wx^2}{2}$$

Integrating, $\quad EI\dfrac{dy}{dx} = \dfrac{wlx^2}{4} - \dfrac{wx^3}{6} + C_1$

$\dfrac{dy}{dx} = 0$ at mid-span, when $x = l/2$

$$\therefore \quad EI \times 0 = \frac{wl}{4} \times \left(\frac{l}{2}\right)^2 - \frac{w}{6}\left(\frac{l}{2}\right)^3 + C_1$$

$$\therefore \quad C_1 = -\frac{wl^3}{24}$$

$$\therefore EI\frac{dy}{dx} = \frac{wlx^2}{4} - \frac{wx^3}{6} - \frac{wl^3}{24}$$

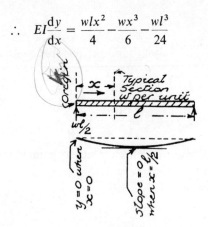

Fig. 8.17.

Integrating again,

$$EIy = \frac{wlx^3}{12} - \frac{wx^4}{24} - \frac{wl^3 x}{24} + C_2$$

$y = 0$, when $x = 0$. $\quad \therefore \quad C_2 = 0$

$$\therefore \quad EIy = \frac{wlx^3}{12} - \frac{wx^4}{24} - \frac{wl^3 x}{24}$$

For maximum deflection, put $x = l/2$.

$$EIy = \left(\frac{wl}{12} \times \frac{l^3}{8}\right) - \left(\frac{w}{24} \times \frac{l^4}{16}\right) - \left(\frac{wl^3}{24} \times \frac{l}{2}\right)$$

$$= wl^4 \left(\frac{1}{96} - \frac{1}{384} - \frac{1}{48}\right)$$

$$= -\frac{5}{384} wl^4$$

$$y_{\text{maximum}} = -\frac{5}{384} \frac{Wl^3}{EI}$$

The remaining standard cases already considered by the *area-moment* method may be taken by the reader as exercises in the mathematical method.

Beams with several load systems

Resolve the loading into simple systems, and deal with each system separately, by graphical or analytical methods. The net deflection at any given point will be the algebraic sum of the component deflections.

Exercises 8

(All beams are assumed to be simply supported at the ends. E is to be taken as 210 kN/mm^2 in the case of steel beams. Efficient lateral restraint of beams is assumed.)

1 A $381 \times 152 \times 52$ kg UB has $I_{max.} = 16046 \text{ cm}^4$. Calculate the maximum deflection for a U.D. load of 180 kN, the span being 6 m.

2 A $254 \times 102 \times 22$ kg UB projects 1·2 m horizontally from its support, and carries at its end a concentrated load of 30 kN. Taking $I_{max.} = 2863 \text{ cm}^4$, calculate:
(a) the maximum stress in the steel,
(b) the maximum deflection in the cantilever.

3 Draw the B.M. diagram for the cantilever of question 2, and check the value of the maximum deflection by applying Mohr's theorem. Find also the maximum slope.

4 In an experiment to determine Young's modulus for a given beam material, by a deflection experiment, it was found that a single central load of 2 kN produced a deflection of 1·5 mm. The span was 1 m, and I for the beam section 420 cm^4. Find E from these results.

5 A $356 \times 171 \times 51$ kg mild-steel UB carries a U.D. load. The ratio $W_D/W_T = 0.6$. What is the maximum permissible span of this beam if it is to comply with Clause 15. What maximum stress is assumed in this computation? What is the value of the total load?

$$I_{maximum} \text{ for the section} = 14118 \text{ cm}^4.$$

6 A grade 43 steel beam of 6 m span carries a single central load of 60 kN. $I_{maximum}$ for the beam section $= 12500 \text{ cm}^4$.

Determine the maximum deflection of the beam (a) by the standard formula, (b) by means of the secondary-bending-moment method.

7 Obtain the maximum deflection for the beam of question 6, by graphical construction.

(Divide the span into 0·6 m bays. The area of B.M. diagram will be, in order from the left, up to the mid-point of the beam: 5·4, 16·2, 27, 37·8 and 48·6 kN m^2 respectively.)

8 Find the maximum deflection for the compound girder given in fig. 7.12, assuming an effective span of 6 m and a total U.D. load (including the self-weight of the girder) of 1150 kN.

9 Treating the half-span of a simply supported beam as an inverted cantilever, and using the expression A/EI, show that the maximum slope (i.e. the slope at the ends) of a simply supported beam is (a) $Wl^2/16EI$ for a single central load W, and (b) $Wl^2/24EI$ for a U.D. load W.

10 A mild-steel beam has to carry a total U.D. load of 240 kN over a span of 8 m. The weight of the structural floor, steelwork and casing is 72 kN. What is the lightest UB which will carry this load and, at the same time, conform with the requirements of Clause 15? The maximum permissible stress is 165 N/mm^2.

9

Shear and its Applications

Relationship between shear force and bending moment in beams

In this chapter will be explained some of the important applications of shear in structural calculations.

Fig. 9.1 shows a simply supported beam, carrying a load system of a

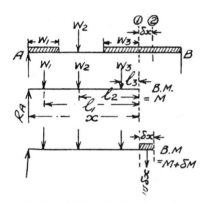

Fig. 9.1 *B.M. and S.F. relationship.*

general type. The B.M. at section 1 is assumed to be M, and to have increased by a small amount δM to a value $M + \delta M$ at section 2, the two sections being a small distance δx apart. A possible load w per unit run is shown as acting on this element of beam length.

$$M = \text{reaction moment} - \text{load moments}$$
$$= R_A x - W_1 l_1 - W_2 l_2 - W_3 l_3$$
$$M + \delta M = R_A(x + \delta x) - W_1(l_1 + \delta x) - W_2(l_2 + \delta x)$$
$$\quad - W_3(l_3 + \delta x) - \tfrac{1}{2} w \delta x^2$$

By subtraction, and dividing by δx,

186 *Structural Steelwork*

$$\frac{\delta M}{\delta x} = R_A - W_1 - W_2 - W_3 - \tfrac{1}{2}w\delta x$$

If δx be taken extremely small, $\tfrac{1}{2}w\delta x$ vanishes, so that, using the calculus notation, we have

$$\frac{dM}{dx} = R_A - W_1 - W_2 - W_3 = S,$$

where S represents the shear force at section 1.

$$\therefore \quad S = \frac{dM}{dx}$$

Integrating both sides, and changing over,

$$M = \int S dx$$

In simple language, these two important results may be expressed as follows:
i) the value of the shear force at any section of a beam is given by the slope of the B.M. diagram at that section;
ii) the difference in bending-moment values, for any two given sections of a beam, equals the area of the shear-force diagram between these two sections.

The case of a simple beam has been taken to investigate the foregoing relationships, but a similar analysis, in the cases of the beam types referred to later in the book, will show that the laws enunciated are true for all types of beams.

Illustrative example

As an overhanging beam involves both positive and negative bending moments, this type has been chosen to exemplify the relationships between B.M. and S.F. in beams.

In fig. 9.2, the B.M. and S.F. diagrams have been constructed in the usual way.

Portion CA From C to A, the slope of the B.M. diagram is uniform, negative, and equals $4\,\text{kN m}/2\,\text{m} = 2\,\text{kN}$. Therefore, by law (i), the shear force in this portion is constant, negative, and equals 2 kN.

Portion AD Slope of B.M. diagram $= +\{(12\cdot8+4)/6\}\,\text{kN} = +2\cdot8\,\text{kN}$, which equals the S.F. over this portion of the beam. The reader can now trace the remaining slopes, and show that they give the true S.F.

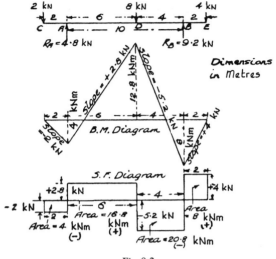

Fig. 9.2.

values. We will now reverse the process, and find B.M. values from the S.F. diagram.

Portion CA Area of S.F. diagram $= -(2\,\text{kN} \times 2\,\text{m}) = -4\,\text{kN m}$. By law (ii), this must be the difference in B.M. values between the sections at C and A. But B.M. at $C = 0$, $\therefore$ B.M. at $A = -4\,\text{kN m}$.

Portion CD Total net area of S.F. diagram $= \{(-2 \times 2) + (2\cdot 8 \times 6)\} = +12\cdot 8\,\text{kN m}$, which gives the B.M. at D. The B.M. at B may similarly be found, and it can be verified easily that the net area of the S.F. diagram from C to E equals zero, giving B.M. at $E = $ zero.

Applications of S.F.–B.M. relationships

The importance of the laws referred to does not lie in the mere deduction of B.M. from S.F., or vice versa. One important application gives a method of determining the *position of maximum B.M. in a beam*.

The slope of the B.M. diagram (shown in fig. 9.3) is zero at the point where the B.M. reaches its maximum value. If the diagram were made up of straight lines, the slope would suddenly change from a positive to a negative value, at the section corresponding to maximum B.M. But the slope of the B.M. diagram is given by the S.F. We have therefore the following very important rules:

i) the B.M. will be a maximum at the beam section at which the S.F. diagram crosses its base, i.e. passes through a zero value;

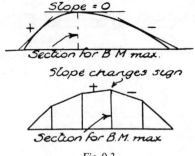

Fig. 9.3.

ii) if we proceed across the beam from the left end, just sufficiently far enough to take up load of equal value to the left-end support reaction, the point arrived at will be that of maximum B.M.

Example Find the position, and value, of maximum B.M. for the example given in fig. 9.4.

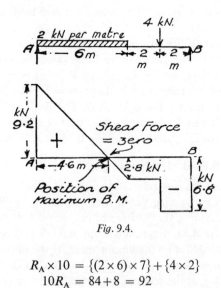

Fig. 9.4.

$$R_A \times 10 = \{(2 \times 6) \times 7\} + \{4 \times 2\}$$
$$10R_A = 84 + 8 = 92$$
$$R_A = 9 \cdot 2 \text{ kN}$$

To make up 9·2 kN, we must proceed 4·6 m across the beam from A; B.M. maximum is therefore at 4·6 m from A.

$$\text{B.M. maximum} = (9.2 \times 4.6) - \left(9.2 \times \frac{4.6}{2}\right) \text{ kNm}$$

$$= 21.16 \text{ kNm}$$

If, on arriving at a concentrated load by this method, it is found that its inclusion makes the load total too great, and that its exclusion gives too small a value, the maximum B.M. will actually occur at the load point. Fig. 9.2 shows an example of a S.F. diagram cutting its base line three times, each position corresponding to a *local* B.M. maximum value.

Rules for constructing S.F. diagrams

The following rules will be found helpful in constructing S.F. diagrams for simply supported beams.
i) Where there is no load on the beam, the diagram will be horizontal.
ii) A vertical load will cause a corresponding vertical jump in the diagram—vertically upwards for a reaction, and vertically downwards for a concentrated load.
iii) The diagram will slope uniformly for U.D. load. Throughout the diagram, the slope will remain constant for constant load per unit run. If the rate of load per unit run increases the slope will correspondingly increase. The slope is always downwards towards the right.

These rules will be found to apply in the cases of 'fixed' beams and 'continuous' beams respectively (see Chapter 10).

Example Draw the S.F. diagram for the case given in fig. 9.5.

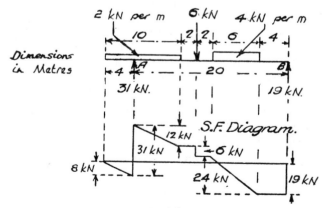

Fig. 9.5.

$$R_A \times 20 = (20 \times 19) + (6 \times 12) + (24 \times 7)$$
$$20R_A = 380 + 72 + 168$$
$$= 620$$
$$R_A = 31 \text{ kN}$$
$$R_B = 19 \text{ kN}$$

From the left end of the beam up to the point A, the slope of the S.F. diagram will be uniform, and the total drop will be (2 kN per metre × 4 metres) = 8 kN. The diagram will then jump vertically upwards a distance equivalent to 31 kN, owing to the vertical reaction at A. The remainder of the diagram may be similarly followed through.

Complementary shear stress

ABCD (fig. 9.6) is a very small square block of metal forming part of the web of a beam, the sides AB and CD being vertical. Assuming the block to be situated near the left end of the beam, the sides AB and CD

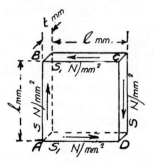

Fig. 9.6.

will be subjected to the type of shear stress indicated in the figure. The total load carried by each face $= s \, (\text{N/mm}^2) \times (l \times t) \, \text{mm}^2 = slt \, \text{N}$. These two forces constitute a couple tending to rotate the block in a clockwise manner with a moment of magnitude $slt \, \text{N} \times l \, \text{mm} = sl^2t$ N mm. It is clear that an equal and opposite couple must be acting on the block, to maintain equilibrium. The forces of this balancing couple are brought about by the stress induced in the fibres of metal along the horizontal faces BC and AD of the block. If $s_1 \, \text{N/mm}^2$ be this stress value, the corresponding couple will have a moment

$$s_1 lt \, \text{N} \times l \, \text{mm} = s_1 l^2 t \, \text{N mm}$$
$$\therefore \quad s_1 l^2 t = s l^2 t \text{ for equilibrium,}$$
$$\text{i.e. } s_1 = s$$

A vertical shear stress of a certain value at a given point in the web is, therefore, accompanied by a horizontal shear stress of equal intensity, at the point.

Further investigation shows that other complementary stresses are involved. Fig. 9.7 (a) indicates how the shear stresses, already referred

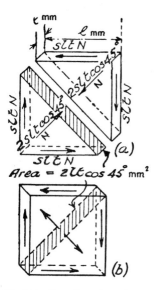

Fig. 9.7   Complementary stresses.

to, result in a *compressive stress* in the material. Resolving the shear loads at right angles to the internal diagonal plane of the block, and dividing by the area of the plane, to obtain the intensity of stress, we get

$$\text{compressive stress} = \frac{\text{load}}{\text{area}} = \frac{2slt \cos 45° \text{ N}}{2lt \cos 45° \text{ mm}^2}$$

$$= s \text{ N/mm}^2$$

Similarly, the *tensile stress* induced $= s$ N/mm² (fig. 9.7 (b)).

The original vertical shear stress is thus accompanied by both compressive and tensile stresses of equal intensity to its own—across planes at 45° to the horizontal. The theory would, of course, hold for the case of any stress units and may be extended to include any elastic material subjected to shear stress, not necessarily forming part of a beam web.

Variation of shear stress in a beam web

The result obtained by dividing the shear load (as obtained from the shear-force diagram) by the area of the web, gives, for any given beam section, the average shear stress in the web. The shear stress is not in fact a constant value for all points in the web depth, and the nature of the variation will now be discussed.

In fig. 9.8, AB and CD are two vertical sections of a beam, assumed

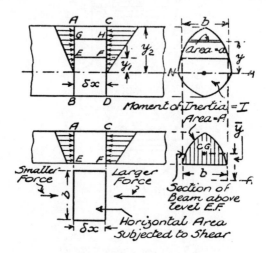

Fig. 9.8 *Horizontal shear stress.*

to be very close together. The bending moment at AB is M, and that at CD is $M+\delta M$. EF is a horizontal layer of material lying between the two vertical sections, its distance from the neutral layer of the beam being y_1. Consider the forces acting on the portion of beam from EF to the top (shown in the lower diagrams of fig. 9.8). Owing to the difference of B.M. at sections AB and CD, the end faces of this little piece of beam will be subjected to different stress intensities—and therefore to different resultant thrusts. The difference of these two end thrusts represents a force tending to slide the portion of beam over its base at EF. We can obtain the corresponding horizontal shear stress by dividing this force by the horizontal area of the base.

GH represents a typical layer of beam situated between level EF and the top of the beam. At the level of GH in section AB the compressive stress will be My/I (obtained from the standard formula $M = fI/y$).

The load on the corresponding elemental strip of cross-section of area a—at AB—will therefore be Mya/I.

The compressive stress at the level GH for section CD will be $(M+\delta M)y/I$, and the load on the corresponding cross-sectional strip $= (M+\delta M)ya/I$. For this one little strip the difference of the end loads will therefore be

$$\frac{(M+\delta M)ya}{I} - \frac{Mya}{I} = \frac{\delta M ya}{I}$$

To obtain the total differences of thrusts referred to, we must add up all these little differences—from level EF to the top of the beam, i.e. from $y = y_1$ to $y = y_2$.

Net resultant thrust $= \Sigma \dfrac{\delta May}{I}$

$$= \frac{\delta M}{I} \times \Sigma ay \quad \text{(between the stated levels)}$$

But $\Sigma ay = A\bar{y}$, where $A =$ the total area of beam section above EF and $\bar{y}$ is its c.g. distance from the *neutral axis* of the section.

$$\therefore \quad \text{Net resultant thrust} = \frac{\delta M}{I} A\bar{y}$$

$$\text{Shear stress at level EF} = \frac{\text{shear load}}{\text{area}} = \frac{\delta M}{\delta x \times b} \times \frac{A\bar{y}}{I}$$

As we take δx smaller and smaller, the value of $\delta M/\delta x$ becomes dM/dx, i.e. S, the shear force at the section of beam under consideration.

If $s =$ the intensity of horizontal (or vertical) shear stress at the level EF,

$$s = \frac{S}{bI} A\bar{y}$$

Application to a rectangular beam section

For the shear stress at the section EF (fig. 9.9), we have

$$b \text{ (in formula)} = b$$

$$I \quad \text{,,} \quad = \frac{bd^3}{12}$$

A (in formula) $= bx$

$\bar{y}$,, $= \left(\dfrac{d}{2} - \dfrac{x}{2}\right)$

$$s = \dfrac{S}{bI} A\bar{y}$$

$$\therefore \quad s = \dfrac{S}{b \times bd^3/12} \times bx \times \left(\dfrac{d}{2} - \dfrac{x}{2}\right)$$

$$= \dfrac{12S}{bd^3} \times \dfrac{x(d-x)}{2} = \dfrac{6S}{bd^3} \times x(d-x)$$

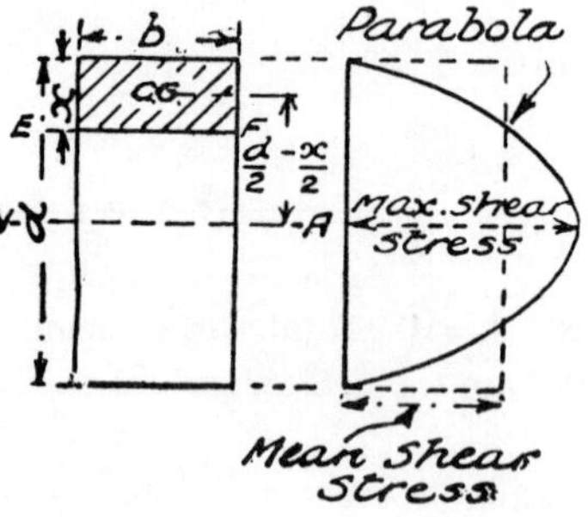

Fig. 9.9 Shear stress in rectangular beam.

If we plotted a diagram showing the shear variation for different values of x, the graph would be parabolic.

When $x = 0$, $s = 0$.

When $x = \dfrac{d}{2}$, $s = \dfrac{6S}{bd^3} \times \dfrac{d}{2} \times \dfrac{d}{2} = \dfrac{3S}{2bd}$.

But the average shear stress would be load/area = S/bd, so that our result shows that *the maximum shear stress, in the case of a rectangular beam section, is $1\frac{1}{2}$ times the mean value.*

Example A rectangular steel beam (fig. 9.10) is used to carry a total U.D. load of 60 kN, the section being 150 mm deep × 75 mm wide, and the span 4 m. Draw a diagram showing the variation of shear (horizontal and vertical) stress down the beam section, and write down the maximum shear stress in the steel.

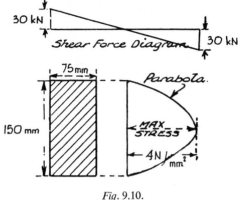

Fig. 9.10.

$$\text{Maximum shear load} = \frac{W}{2} = 30 \text{ kN}$$

$$\text{Maximum shear stress} = \tfrac{3}{2} \times \text{mean stress}$$

$$\text{Mean stress} = \frac{\text{load}}{\text{area}} = \frac{30 \times 1000}{150 \times 75} = \tfrac{8}{3} \text{ N/mm}^2$$

$$\therefore \text{Maximum shear stress} = \tfrac{3}{2} \times \tfrac{8}{3} \text{ N/mm}^2 = 4 \text{ N/mm}^2.$$

The maximum tensile and compressive bending stresses in this example will be found to be 105 N/mm². The steel is only required to develop a maximum shear stress value of 4 N/mm², much below what is permissible in the usual case. As we have seen, the UB form of beam section is more economical from the point of view of flexural stress, and the heavier loads carried will cause the maximum shear stress in the thin web to approach nearer its safe value, so that it requires to be checked in design.

Distribution of shear stress in a UB type of section

Example Illustrate the shear stress distribution at the given beam section (fig. 9.11), for a shear load of 60 kN. Compare the maximum shear stress with the mean value, as usually computed.

$$I \text{ for section} = \frac{BD^3}{12} - \frac{bd^3}{12}$$

$$I = \frac{100 \times 200^3}{12} - \frac{90 \times 160^3}{12} = 35\cdot 95 \times 10^6 \text{ mm}^4$$

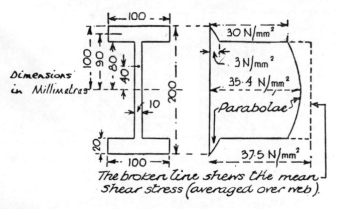

Fig. 9.11 Shear stress distribution.

Shear stress values
i) At level of flange and web junction
(a) Just inside flange:

$$s = \frac{S}{bI} A\bar{y}$$

$S = 60$ kN; $b = 100$ mm; $I = 35\cdot 95 \times 10^6$ mm^4; $A = 2000$ mm^2
$\bar{y} = 90$ mm

$$s = \left(\frac{60 \times 1000}{100 \times 35\cdot 95 \times 10^6} \times 2000 \times 90 \right) \text{N/mm}^2 = 3 \text{N/mm}^2$$

(b) Just inside web:

$$s = \left(\frac{60 \times 1000}{10 \times 35\cdot 95 \times 10^6} \times 2000 \times 90 \right) \text{N/mm}^2 = 30 \text{ N/mm}^2$$

ii) At neutral axis of beam

$b = 10$ mm
$A\bar{y} = [2000 \times 90 + \{(80 \times 10) \times 40\}] = 212\,000$ mm^3

$$s = \frac{S}{bI} A\bar{y}$$

$$s = \left(\frac{60 \times 1000}{10 \times 35\cdot 95 \times 10^6} \times 212\,000\right) \text{N/mm}^2$$

$$= 35\cdot 4 \text{ N/mm}^2$$

Maximum shear stress $= 35\cdot 4$ N/mm^2

$$\text{Mean shear stress} = \frac{\text{shear load}}{\text{area of web}} = \frac{60 \times 1000}{160 \times 10}$$

$= 37\cdot 5$ N/mm^2, or 30 N/mm^2 if the web is taken as 200 mm deep.

The maximum and mean values approximately agree. In most cases of standard sections, it will be sufficiently accurate to compute the shear stress by averaging over the web sectional area. The full beam depth is taken, in practice, in calculating web area for shear calculations. For web buckling see page 296.

Shear strain

The method of measurement of shear strain has been referred to in Chapter 1. The proportional law of elasticity applies to shear stresses and shear strains. The modulus of elasticity—corresponding to Young's modulus for tension and compression—is termed the *shear modulus*. It is sometimes referred to as the *modulus of rigidity*. In the case of mild steel, the value of the shear modulus is about 100 kN/mm^2. Shear strains in beams are very small, and may be neglected in calculations involving deflection.

Application of theory of shear to built-up beams

Stiffening of plate-girder webs

Fig. 9.12 shows a portion of a plate girder. The shear force in this case

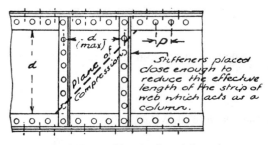

Fig. 9.12 *Buckling in plate girder web.*

is assumed to be negative, so that the compressive stresses in the web act at 45°, across the plane of compression indicated. If the web stiffeners shown are placed close enough together to both cut a plane of compression, i.e. if they are not farther apart than the depth d in the diagram, the lengths of inclined strips of web which have to act as virtual columns will be lessened. As will be seen later, in Chapter 11, the length of a column is a vital factor in its strength, so that the employment of stiffeners greatly increases the resistance of the web to 'buckling' or failure as a column. Various formulae of the column type, are in use, relating the spacing of stiffeners to web thickness. Practically, the spacing is a combination of theoretical principles and constructional requirements (see Chapter 15).

Connection in compound and plate girders (e.g. bolting)

The detail of the connection of the flange angles to the web, and to the flange plates, is concerned with the horizontal shear which accompanies the vertical shear. The shear load per metre length of girder flange has to be resisted by the bolt strength provided per metre of length.

Consider the bolts connecting the flange angles to the web in fig. 9.12. Expressing the equality of horizontal and vertical shear stress for the faces of a rectangular block of web one metre deep, one metre long, and t metres thick, we have (assuming S_V and S_H to be the shear loads)

$$\frac{S_V}{1 \times t} = \frac{S_H}{1 \times t} \quad \text{or} \quad S_V = S_H$$

i.e. the *horizontal shear load per metre length of girder = the vertical shear load per metre of depth*. This method of investigation is necessarily approximate, as variation of web stress has been neglected, but a closer analysis, based on the stress-variation theory, leads to the same result for the position in which the bolt line, or lines, are fixed.*

Let V N = value of one bolt = maximum shear load per pitch length.

Let p mm = pitch of bolts (single bolting, as in figure).

Let D mm = depth of web (sometimes taken as depth between bolt lines).

Let S N = shear force at the portion of the girder where the bolting is being considered.

Applying the above result we get

---

* See *Structural Engineering*, by J. Husband and W. Harby (Longman).

$$\frac{V}{p} = \frac{S}{D} \quad \text{or} \quad p = \frac{VD}{S}$$

In double bolting, as in 125 mm × 125 mm and larger angles, V = value of two bolts, and 'p' represents the straight-line pitch.

Example A plate girder of depth (web) 1 m carries a total U.D. load of 1000 kN. The web is 10 mm thick and 22 mm diameter bolts are used. Find the maximum permissible bolt pitch at the girder ends.

The value of one bolt in this case = dtf_b = (22 × 10 × 200) N = 44·0 kN.

Method 1   S.F. maximum = $\dfrac{W}{2} = \dfrac{1000 \text{ kN}}{2}$ = 500 kN.

∴ Shear load per metre of depth = 500 kN
∴ Shear load per metre length of girder = 500 kN

Number of bolts required per metre = 500/44·0 = 12, i.e. the maximum pitch = 80 mm.

Method 2 $p = \dfrac{VD}{S} = \dfrac{44 \cdot 0 \times 1000 \text{ mm}}{500}$ = 88 mm.

The usual bolt pitches are 75, 100, and 150 mm, and consequently a pitch of 75 mm would be used.

The bolt pitch may be changed to 150 mm, if desirable, at the section of the beam where the shear force has fallen to a value

$$S = \frac{VD}{150 \text{ mm}}, \quad \text{i.e.} \quad \frac{44 \cdot 0 \times 1000}{150} \text{ kN} = 293 \text{ kN}.$$

This position may be fixed by means of the S.F. diagram, or by calculation.

The bolt pitch in the flange plates will normally correspond with that in the 'angle-to-web' connection.

Shear reinforcement in R.C. beams

Fig. 9.13 illustrates how the induced tensile and compressive forces are resisted in the case of a reinforced-concrete beam. The weakness of concrete in tension necessitates steel reinforcement. As the B.M. falls away, some of the main reinforcing bars may be dispensed with, as far

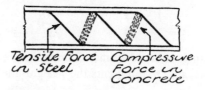

Fig. 9.13 *Shear reinforcement in R.C. beam.*

as their employment in providing moment of resistance is concerned. These are turned up to provide the necessary resistance in tension. In the diagram, the inclined steel cuts the tension planes (as in fig. 9.7 (b)) at right angles, thus taking up the stress brought about by the positive vertical shear force in the beam.

Exercises 9

1 Draw the shear-force diagram for the beam given in fig. 9.14. Show that the area (in kN m units) above the base line equals that below. Why is this?

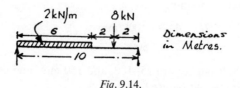

Fig. 9.14.

2 The maximum B.M. for the beam given in fig. 9.15 occurs at 9 m from the left end. Verify this by drawing a shear-force diagram, and calculate the value of B.M. maximum.

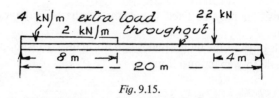

Fig. 9.15.

3 Deduce the load system which will result in the B.M. diagram given in fig. 9.16. (Construct the S.F. diagram by the *slope of B.M. diagram* method, and determine the loads from the vertical jumps in this diagram.)

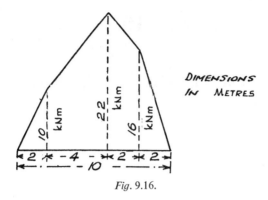

Fig. 9.16.

4 Obtain (without drawing a S.F. diagram) the position, and value, of the maximum B.M. for a beam of 12 m span, which carries two U.D. load systems, viz. 2 kN per metre for the whole span and 4 kN per metre in addition for the first 3 m of span, measured from the left end.

5 A rectangular beam section, 50 mm wide × 150 mm deep, is subjected to a vertical shear load of 240 kN. Calculate the intensity of shear stress at a level 50 mm below the upper surface. Obtain the value of maximum shear stress, and construct a diagram showing the variation of shear stress down the section.

6 A steel joist section has the following dimensions: flange width = 200 mm, flange thickness = 20 mm, overall depth = 400 mm, and web thickness = 10 mm. Construct a diagram showing the shear stress distribution across the section, for a vertical shear load of 200 kN.

7 A plate girder carries a total U.D. load of 2500 kN. Taking the particulars given, determine a suitable bolt pitch for the girder near the supports.

Depth of web = 1·2 m.
Thickness of web = 12 mm.
Bolt diameter = 22 mm (f_b = 200 N/mm², f_s = 110 N/mm²).
Hole diameter = 24 mm.

The flange angles are 150 × 150 × 12, requiring two lines of bolts.

10

Fixed and Continuous Beams

Fixed beams

The bending and shear effects produced in a beam by a given load system depend upon the way in which the ends of the beam are held in position. In fig. 10.1 (a), the ends are subjected to no end restraint, and the beam is said to be *simply supported*. In fig. 10.1 (b), the ends are

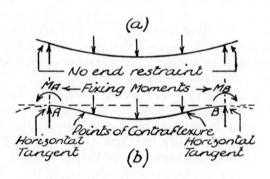

Fig. 10.1.

constrained to bend *until the tangent to the beam at each support is horizontal*. Such constraint represents *perfect fixture*, and the beam is termed a *fixed beam*. It is important to note that a beam is only partially fixed if it does not fulfil the qualification of zero slope at the supports. A beam rigidly fixed at its ends by top, web, and base cleats (or by welding) to a stanchion transmits a moment to the stanchion which the latter must be capable of resisting.

Clause 9b sets out three methods which may be employed in the design of a steel frame. These are the methods of *simple design, semi-rigid design*, and *fully-rigid design*.

Many designs are based on the *simple* method. This is probably because of the ease with which analysis can be made by this method, and the doubtful economy of other methods, in that any savings effected by a more detailed analysis could well be offset by the cost of the additional time spent in the analysis.

In practice, most beams are partially fixed at the ends. The *semi-rigid* method sets down some empirical rules which take account of this partial fixity.

It is possible that, with the increasing use of computers for making complex detailed analyses, the *fully-rigid* method may become more extensively used in the future. Prior to the use of the computer, the analysis of a fully-rigid, multi-bay, multi-storey steel frame by a manual process would have been very uneconomic from the point of view of the time spent.

The nature of the bending moments set up in a fixed beam will now be considered.

Relationship between fixed and overhanging beams

The negative support moments necessary to bring about the condition of fixed ends may be produced by overhanging the ends of the beam, as shown in fig. 10.2. In (a) we have the ordinary B.M. diagram, as for free

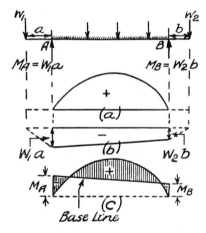

Fig. 10.2 *B.M. diagram for fixed beam.*

ends. In (b) is shown the B.M. diagram (negative) for loads W_1 and W_2—the loads introduced to create the required 'end-fixing' moments.

The positive and negative diagrams are superimposed in (c), the final net B.M. diagram being shown hatched. It should be noted that, in all such diagrams with a sloping base line, *the B.M. values are obtained by scaling vertically*—and not at right angles to the base. *The B.M. diagram for a fixed beam is thus the summation of two diagrams:* (i) *the ordinary B.M. diagram as for free ends* (*a positive diagram*), *and* (ii) *a negative B.M. diagram in the form of a trapezium*. The problem in any given case is, therefore, to obtain the dimensions of the *negative fixing trapezium* in order to superimpose it on the *free-end* B.M. diagram which is first drawn by the usual methods.

Properties of the fixing trapezium

The negative fixing trapezium referred to above has two important relationships with the 'free-end' B.M. diagram. These are determined by an application of Mohr's theorem for the deflection of a cantilever. It is assumed in fig. 10.3 that the ends A and B are at the same level, and that the beam is of constant section. We may regard the beam as being a cantilever loaded with a positive and with a negative B.M. diagram, in such a way that, with respect to the support end A, the deflection at the end B is zero.

Similarly, the support end may be assumed to be at B, and the deflection at A taken as zero. Using the symbols given in fig. 10.3, and

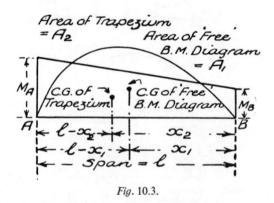

Fig. 10.3.

expressing Mohr's theorem for the case of the support at A, we get

$$\frac{A_1 x_1}{EI} - \frac{A_2 x_2}{EI} = 0$$

$$\therefore A_1 x_1 = A_2 x_2$$

For the support at B, and zero deflection at A, the theorem gives

$$\frac{A_1(l-x_1)}{EI} - \frac{A_2(l-x_2)}{EI} = 0$$

$$\therefore A_1(l-x_1) = A_2(l-x_2)$$
i.e. $A_1 l - A_1 x_1 = A_2 l - A_2 x_2$

But, as $A_1 x_1 = A_2 x_2$,
$$A_1 l = A_2 l$$
i.e. $A_1 = A_2$
and $x_1 = x_2$

The following relationships are thus established:
(i) the area of the fixing trapezium is equal to that of the 'free' B.M. diagram;
(ii) the centres of gravity of the two diagrams lie in the same vertical line, i.e. are equidistant from a given end of the beam.

Fixed beam with central concentrated load

The fixing trapezium is in this case a rectangle, as its centre of gravity has to be at mid-span. It must clearly also have half the height of the triangular free B.M. diagram in order to be of equal area. The net diagram for this case is, therefore, as shown hatched in fig. 10.4. The

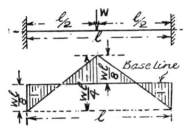

Fig. 10.4 *Fixed beam with single concentrated central load.*

maximum B.M. is halved by fixing the ends of the beam, but the supports are required to resist a B.M. of $Wl/8$, which is equal to that at the centre of the fixed beam.

Fixed beam with uniformly distributed load

The fixing trapezium will again be a rectangle (fig. 10.5). The area of the parabolic free B.M. diagram = $\frac{2}{3}$ base × height, so that the height

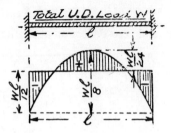

Fig. 10.5 *Fixed beam with U.D. load.*

of the rectangle for equal area must be $\frac{2}{3}$ × height of parabola, $= \frac{2}{3} \times Wl/8 = Wl/12$. In this case it will be seen that the maximum B.M. is not at the centre of the beam, but at the supports. This illustrates the importance of the consideration of the bending moments transmitted by such beams to members with which they are connected.

It will be seen from fig. 10.5 that the maximum positive B.M. $= Wl/24$. By the use of the *semi-rigid* method, as set down in Clause 9.b.2 of BS 449, the negative moments at the ends of the beam would each be $Wl/80$, while the positive B.M. at the centre would be $9Wl/80$.

Comparing the two methods, it will be seen that the maximum B.M. in the beam is $0.1125\ Wl$ using the semi-rigid method, compared with $0.0833\ Wl$ using the fully-rigid method, but that the fixing moments to be transmitted by the end connections to the stanchions are increased from $0.0125\ Wl$ with the semi-rigid method, to $0.0833\ Wl$ with the fully-rigid.

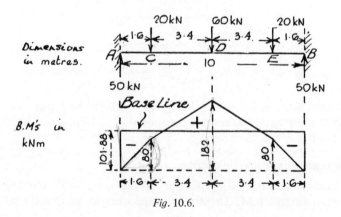

Fig. 10.6.

Fixed beam with a symmetrical load system

In all such cases, the fixing trapezium will clearly be a rectangle, as both c.g.'s will be at mid-span. The height of the rectangle is found by equating its area to that of the free B.M. diagram.

Example Construct the B.M. diagram for the fixed beam given in fig. 10.6.

Free B.M. diagram:

$$R_A = R_B = \frac{100 \text{ kN}}{2} = 50 \text{ kN}$$

$$\text{B.M.}_C = (50 \times 1 \cdot 6) \text{ kNm} = 80 \text{ kNm}$$
$$\text{B.M.}_D = \{(50 \times 5) - (20 \times 3 \cdot 4)\} \text{ kNm} = 182 \text{ kNm}$$

$$\text{Area of diagram} = 2\left(\frac{1 \cdot 6 \times 80}{2}\right) + 2\left(\frac{80 + 182}{2} \times 3 \cdot 4\right) \text{ kN m}^2$$

$$= (128 + 890 \cdot 8) \text{ kNm}^2$$
$$= 1018 \cdot 8 \text{ kNm}^2$$

Fixing trapezium:
The height of the rectangle will be $1018 \cdot 8 \text{ kNm}^2 / 10 \text{ m}$
$$= 101 \cdot 8 \text{ kNm}$$

The top of the rectangle forms the base line of the 'fixed beam' diagram.

Shear-force diagrams for fixed beams

In Chapter 9 it was explained that the shear force at any section of a beam is given by the slope of the B.M. diagram at the given section. The slope at any section will remain unaltered, if the base of the B.M. diagram remains horizontal. In all cases of symmetrical loading, therefore, the shear force values will be identical for both 'free' and 'fixed ended' beams, and the shear force diagrams will be the same for both cases.

If M_A, the fixing moment at A, is not equal to M_B, the fixing moment at B, the base line for the fixed beam B.M. diagram will be inclined, so that all slopes will be altered—but by the same amount, i.e. by the increase in slope. The *positive* increase in slope $= (M_A - M_B)/l$, so that it will be necessary to *lower* the base line of the original 'free' S.F. diagram by this amount, for the fixed-end condition. If the value is negative, i.e. if M_B is greater than M_A, then the base line must be *raised*.

Fixed beam with a single non-central concentrated load

This case lends itself to a simple solution derived from the properties of a trapezium.

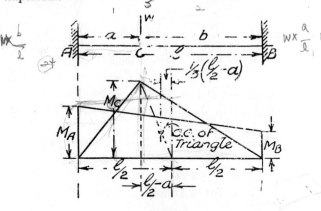

Fig. 10.7.

In fig. 10.7, $M_C =$ the free B.M. at C

$$= \frac{Wab}{l}$$

Equating the areas of the triangle (representing the 'free' diagram) and the 'fixing trapezium',

$$\frac{M_C \times l}{2} = \frac{M_A + M_B}{2} \times l$$

i.e. $M_C = M_A + M_B$

Distance of c.g. of triangle from B $= \dfrac{l}{2} + \tfrac{1}{3}\left(\dfrac{l}{2} - a\right)$

$$= \frac{l}{2} + \frac{l}{6} - \frac{a}{3} = \frac{2l-a}{3}$$

Fixed and Continuous Beams 209

Distance of c.g. of trapezium from B (using the standard expression for the c.g. position in a trapezium)

$$= \frac{2M_A + M_B}{3(M_A + M_B)} \times l$$

Equating these distances,

$$\frac{2M_A + M_B}{3(M_A + M_B)} \times l = \frac{2l - a}{3}$$

Let $M_A = xM_B$,

$$\therefore \quad \frac{(2xM_B + M_B)l}{3(xM_B + M_B)} = \frac{2l - a}{3}$$

$$\therefore \quad \frac{(2x + 1)l}{3(x + 1)} = \frac{2l - a}{3}$$

i.e. $6xl + 3l = 3(x + 1)(2l - a)$
$= 6lx - 3ax + 6l - 3a$
$\therefore \quad 3ax + 3a = 3l = 3a + 3b$

$$\therefore \quad ax = b, \quad \text{i.e. } x = \frac{b}{a}$$

$$\therefore \quad \frac{M_A}{M_B} = \frac{b}{a}$$

$$\therefore \quad \frac{M_A}{M_A + M_B} = \frac{b}{a + b} = \frac{b}{l}$$

But $M_A + M_B = M_C$

$$\therefore \quad M_A = \frac{b}{l} \times M_C$$

$$\text{and } M_B = \frac{a}{l} \times M_C$$

M_A and M_B having been obtained, the base line for the fixed beam can be constructed. The final diagrams are as shown in fig. 10.8. The base line of the S.F. diagram is *lowered*, the broken line representing the base for 'ends free'.

210 Structural Steelwork

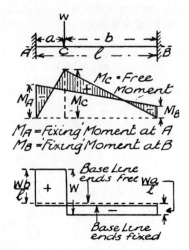

Fig. 10.8 Fixed beam with single non-central load.

Example Construct the B.M. and S.F. diagrams for the fixed beam given in fig. 10.9.

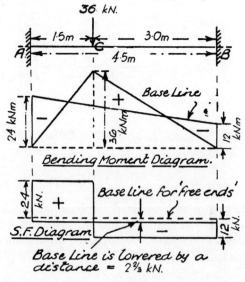

Fig. 10.9.

$$M_C \text{ (free B.M. at C)} = \frac{36 \times 3 \times 1 \cdot 5}{4 \cdot 5} \text{ kNm} = 36 \text{ kNm}$$

$$M_A = \frac{b}{l} \times M_C = \left(\frac{3 \cdot 0}{4 \cdot 5} \times 36\right) \text{ kNm} = 24 \text{ kNm}$$

$$M_B = \frac{a}{l} \times M_C = \left(\frac{1 \cdot 5}{4 \cdot 5} \times 36\right) \text{ kNm} = 12 \text{ kNm}$$

The diagram is constructed as shown.

$$R_A \text{ (for free ends)} = \frac{36 \times 3 \cdot 0}{4 \cdot 5} \text{ kN} = 24 \text{ kN}$$

$$R_B \text{ (for free ends)} = \frac{36 \times 1 \cdot 5}{4 \cdot 5} \text{ kN} = 12 \text{ kN}$$

The base line for the S.F. diagram (free ends) is shown by the broken line. For fixed end condition this must be *lowered* by

$$\frac{M_A - M_B}{l} = \frac{24 - 12}{4 \cdot 5} \text{ kN} = \frac{12}{4 \cdot 5} \text{ kN} = 2\tfrac{2}{3} \text{ kN}$$

Fixed beam with several concentrated loads

The B.M. diagram for free ends is first drawn. The fixing moment at the left end, corresponding to each load taken separately, is computed, and the total fixing moment is obtained by addition. Similarly, the total fixing moment at the right end is determined.

Example Fig. 10.10 shows a fixed beam carrying two concentrated loads. Construct B.M. and S.F. diagrams for the beam.

Free B.M. diagram:

$$R_A \times 5 = (60 \times 1) + (40 \times 3) = 180$$
$$R_A = 36 \text{ kN}, \; R_B = 64 \text{ kN}$$
$$M_C = (R_A \times 2) \text{ kNm} = 72 \text{ kNm}$$
$$M_D = (R_B \times 1) \text{ kNm} = 64 \text{ kNm}$$

NOTE: M_C and M_D are 'free' moments throughout.

$$M_C \text{ (due to 40 kN load alone)} = \frac{40 \times 3 \times 2}{5} \text{ kNm} = 48 \text{ kNm}$$

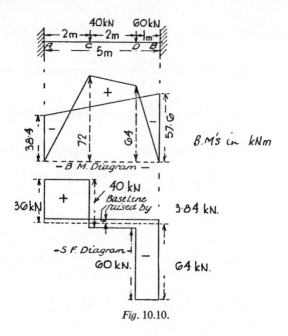

Fig. 10.10.

$\therefore$ M_A (due to 40 kN load alone) $= \dfrac{48 \times 3}{5}$ kNm $= 28\cdot8$ kNm

$\therefore$ M_B (due to 40 kN load alone) $= \dfrac{48 \times 2}{5}$ kNm $= 19\cdot2$ kNm

M_D (due to 60 kN load alone) $= \dfrac{60 \times 4 \times 1}{5}$ kNm $= 48$ kNm

$\therefore$ M_A (due to 60 kN load alone) $= \dfrac{48 \times 1}{5}$ kNm $= 9\cdot6$ kNm

$\therefore$ M_B (due to 60 kN load alone) $= \dfrac{48 \times 4}{5}$ kNm $= 38\cdot4$ kNm

Total $M_A = (28\cdot8 + 9\cdot6)$ kNm $= 38\cdot4$ kNm
Total $M_B = (19\cdot2 + 38\cdot4)$ kNm $= 57\cdot6$ kNm
The diagram is constructed as indicated.

M_B is greater than M_A, therefore the base line of the free-end S.F. diagram must be *raised*.

$$\frac{M_A - M_B}{l} = \frac{38 \cdot 4 - 57 \cdot 6}{5} \text{ kN}$$

$$= -3 \cdot 84 \text{ kN}$$

Deflection of fixed beams

a) Fixed beam with single concentrated central load

Fig. 10.11 illustrates a convenient method of solution. The beam is regarded as being a double inverted cantilever. The cantilever shown to the left of the central section is loaded with two B.M. diagrams, and

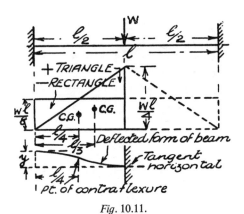

Fig. 10.11.

by applying Mohr's theorem for the maximum upward deflection y we get:

$$y = \frac{A\bar{x}}{EI} = \frac{1}{EI}\left\{\left(\frac{Wl}{4} \times \frac{l}{4} \times \frac{l}{3}\right) - \left(\frac{Wl}{8} \times \frac{l}{2} \times \frac{l}{4}\right)\right\}$$

$$= \frac{1}{EI}\left(\frac{Wl^3}{48} - \frac{Wl^3}{64}\right)$$

$$= \frac{1}{192} \frac{Wl^3}{EI}$$

b) Fixed beam with uniformly distributed load

Adopting the same method as in the last case, the expression for

214 *Structural Steelwork*

$$y \text{ (fig. 10.12)} = \frac{A\bar{x}}{EI}$$

$$= \frac{1}{EI}\left\{\left(\tfrac{2}{3} \times \frac{l}{2} \times \frac{Wl}{8} \times \frac{5}{16}l\right) - \left(\frac{Wl}{12} \times \frac{l}{2} \times \frac{l}{4}\right)\right\}$$

$$= \frac{1}{EI}\left(\frac{5Wl^3}{384} - \frac{Wl^3}{96}\right)$$

$$= \frac{1}{384}\frac{Wl^3}{EI}$$

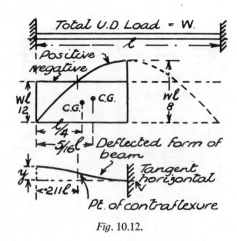

Fig. 10.12.

For other cases of loading, the graphical method given later may be employed.

Points of contraflexure

These are the points in a beam at which the character of the bending changes from positive to negative, or vice versa. In the case of a single central load (fig. 10.11) the points are clearly situated at one quarter of the span, respectively, from each support. In fig. 10.12, in which the load is uniformly distributed, the points can be fixed by equating the general expression for the 'free B.M.' at a given point in the span to $Wl/12$, the end fixing moment. If w be the load per unit run, the 'free' B.M. at x from the left end $= wlx/2 - wx^2/2$.

$$\therefore \quad \frac{wlx}{2} - \frac{wx^2}{2} = \frac{wl^2}{12}$$

$$6x^2 - 6lx + l^2 = 0$$

$$x = \frac{6l \pm \sqrt{(36l^2 - 24l^2)}}{12} = \frac{l}{2} \pm \sqrt{\frac{l^2}{12}}$$

$$= 0.5l \pm 0.289l$$
$$= 0.211l \quad \text{or} \quad 0.789l$$

i.e. the points of contraflexure are $0.211l$ from each support.

Continuous beams

A continuous beam is one which covers more than one span, so that it has at least three supports. Continuous beams are <u>not so common</u> in steelwork <u>as</u> in reinforced concrete. The calculation of B.M. values for a beam of this type is a reversal of the normal procedure in B.M. calculations. In the case of continuous beams the B.M. values are determined first, and the support reactions deduced therefrom. Evaluations of bending moments are effected by means of a theorem known as the '*theorem of three moments*'. A proof of this theorem will be found in books on the theory of structures.

Theorem of three moments*

In fig. 10.13 we have a continuous beam for which the *free* B.M.

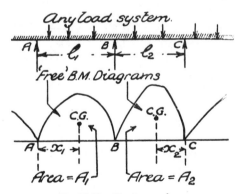

Fig. 10.13 *Continuous beam.*

---

*The method of solution of continuous beams by 'moment distribution' will be found in text-books on 'Theory of Structures' or 'Structural Analysis'.

diagrams are shown for two adjacent spans. These diagrams are drawn as if AB and BC were separate beams, having freely supported ends. Owing to continuity, there will be bending moments at the supports A, B, and C—similar in character to the fixing moments in fixed beams. The theorem connects up the values of these three 'moments' with quantities derived from the free B.M. diagrams.

If A_1 is the area of the diagram on span AB, and x_1 its c.g. distance from end A, and if A_2 is the area for span BC, and x_2 the c.g. distance from end C, the theorem is expressed as follows:

$$M_A l_1 + 2M_B(l_1 + l_2) + M_C l_2 = 6\left(\frac{A_1 x_1}{l_1} + \frac{A_2 x_2}{l_2}\right)$$

M_A, M_B, and M_C represent the *numerical* values of the support bending moments at A, B, and C respectively.

The theorem assumes uniform beam section throughout, and is true only provided the supports at A, B, and C are at the same level.

Expression of the theorem for U.D. load

If for each span in the series of spans the loading be uniformly distributed, the theorem may be expressed in simpler form.

In fig. 10.14, $A_1 = \frac{2}{3} \times \text{base} \times \text{height} = \frac{2}{3} \times l_1 \times w_1 l_1^2/8 = w_1 l_1^3/12$ and $x_1 = l_1/2$. Similar values will be obtained for the second span.

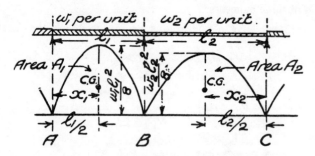

Fig. 10.14 *Uniformly distributed load.*

Inserting these special values in the general expression of the theorem, we get

$$M_A l_1 + 2M_B(l_1 + l_2) + M_C l_2 = 6\left(\frac{w_1 l_1^3/12 \times l_1/2}{l_1} + \frac{w_2 l_2^3/12 \times l_2/2}{l_2}\right)$$
$$= \tfrac{1}{4}(w_1 l_1^3 + w_2 l_2^3).$$

Example 1 Draw the B.M. and S.F. diagrams for the continuous beam given in fig. 10.15. The ends A and C are freely supported.

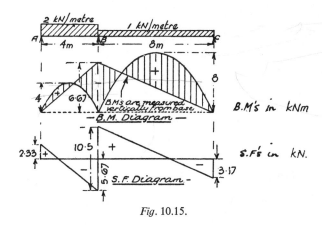

Fig. 10.15.

Free B.M. diagrams

Span AB, B.M. maximum $= \dfrac{Wl}{8} = \dfrac{8 \times 4}{8}$ kNm $= 4$ kNm ✓

Span BC, B.M. maximum $= \dfrac{Wl}{8} = \dfrac{8 \times 8}{8}$ kNm $= 8$ kNm

The free B.M. diagrams are parabolae, as shown.

Expressing the theorem of three moments,

$$M_A l_1 + 2M_B(l_1 + l_2) + M_C l_2 = \tfrac{1}{4}(w_1 l_1^3 + w_2 l_2^3)$$

$M_A = M_C = 0$, as the extreme ends are 'free',

$$\therefore \quad 2M_B(4+8) = \tfrac{1}{4}(2 \times 4^3 + 1 \times 8^3)$$
$$24\,M_B = 160$$
$$\therefore \quad M_B = 6{\cdot}67 \text{ kNm}$$

The negative support B.M. at B is therefore 6.67 kNm. This is set up to scale at B, and the B.M. diagram is completed by drawing in the base line. It is not necessary to reduce the diagram to a horizontal base, provided B.M. values are measured vertically from the jointed, inclined base shown.

Support reactions

These are obtained by the ordinary methods of moments, *but care must be taken with the sign of a support moment.*

Taking moments about B,

$$(R_A \times 4) - (8 \times 2) = -6.67 \quad \text{(note the sign)}$$
$$4R_A = 16 - 6.67 = 9.33$$
$$R_A = 2.33 \text{ kN}$$

Moments about B (to obtain R_C),

$$(R_C \times 8) - (8 \times 4) = -6.67$$
$$8R_C = 32 - 6.67 = 25.33$$
$$R_C = 3.17 \text{ kN}$$

Moments about C,

$$(R_A \times 12) + (R_B \times 8) - (8 \times 10) - (8 \times 4) = 0$$
$$8R_B = 80 + 32 - 28 = 84$$
$$R_B = 10.5 \text{ kN}$$

Check: $R_A + R_B + R_C = (2.33 + 10.5 + 3.17)$ kN
$$= 16 \text{ kN} \quad \text{(load on beam)}$$

The S.F. diagram is constructed by the rules given in Chapter 9, page 189.

Example 2 Fig. 10.16 shows a beam continuous over three spans, the extreme ends being freely supported. Construct the B.M. and S.F. diagrams for the beam.

Free B.M. maximum values

Span AB, B.M. maximum $= \dfrac{Wl}{8} = \dfrac{120 \times 15}{8} = 225$ kNm

Span BC, B.M. maximum $= \dfrac{Wl}{8} = \dfrac{80 \times 20}{8} = 200$ kNm

Span CD, B.M. maximum $= \dfrac{Wl}{8} = \dfrac{120 \times 10}{8} = 150$ kNm

Expressing the 'theorem of three moments' for the first two spans, viz. AB and BC, we have

$$M_A l_1 + 2M_B(l_1 + l_2) + M_C l_2 = \tfrac{1}{4}(w_1 l_1^3 + w_2 l_2^3)$$

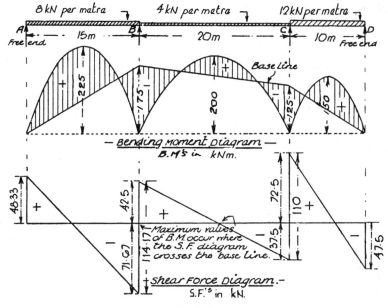

Fig. 10.16.

But $M_A = 0$,

$$\therefore 2M_B(15+20) + (M_C \times 20) = \tfrac{1}{4}(8 \times 15^3 + 4 \times 20^3)$$
$$70M_B + 20M_C = 14\,750$$
$$\therefore 7M_B + 2M_C = 1475 \quad \text{(i)}$$

In order to find M_B and M_C we require another simultaneous equation. This is obtained by considering spans BC and CD:

$$M_B l_2 + 2M_C(l_2 + l_3) + M_D l_3 = \tfrac{1}{4}(w_2 l_2^3 + w_3 l_3^3)$$

But $M_D = 0$,

$$\therefore 20M_B + 2M_C(20+10) = \tfrac{1}{4}(4 \times 20^3 + 12 \times 10^3)$$
$$20M_B + 60M_C = 11\,000$$
$$M_B + 3M_C = 550 \quad \text{(ii)}$$

Combining these equations,

$$7M_B + 2M_C = 1475 \quad \text{(i)}$$
$$M_B + 3M_C = 550 \quad \text{(ii)}$$

Multiplying (ii) by 7, and subtracting (i),

220 *Structural Steelwork*

$$19 M_C = 2375$$
$$M_C = 125 \text{ kNm}$$
$$\text{and } M_B = 175 \text{ kNm}$$

Support reactions:
$(R_A \times 15) - (120 \times 7 \cdot 5) = -175$
$\qquad R_A = 48 \cdot 33 \text{ kN}$
$(R_B \times 20) + (48 \cdot 33 \times 35) - (120 \times 27 \cdot 5) - (80 \times 10) = -125$
$\qquad R_B = 114 \cdot 17 \text{ kN}$
$(R_D \times 10) - (120 \times 5) = -125$
$\qquad R_D = 47 \cdot 5 \text{ kN}$
$(R_C \times 20) + (47 \cdot 5 \times 30) - (120 \times 25) - (80 \times 10) = -175$
$\qquad R_C = 110 \text{ kN}$

The sum of the reactions is 320 kN, which checks the numerical working.

Example 3 AC (fig. 10.17) is a continuous beam, freely supported at A and C. Construct the B.M. and S.F. diagrams for the beam, which carries the two given concentrated loads.

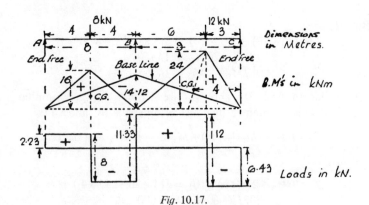

Fig. 10.17.

Area of free B.M. diagram for AB $= A_1 = (\tfrac{1}{2} \times 16 \times 8) \text{ kNm}^2$
$\qquad\qquad\qquad\qquad\qquad\qquad = 64 \text{ kNm}^2$
Area of free B.M. diagram for BC $= (\tfrac{1}{2} \times 24 \times 9) \text{ kNm}^2$
$\qquad\qquad\qquad\qquad \therefore \quad A_2 = 108 \text{ kNm}^2.$

$x_1 = 4 \text{ m}$ (measured to A), $x_2 = 4 \text{ m}$ (measured to C).

$$M_A l_1 + 2M_B(l_1+l_2) + M_C l_2 = 6\left(\frac{A_1 x_1}{l_1} + \frac{A_2 x_2}{l_2}\right)$$

$M_A = M_C = 0$,

$$\therefore 2M_B(8+9) = 6\left(\frac{64\times 4}{8} + \frac{108\times 4}{9}\right)$$

$$34M_B = 6\times 80$$
$$M_B = 14\cdot 12 \text{ kNm}$$

Reactions at supports:

$$(R_A \times 8) - (4\times 8) = -14\cdot 12$$
$$8R_A = 17\cdot 88$$
$$R_A = 2\cdot 23 \text{ kN}$$
$$(R_B \times 9) + (R_A \times 17) - (8\times 13) - (12\times 3) = 0$$
$$9R_B = 36 + 104 - (17\times 2\cdot 235)$$
$$9R_B = 101\cdot 92$$
$$R_B = 11\cdot 33 \text{ kN}$$
$$(R_C \times 9) - (6\times 12) = -14\cdot 12$$
$$9R_C = 72 - 14\cdot 12 = 57\cdot 88$$
$$R_C = 6\cdot 43 \text{ kN}$$

The diagrams are completed as shown in fig. 10.17.

Graphical method for deflection

The method shown in Chapter 8 for obtaining the deflection of a simply supported beam graphically, may be extended to fixed and continuous beams. Fig. 10.18 illustrates the method applied to one span

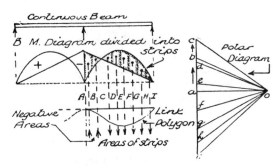

Fig. 10.18.

of a continuous beam. Negative areas are drawn upwards in the polar diagram, otherwise the procedure is as for simple beams. As the slope in space A = zero, in the example chosen, the pole *o* is conveniently chosen on a level with point *a* in the polar diagram. This is done to obtain a horizontal base for the deflection diagram.

Characteristic points

Bending-moment and shear-force diagrams for fixed and continuous beams are sometimes constructed by the aid of points termed *characteristic points*. The method is particularly useful when the conditions are rather complicated for the use of the 'theorem of three moments'. The theory is due to Professor Claxton Fidler. Readers interested in the application of the method may with advantage consult 'Selected Engineering Paper No. 46' of the Institution of Civil Engineers. This paper, entitled 'Characteristic Points', is written by Dr E. H. Salmon, M.I.C.E., and in it the author shows how the method may be developed and extended. It will be possible here to refer only briefly to the use of the theory in a few of the examples already considered.

Each free B.M. diagram for a span in a continuous beam (or for a given fixed-beam span) will have two characteristic points. To obtain these points the span is divided into three equal parts, and, at the third points in the span, ordinates are erected of a certain height. The tops of these ordinates are the characteristic points required.

For a given span AB, the ordinate nearer A must have such a height x as to satisfy the following equation:

$x \times \text{span}^2 =$ twice the moment of the free B.M. diagram about B.

Note that for the ordinate nearer A, the moment is taken about B, and vice versa.

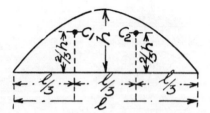

Fig. 10.19 *Characteristic points.*

Applying the rule to a parabolic B.M. diagram (fig. 10.19) of height h, we have

$$x \times l^2 = 2 \times (\tfrac{2}{3} l \times h) \times \left(\frac{l}{2}\right)$$

$$x = \tfrac{2}{3} h$$

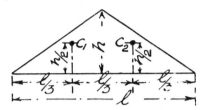

Fig. 10.20   *Characteristic points.*

The points are shown as C_1 and C_2 in fig. 10.19. For the case of fig. 10.20 (representing the B.M. diagram for a single central load):

$$x \times l^2 = 2 \times (\tfrac{1}{2} \times l \times h) \times \frac{l}{2}$$

$$x = \frac{h}{2}$$

Having constructed the free B.M. diagrams in the usual manner, the base line is drawn in—using the characteristic points—with the aid of the following rules.

Fixed beams

Simply draw the base line through the two characteristic points. This clearly gives the required B.M. diagram in the cases given in figs 10.19 and 10.20, and should be tested for the other fixed beam diagrams already taken.

Continuous beams

The method is a little more complicated, but will be set out as briefly as possible.

Take the example of fig. 10.21 (previously solved in fig. 10.16). The point C_1 (being adjacent to a free end) is not required. The base line must obviously pass above C_2, through C_2, or below C_2. If above C_2 it must be below C_3; if below C_2, it must be above C_3, i.e. it must alternate, above and below, for either side of a support. For two points, such as C_3 and C_4 *in the same span*, there is no such required relationship. If the base line passes through C_2, it must pass through C_3 (but not necessarily through C_4). There is one further relationship, governing the base-line position with respect to C_2 and C_3 (or C_4 and C_5). *The respective vertical distances between the base line and a pair of such points must be inversely as the spans in which they are situated.* Thus the distance below C_2 (in the example of fig. 10.21) is to that above C_3, in

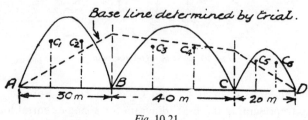

Fig. 10.21.

the same ratio, as span BC is to span AB, i.e. 40:30. The method is not so involved as it appears in a description, and after a few trial lines the correct base line can usually be fitted in. The reader is advised to draw out the given example to a fairly big scale, and to check the base-line position by the support moments already computed for this case.

For the example given in fig. 10.22 (see also fig. 10.17) the charac-

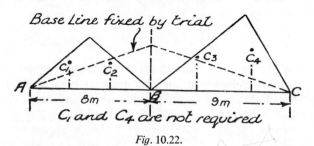

Fig. 10.22.

teristic point is just below the base line for span AB and slightly less above span BC, the ratio of distances being 9:8. Its position in AB will be 8/3 m ($= 2\frac{2}{3}$ m) towards A from B, and the height above base AB $= 16$ kNm/2 $= 8$ kNm (to scale). In BC it will be 9/3 m ($= 3$ m) from B, and x above base, where

$$x \times 9^2 = 2 \times \frac{9 \times 24}{2} \times 4 \quad \left[= \frac{\text{twice moment}}{\text{of area about C}} \right]$$

i.e. $x = 10\tfrac{2}{3}$ kN m   (to scale)

The characteristic points for an unsymmetrical B.M. diagram—as that for span BC in the last example—will not, of course, both be at the same height above the base line of the free B.M. diagram. To find the height of C_4, if it were actually required, the moment of the B.M. diagram area would be taken about B.

Exercises 10

1 A steel beam of 6 m span carries a total U.D. load of 120 kN. Taking a working stress of 165 N/mm², calculate the necessary section modulus for the beam, assuming (a) ends simply supported, (b) ends fixed, (c) semi-rigid design, taking the fixing moment as 10% of the free B.M. Draw the B.M. and S.F. diagrams for cases (b) and (c).

2 A 203 × 133 × 25 kg UB ($Z = 231$ cm³) is securely held by its end connections to columns, so that the ends may be regarded as being completely fixed. The span is 2·4 m and the loads carried are 72 kN (central) plus 36 kN (uniformly distributed). Find the B.M. transmitted to each column, and deduce the maximum stress in the steel of the beam.

3 A beam, fixed at both ends, has a span of 6 m. A concentrated load of 90 kN is carried at 2 m from the left end. Calculate the *fixing moment* at each support, and draw the B.M. and S.F. diagrams for the beam.

4 Draw the B.M. diagram for the example given in question 3 by means of *characteristic points*.

5 A continuous girder of 10 m total length, and of constant section, carries a uniform load of 60 kN per metre run. It rests freely on three supports, at the same level, one at each end and one 4 m from the left end. Calculate the B.M. at the intermediate support and the reaction at each support. Draw the B.M. and S.F. diagrams for the beam.

6 A beam, fixed at one end, is propped at the other end so that both ends of the beam are at the same level. Assuming the beam to carry a total U.D. load W, obtain an expression for the value of the pressure on the prop. (Assume the beam to be half a continuous beam of two equal spans, with a load W on each span.)

7 Fig. 10.23 shows a continuous beam carrying a wall of uniform thickness. Calculate the support bending moments and reactions, and

226 Structural Steelwork

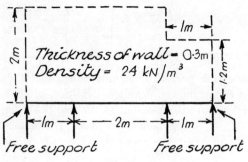

Fig. 10.23.

draw the B.M. and S.F. diagrams for the beam.

8 A fixed beam of 4 m span carries two concentrated loads, 80 kN at 1 m from the left end and 40 kN at 1 m from the right end. Calculate the positions of the characteristic points and complete the B.M. diagram. Obtain the end fixing moments and draw the S.F. diagram for the beam.

11

Practical Design of Compression Members

Introduction

In the computation of the strength of a compression member, several factors have to be considered which do not influence the calculations in the corresponding example of a member in tension. In the latter case, the question of length does not usually arise. Further, the methods by which the ends of a tie are fixed are important only in so far as they may influence the axiality of the load. In a compression member, these considerations are of vital importance. The word *strut* will be used in a general sense throughout the chapter to include all members in compression, such as *columns* or *stanchions*.

Length of a strut

The terms *long* and *short* are of a special relative character, when used in connection with 'struts'. A *long strut* may be actually shorter than a *short strut*. The terms have reference to the relationship between the actual length of the member and its cross-sectional dimensions. Thus a concrete cube, 150 mm high, would be a *short* strut, but a needle 50 mm long would be a *long* strut.

Long struts are liable to failure by side-bending or *buckling*, as well as by direct crushing, and the 'longer' the member is, the greater is the importance of the buckling tendency.

Classification of struts

(a) *Short struts*—in which failure is due to the direct crushing of the material, without the complication of buckling. The design of such members depends simply upon the permissible working stress in compression for the material. BS 449 gives 155 N/mm^2 for grade 43 steel as the maximum permissible working stress for uncased struts.

(b) *Medium struts*—in which failure is a combination of crushing (direct stress) and buckling. This group includes the majority of practical struts, and is discussed in detail later.

(c) *Long struts*—in which the direct stress plays an unimportant part in comparison with that due to buckling.

The actual numerical limiting values to be given to each of these groups cannot be assigned until we have considered the exact way in which the cross-sectional dimensions enter into the question of *length*.

Radius of gyration

In Chapter 6 two properties of section were considered. These had important implications in the design of beams. The property of section, now to be considered, is of equal importance in the design of 'struts'. It involves the 'moment of inertia' of the strut cross-section, and also its 'sectional area'. Experiment has shown that the association of these two quantities is in the form of the square root of their 'ratio'.

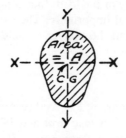

Fig. 11.1.

If I_{XX} = moment of inertia of the section given in fig. 11.1 about the axis XX, and A = its sectional area, the radius of gyration of the section with respect to the axis XX is given by the expression $\sqrt{(I_{XX}/A)}$.

Various symbols are employed to denote this property. The letter r is commonly used in regulations. Some formulae use the letter k. The letter g is also sometimes employed.

$$r_{XX} = \sqrt{\frac{I_{XX}}{A}}$$

Similarly,
$$r_{YY} = \sqrt{\frac{I_{YY}}{A}}$$

It is possible to constrain a strut so that its tendency to bend must be about a particular axis of its section. In the usual case no such restraint is present, and the radius of gyration has to be evaluated for the principal axis of section for which it has the lesser value. This is termed *least r* (or *least k*, etc.).

Calculation of radius of gyration values

The following examples illustrate the nature of the calculations involved in the derivation of this property. Bolt holes are not allowed for in obtaining r values, the gross section being always taken.

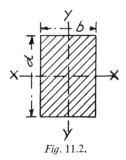

Fig. 11.2.

Example 1 Find r_{XX} for a rectangle b (wide) $\times$ d (deep) (fig. 11.2).

$I_{XX} = bd^3/12, \quad A = b \times d$:

$$r_{XX} = \sqrt{\frac{I_{XX}}{A}} = \sqrt{\frac{bd^3/12}{b \times d}} = \frac{d}{\sqrt{12}}$$

Similarly, $\quad r_{YY} = \dfrac{b}{\sqrt{12}}$

Example 2 Obtain an expression for r_{XX} (i.e. 'r' about a diameter) for a solid circular section of diameter D (fig. 11.3).

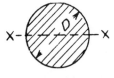

Fig. 11.3.

$$r_{XX} = \sqrt{\frac{I_{XX}}{A}} = \sqrt{\frac{\pi D^4/64}{\pi D^2/4}} = \sqrt{\frac{D^2}{16}} = \frac{D}{4}$$

Example 3 Calculate the value of least r for the universal column section given in fig. 11.4.

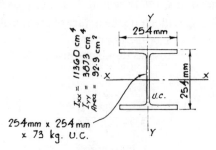

Fig. 11.4.

From tables, $I_{XX} = 11\,360$ cm^4, $I_{YY} = 3873$ cm^4, $A = 92 \cdot 9$ cm^2

$$r_{XX} = \sqrt{\frac{I_{XX}}{A}} = \sqrt{\frac{11\,360}{92 \cdot 9}} = 11 \cdot 05 \text{ cm}$$

$$r_{YY} = \sqrt{\frac{I_{YY}}{A}} = \sqrt{\frac{3873}{92 \cdot 9}} = 6 \cdot 45 \text{ cm}$$

Least r is therefore 6·45 cm. It is clear that *the least r value must be associated with the least I value.*

The values above may be checked by means of the tables given on page 127.

Example 4 Find the values of r_{XX} and r_{YY} for the compound column section of fig. 11.5.

I_{XX} for UC section = 11 360 cm^4

I_{XX} for plates = $2\{I_{CG} + AD^2\}$

$$= 2\left\{\frac{30 \times 2^3}{12} + (30 \times 2 \times 13 \cdot 7^2)\right\} = 22\,563 \text{ cm}^4$$

Total $I_{XX} = (11\,360 + 22\,563)$ cm^4 = 33 923 cm^4

Total area = $(92 \cdot 9 + 120)$ cm^2 = 212·9 cm^2

Practical Design of Compression Members   231

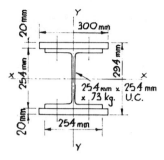

Fig. 11.5.

$$r_{XX} = \sqrt{\frac{I_{XX}}{A}} = \sqrt{\frac{33923}{212 \cdot 9}} = 12 \cdot 6 \text{ cm}$$

I_{YY} for UC section = 3873 cm^4

$$I_{YY} \text{ for plates} = 2\left(\frac{2 \times 30^3}{12}\right) = 9000 \text{ cm}^4$$

$$\text{Total } I_{YY} = (9000 + 3873) \text{ cm}^4 = 12873 \text{ cm}^4$$

$$r_{YY} = \sqrt{\frac{I_{YY}}{A}} = \sqrt{\frac{12873}{212 \cdot 9}} = 7 \cdot 77 \text{ cm}$$

Example 5   Calculate the values of the greatest and the least radii of gyration respectively for the compound column section given in fig. 11.6.

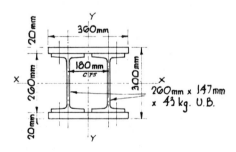

Fig. 11.6.

From section tables (page 124) the following properties of a 254 × 146 × 43 kg UB are obtained.

$I_{XX} = 6546$ cm^4, $I_{YY} = 633$ cm^4, $A = 55$ cm^2

Total I_{XX} for UB's $= 2 \times 6546 = 13092$ cm^4

$$\text{Total } I_{XX} \text{ for plates} = 2\left\{\frac{36 \times 2^3}{12} + (36 \times 2 \times 14^2)\right\}$$

$= 28272$ cm^4

Total I_{XX} for section $= (13092 + 28272)$ cm^4
$= 41364$ cm^4

Total area of section $= \{(2 \times 55) + (2 \times 36 \times 2)\}$ cm^2
$= 254$ cm^2

$$r_{XX} = \sqrt{\frac{I_{XX}}{A}} = \sqrt{\frac{41364}{254}} = 12.7 \text{ cm}$$

Total I_{YY} for UB's $= 2\{I_{CG} + AD^2\}$
$= 2\{633 + (55 \times 9^2)\}$ cm^4
$= 10176$ cm^4

$$\text{Total } I_{YY} \text{ for plates} = \frac{2 \times 2 \times 36^3}{12} \text{ cm}^4$$

$= 15552$ cm^4

Total I_{YY} for section $= (10176 + 15552)$ cm$^4 = 25728$ cm^4

$$r_{YY} = \sqrt{\frac{I_{YY}}{A}} = \sqrt{\frac{25728}{254}} = 10.1 \text{ cm}.$$

End fixture of struts

It is important to distinguish between two terms which are used in connection with the end fixture of a strut.

a) *Position fixed* This mode of fixture implies that the end is unable to move its position but also that the strut has freedom of bending, just as if the end were held by a pin joint (fig. 11.7 (a)).

Fig. 11.7.

b) *Direction fixed* Fig. 11.7 (b) illustrates a method of end fixture whereby the end of the strut is constrained to deflect in the manner of a fixed beam end. Such end fixture is termed *direction fixed*.

In the illustrations given later it will be seen that the form of the curve into which a strut tends to deflect depends upon the mode of end fixture. In each case there is a portion of the length of the strut which bends as if this part had pin-jointed ends. The length of this portion is known as the *equivalent* or *effective* length of the strut. The precise value to be taken for this length depends upon certain practical considerations (noted later). It will be convenient to explain here, in summary form, the incidence of *equivalent lengths* in the various methods used for strut calculations:

a) In formulae of the 'Euler' and 'Rankine' type (see later) the 'equivalent length' of the strut must be inserted.

b) BS 449 generally uses 'effective lengths' as do Codes of Practice and building regulations. However, the BS uses 'actual strut lengths' for 'angle struts' in some cases.

The values given below may be used by students in examination questions which involve the use of the formulae of 'Euler' or 'Rankine'.

Equivalent length of a strut

a) *Hinged ends*, i.e. ends 'position fixed' only. The strut bends freely from end to end (fig. 11.8), and the 'equivalent length' is the actual length of the strut.

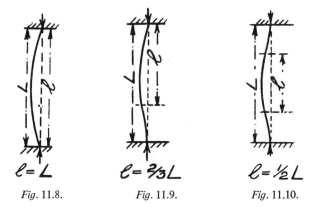

Fig. 11.8.   Fig. 11.9.   Fig. 11.10.

b) *One end hinged, one end fixed*, i.e. 'position fixed' only at one end, 'position and direction fixed' at the other (fig. 11.9). The value '$\frac{2}{3} \times$

actual length' is commonly taken for the equivalent length in this case.
c) *Both ends fixed*, i.e. 'position and direction fixed' at both ends. The equivalent length in this case is half the actual length—as indicated in fig. 11.10.
d) *One end fixed, one end free*, i.e. 'position and direction fixed' at one end, and no effective restraint at the other. This is the extreme case, and the equivalent length, as shown in fig. 11.11, is twice the actual length of the strut.

$\ell = 2L$

Fig. 11.11.

It is not easy to decide to which of these four sections any practical strut in a steel frame should be assigned, but commonly adopted standards will be found later in this chapter.

The strength of struts

The methods used for determining the strength of struts may be divided into three groups:

i) theoretical methods, based on certain ideal conditions—exemplified in Euler's theory;

ii) methods involving the use of formulae having a theoretical background, but which are rendered *empirical* (or 'practical') by the insertion of 'constants', which are found by the actual testing of struts—Rankine's formula is an example of this group;

iii) methods based on lists of working stresses, which are laid down in regulations issued by local authorities, or in standard specifications representing the results of research.

Euler's theory

Euler neglected direct stress, and derived a formula on the assumption that the strut would fail by 'buckling'. He assumed that the loading was perfectly axial and the material homogeneous throughout, and that the length was such that the assumption as to buckling failure was permissible. It is of interest to test Euler's expression for the ultimate load for a strut by a series of approximations.

In fig. 11.12, as a first approximation, assume that the strut bends to the arc of a circle.

Fig. 11.12.

$$y = \frac{Ml^2}{8EI} \quad \text{(for circular deflection)}$$

But $M = P \times y$ in this case,

$$\therefore \quad y = \frac{Pyl^2}{8EI}$$

$$\text{or } P = \frac{8}{l^2} \times EI$$

As a nearer approximation, assume that the B.M. does not remain constant (as is assumed in circular deflection with a constant section of the member), but varies, as in the case of a beam with U.D. load.

$$y = \frac{5}{384} \frac{Wl^3}{EI} = \frac{5}{48} \times \frac{Wl}{8} \times \frac{l^2}{EI}$$

$$= \frac{5}{48} \times M \times \frac{l^2}{EI} = \frac{5}{48} \times Py \times \frac{l^2}{EI}$$

$$\therefore P = \frac{9 \cdot 6}{l^2} EI$$

Euler in his theory obtained π^2 instead of 9·6, and defined P as the *least load required to produce instability*. He assumes that the strut remains perfectly straight until this critical load is reached, and that collapse takes place without any intermediate state of equilibrium.

$$P = \frac{\pi^2}{l^2} EI$$

Where l = the equivalent length of the strut in mm, E = Young's modulus (kN/mm^2), I = least moment of inertia (mm^4), and P = crippling load (axial) in kN.

A factor of safety of 4 is usually used in this theory for mild steel. The theory gives inadmissible results if the ratio of the length of the strut to its least radius of gyration is less than about 110 to 120.

Example A 120 × 120 × 15 BS angle is used as a strut in a truss. It may be assumed to have 'one end hinged' and 'one end fixed'. Its actual length is 3 m. Calculate the safe axial thrust.

Least I (from tables) = 185 cm^4; E = 210 kN/mm^2; factor of safety = 4.

Equivalent length of strut = $\frac{2}{3} \times 3000$ mm = 2000 mm

$$P = \frac{\pi^2}{l^2} EI = \left(\frac{\pi^2}{2000^2} \times 210 \times 185 \times 10000 \right) \text{kN}$$

$$= 937 \text{ kN}$$

Safe axial load = $\dfrac{P}{4}$ = 234 kN

Rankine's formula

Rankine gave the following formula for the crippling load of a strut, axially loaded:

$$P = \frac{A f_c}{1 + a(l/k)^2}$$

Where A = sectional area of strut, l = equivalent length of strut, k = least radius of gyration, f_c = a practical constant (associated with the yield point of the strut material) commonly taken as 330 N/mm² for mild steel, a = a constant, usually taken as 1/7500 for mild steel.

Taking the case of hinged ends (the case for which these formulae are originally considered) it can be shown that for low values of l/k the formula becomes $P = A \times f_c$, and for high values P = the corresponding Eulerian value, so that it agrees with accepted theory at the extreme values of the l/k range.

Example Using Rankine's formula, find the safe axial load for a 308 × 305 × 97 kg universal column 3·3 m high—to be regarded as having ends fixed—using the usual constants for mild steel and adopting a factor of safety of 4. The least radius of gyration for the given standard section is 7·67 cm, and the sectional area 123·3 cm².

$$P = \frac{Af_c}{1 + a(l/k)^2}$$

l = equivalent length of column = $\frac{1}{2} \times 3300$ = 1650 mm

$$a\left(\frac{l}{k}\right)^2 = \frac{1}{7500} \times \left(\frac{1650}{76\cdot7}\right)^2 = 0\cdot062$$

$$P = \frac{123\cdot3 \times 100 \times 0\cdot33}{1 + 0\cdot062} \text{ kN} = 4069 \text{ kN}$$

$$\text{Safe axial load} = \frac{4069}{4} \text{ kN} = 1017 \text{ kN}$$

The *Euler* and *Rankine* formulae are based on struts which are initially straight, and to which the load is applied concentrically.

Early in the nineteenth century, Young deduced a formula which made an allowance for an initial lack of straightness. He assumed that a strut was initially bent in the form of a sine curve.

In 1886, Perry and Ayrton produced a paper based on Young's hypothesis, but including, in addition, an allowance for eccentricity of loading.

In 1925, Robertson wrote a paper in which he stated that the combined effects of initial lack of straightness and eccentric loading are most conveniently dealt with by adopting the suggestion of Perry and Ayrton. However, Robertson made suggestions for some slight modifications to certain factors, and this resulted in the *Perry–Robertson* formula. This formula is given in Appendix B of BS 449, and is the formula from which the allowable stresses, set down in Table 17, are derived.

Practical interpretation of end fixture*

We have still to decide what constitutes adequate restraint in 'position', or in 'direction', or in both 'position and direction'. The following will be found to be common practice.

Columns of one storey or bottom length of a continuous column.

Let L = height from floor level to floor level.
With one-way connection, effective length = $1 \cdot 50L$.
With two-way connections, effective length = $1 \cdot 00L$.
With three- or four-way connections, effective length = $0 \cdot 85L$.

The same values are taken for the topmost length in a continuous column through several storeys, L being then the height from top floor level to roof.

Columns continuing through two or more storeys or top length of a continuous column.

L = height from floor level to floor level.
With one-way connection, effective length = $1 \cdot 00L$.
With two-way connections, effective length = $0 \cdot 85L$.
With three- or four-way connections, effective length = $0 \cdot 70L$.

Effect of eccentric loading†

In the case of a column continuing through several floors of a building, the connections of the main floor beams will have to be made to the flange or to the web of the column. The loads transmitted by these beams will therefore not coincide with the axis of the column. The effect of eccentricity of loading is very marked, and even small eccentricities may considerably reduce the carrying capacity of a column.

Consider the eccentric load W in the case of the *short* column shown in fig. 11.13. Its effect on the column is really two-fold: this may be demonstrated by introducing two vertical forces, as shown, each equal to W, at the centroid of the section. We have now an axial load, W, producing direct compressive stress of a uniform value all over the section, together with a bending moment, of value $W \times$ arm of eccen-

---

*See also page 251. †See also page 252.

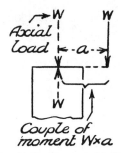

Fig. 11.13 *Eccentric loading.*

tricity, creating bending stresses. The resultant stress in any fibre in the section will be the algebraic sum of these stresses.

Example A short steel column of rectangular section (indicated in fig. 11.14) carries an eccentric load of 150 kN. Taking the dimensions given, calculate the maximum and minimum stresses produced, and draw a diagram showing the stress distribution across the section.

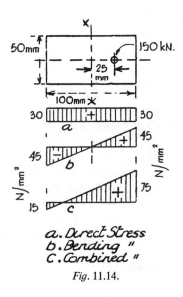

Fig. 11.14.

The eccentric load of 150 kN is equivalent in its effect to a concentric load of 150 kN, together with a bending moment of (150×25) kN mm.

$$\text{Direct stress} = \frac{\text{load}}{\text{area}} = \frac{150 \times 1000}{100 \times 50} = 30 \text{ N/mm}^2$$

Section modulus of column section about axis XX

$$= \frac{bd^2}{6} = \frac{50 \times 100 \times 100}{6} \text{ mm}^3 = \frac{500\,000}{6} \text{ mm}^3$$

$$M = fZ$$

$$\text{or } f = \frac{M}{Z} = \frac{150 \times 1000 \times 25 \times 6}{500\,000} \text{ N/mm}^2 = 45 \text{ N/mm}^2$$

∴ Maximum compressive and tensile bending stresses
  = 45 N/mm²

Total maximum compressive stress = (30+45) = 75 N/mm²

Total minimum compressive stress = (30−45) = −15 N/mm²

i.e. there is a resultant tensile stress of 15 N/mm². As the column is short, the maximum compressive stress is not excessive. We cannot decide, in general case, whether such stress is too great for a practical column, until the length and end fixture are known. *The maximum stress thus computed must not exceed the appropriate 'column stress', determined with reference to the axis of least radius of gyration—in this case YY.* A slight increase in the tabular column stress is permitted when part of the stress is due to eccentric loading.

(NOTE: The letter *r* is used for 'radius of gyration' in the following theory to comply with BS symbols used later.)

Equivalent concentric load

In fig. 11.15, let W_E = the actual eccentric load, and *a* the arm of eccentricity for the principal axis shown, Z = the section modulus for this axis, and y = distance of extreme fibre from the axis, measured towards the side on which the load is situated. If A = the sectional area of the column, the maximum compressive stress will be

$$\text{direct stress} + \text{bending stress} = \frac{W_E}{A} + \frac{W_E \times a}{Z}$$

Writing I/y for Z,

$$\text{maximum compressive stress} = \frac{W_E}{A} + \frac{W_E a y}{I}$$

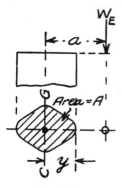

Fig. 11.15.

If r be the radius of gyration of the section for the given axis, $r = \sqrt{(I/A)}$ or $I = Ar^2$.

$$\therefore \text{ maximum compressive stress} = \frac{W_E}{A} + \frac{W_E ay}{Ar^2}$$

$$= \frac{W_E}{A}\left(1 + \frac{ay}{r^2}\right)$$

Let W_C = the concentric load which would produce, as a uniform stress all over the section, a stress intensity equal to the maximum actually created by the eccentric load.

$$\frac{W_C}{A} = \frac{W_E}{A}\left(1 + \frac{ay}{r^2}\right)$$

$$\therefore W_C = W_E\left(1 + \frac{ay}{r^2}\right)$$

Equivalent concentric load = actual eccentric load $\times \left(1 + \frac{ay}{r^2}\right)$

Equivalent concentric load

$$= \left\{\text{actual eccentric load} + \left(\text{eccentric load} \times \frac{ay}{r^2}\right)\right\}$$

The portion ay/r^2 simply depends for its value on the properties of the column section and the 'arm of eccentricity'. It may therefore be evaluated for a given standard section in terms of the 'eccentric arm'. The symbols *ex* and *ey* are often used instead of *a*, and they denote respectively *eccentricity with respect to axis XX* and *with respect to axis YY*. The phrase '*with respect to*' means in this connection '*at right angles to*'.

BS 449 methods

Axial stresses in struts

The working loads in column shafts are contained in Clause 30. Clause 30 (a) deals with *uncased struts* and reads as follows:

> The calculated average stresses on the gross sectional area of struts axially loaded shall not exceed p_c in Table 17, in which l/r is the effective length divided by appropriate radius of gyration.

Maximum slenderness ratio of struts

Clause 33 deals with the maximum permissible ratio of effective column length to radius of gyration, and reads as follows:

> 33. The ratio of the effective length, or of the length centre-to-centre of connections, to the appropriate radius of gyration, shall not exceed the following values:
> 1. For any member carrying loads resulting from dead weights with or without imposed loads and for single bolted or riveted single angle struts.   180
> 2. For any member carrying loads resulting from wind forces only, and provided that the deformation of such member does not cause an increase of stress, in any part of the structure, beyond the permissible stress.   250

Example Calculate the safe axial load for a 254 × 254 × 73 kg uncased universal column in grade 43 steel given:

area of section = 92·9 cm²
least r = 6·45 cm
effective height = 3 m
use the table given in fig. 11.16.

$$\frac{l}{r} = \frac{300 \text{ cm}}{6·45 \text{ cm}} = 46·5$$

TABLE 17a. ALLOWABLE STRESS p_o ON GROSS SECTION FOR AXIAL COMPRESSION

l/r	p_o (N/mm²) for grade 43 steel									
	0	**1**	**2**	**3**	**4**	**5**	**6**	**7**	**8**	**9**
0	155	155	154	154	153	153	153	152	152	151
10	151	151	150	150	149	149	148	148	148	147
20	147	146	146	146	145	145	144	144	144	143
30	143	142	142	142	141	141	141	140	140	139
40	139	138	138	137	137	136	136	136	135	134
50	133	133	132	131	130	130	129	128	127	126
60	126	125	124	123	122	121	120	119	118	117
70	115	114	113	112	111	110	108	107	106	105
80	104	102	101	100	99	97	96	95	94	92
90	91	90	89	87	86	85	84	83	81	80
100	79	78	77	76	75	74	73	72	71	70
110	69	68	67	66	65	64	63	62	61	61
120	60	59	58	57	56	56	55	54	53	53
130	52	51	51	50	49	49	48	48	47	46
140	46	45	45	44	43	43	42	42	41	41
150	40	40	39	39	38	38	38	37	37	36
160	36	35	35	35	34	34	33	33	33	32
170	32	32	31	31	31	30	30	30	29	29
180	29	28	28	28	28	27	27	27	26	26
190	26	26	25	25	25	25	24	24	24	24
200	24	23	23	23	23	22	22	22	22	22
210	21	21	21	21	21	20	20	20	20	20
220	20	19	19	19	19	19	19	18	18	18
230	18	18	18	18	17	17	17	17	17	17
240	17	16	16	16	16	16	16	16	16	15
250	15									
300	11									
350	8									

Intermediate values may be obtained by linear interpolation.

NOTE. For material over 40 mm thick, other than rolled I-beams or channels, and for Universal columns of thicknesses exceeding 40 mm, the limiting stress is 140 N/mm².

Fig. 11.16(a)   *Column stresses.*

TABLE 17b. ALLOWABLE STRESS p_c ON GROSS SECTION FOR AXIAL COMPRESSION

l/r	p_c (N/mm²) for grade 50 steel									
	0	1	2	3	4	5	6	7	8	9
0	215	214	214	213	213	212	212	211	211	210
10	210	209	209	208	208	207	207	206	206	205
20	205	204	204	203	203	202	202	201	201	200
30	200	199	199	198	197	197	196	196	195	194
40	193	193	192	191	190	189	188	187	186	185
50	184	183	181	180	179	177	176	174	173	171
60	169	168	166	164	162	160	158	156	154	152
70	150	148	146	144	142	140	138	135	133	131
80	129	127	125	123	121	119	117	115	113	111
90	109	107	106	104	102	100	99	97	95	94
100	92	91	89	88	86	85	84	82	81	80
110	78	77	76	75	74	72	71	70	69	68
120	67	66	65	64	63	62	61	60	60	59
130	58	57	56	55	55	54	53	52	52	51
140	50	50	49	48	48	47	47	46	45	45
150	44	44	43	43	42	42	41	41	40	40
160	39	39	38	38	37	37	36	36	36	35
170	35	34	34	34	33	33	33	32	32	31
180	31	31	30	30	30	30	29	29	29	28
190	28	28	27	27	27	27	26	26	26	26
200	25	25	25	25	24	24	24	24	23	23
210	23	23	23	22	22	22	22	22	21	21
220	21	21	21	20	20	20	20	20	20	19
230	19	19	19	19	19	18	18	18	18	18
240	18	18	17	17	17	17	17	17	17	16
250	16									
300	11									
350	8									

Intermediate values may be obtained by linear interpolation.

NOTE. For material over 65 mm thick, the allowable stress p_c on gross section for axial compression shall be calculated in accordance with the procedure in Appendix B taking Y_s equal to the value of the yield stress agreed with the manufacturer, with a maximum value of 350 N/mm².

Fig. 11.16(b)

TABLE 17c. ALLOWABLE STRESS p_c ON GROSS SECTION FOR AXIAL COMPRESSION

l/r	p_c (N/mm²) for grade 55 steel									
	0	1	2	3	4	5	6	7	8	9
0	265	264	264	263	262	262	261	260	260	259
10	258	258	257	256	256	255	254	254	253	252
20	252	251	250	250	249	248	248	247	246	246
30	245	244	244	243	242	241	240	239	239	238
40	236	235	234	233	232	230	229	227	226	224
50	222	220	219	217	214	212	210	208	205	203
60	200	197	195	192	189	186	183	180	178	175
70	172	169	166	163	160	157	154	151	148	146
80	143	140	138	135	133	130	128	125	123	121
90	118	116	114	112	110	108	106	104	102	100
100	99	97	95	93	92	90	89	87	86	84
110	83	82	80	79	78	76	75	74	73	72
120	71	69	68	67	66	65	64	63	62	62
130	61	60	59	58	57	56	56	55	54	53
140	53	52	51	50	50	49	49	48	47	47
150	46	45	45	44	44	43	43	42	42	41
160	41	40	40	39	39	38	38	37	37	37
170	36	36	35	35	34	34	34	33	33	33
180	32	32	32	31	31	31	30	30	30	29
190	29	29	28	28	28	28	27	27	27	27
200	26	26	26	25	25	25	25	25	24	24
210	24	24	23	23	23	23	23	22	22	22
220	22	22	21	21	21	21	21	20	20	20
230	20	20	20	19	19	19	19	19	19	18
240	18	18	18	18	18	18	17	17	17	17
250	17									
300	12									
350	9									

Intermediate values may be obtained by linear interpolation.

NOTE. For material over 40 mm thick, other than rolled I-beams or channels, and for Universal columns of thicknesses exceeding 40 mm, the limiting stress is 245 N/mm².

Fig. 11.16(c)

246 *Structural Steelwork*

From the table in fig. 11.16 (a),
$$p_c = 136 \text{ N/mm}^2$$
$$\therefore \text{ Safe axial load} = (136 \times 92 \cdot 9 \times 100) \text{ N}$$
$$= 1259 \text{ kN}$$

This result may be checked on page 267.

Cased struts. Radius of gyration

BS 449 also takes account of the strengthening given to a strut if the strut is cased in concrete, and Clause 30 (b) gives the requirements as to casing and the benefits that may be derived therefrom.

It is worth while to repeat this Clause in full.

> 30*b*. **Cased Struts** Struts of single I section or of two channels back to back in contact or spaced apart not less than 20 mm or more than half their depth and battened or laced in accordance with the requirements of Clauses 35 and 36 may be designed as cased struts when the following conditions are fulfilled:
> 1. The steel strut is unpainted and solidly encased in ordinary dense concrete, with 10 mm aggregate (unless solidity can be obtained with a larger aggregate) and of a works strength not less than 21 N/mm² at 28 days when tested in accordance with BS 1881, 'Methods for testing concrete'.
> 2. The minimum width of solid casing is equal to $b+100$ mm, where b is the width overall of the steel flange or flanges in mm.
> 3. The surface and edges of the steel strut have a concrete cover of not less than 50 mm.
> 4. The casing is effectively reinforced with wire to BS 4449, 'Hot Rolled steel bars for the reinforcement of concrete.' The wire shall be at least 5 mm diameter and the reinforcement shall be in the form of stirrups or binding at not more than 200 mm pitch, so arranged as to pass through the centre of the covering of the edges and outer faces of the flanges and supported by and attached to longitudinal spacing bars not fewer than 4 in number.
>
> The radius of gyration r of the strut section about the axis in the plane of its web or webs may be taken as $0 \cdot 2 \, (b+100)$ mm. The radius of gyration about its other axis shall be taken as that of the uncased section.
>
> In no such case shall the axial load on a cased strut exceed twice that which would be permitted on the uncased section, nor shall

the slenderness ratio of the uncased section, measured over its full length centre-to-centre of connections, exceed 250.

In computing the allowable axial load on the cased strut the concrete shall be taken as assisting in carrying the load over its rectangular cross section, any cover in excess of 75 mm from the overall dimensions of the steel section of the cased strut being ignored. This cross section of concrete shall be taken as assisting in carrying the load on the basis of a stress equal to the allowable stress in the steel (as given in Table 17) divided by 0·19 times the numerical value of p_{bc} given in Table 2 for the grade of steel concerned.

NOTE: This Clause does not apply to steel struts of overall sectional dimensions greater than 1000 mm × 500 mm, the dimension of 1000 mm being measured parallel to the web, or to box sections.

It will be noted that, whereas in previous issues of BS 449 the concrete casing was considered to be effective to the extent of stiffening the steel section in that an increased value of the radius of gyration about the weak axis could be used, it is now permitted to consider the concrete casing as assisting the steel section in the carrying of the axial load.

Example A 254 × 254 × 73 kg cased universal column of grade 43 steel has the minimum edge concrete cover of 50 mm, making the overall size, 354 mm × 354 mm. The section area of the steel joist = 92·9 cm² and the effective height of the stanchion is 3 m. Calculate the safe axial load.

Radius of gyration about YY axis =
$$0·2(b+100) \text{ mm} = 0·2(254+100) \text{ mm} = 70·8 \text{ mm}$$

$$\frac{l}{r} = \frac{3000 \text{ mm}}{70·8 \text{ mm}} = 42·4$$

p_c (from fig. 11.16(a)) = 137 N/mm²

Cross-sectional area of concrete casing = 354 × 354
= 1·253 × 10⁵ mm²

∴ Safe axial load = $137 \times 92·9 \times 100 + \dfrac{137 \times 1·253 \times 10^5}{0·19 \times 165}$

= 1 273 000 + 547 500 = 1 820 500 N
= 1820·5 kN

As in no case shall the axial load on the cased strut exceed twice that which would be permitted on the uncased section, this check will now be made.

By reference to the previous example, twice the load on the uncased strut $= 2 \times 1263 = 2526$ kN,

∴ The safe axial load on the cased strut $= 1820 \cdot 5$ kN.

By comparing this value with that given for the previous example, we can see the effect of the casing.

Angles as struts. Permissible stresses.

The permissible stresses in angle struts are governed by the type of connection at the ends of the struts, in accordance with the following rules:

Case 1

Clause $30c(i)$. For single-angle discontinuous struts connected to gussets or to a section either by riveting or bolting by not less than two rivets or bolts in line along the angle at each end, or by their equivalent in welding, the eccentricity of the connection with respect to the centroid of the strut may be ignored and the strut designed as an axially-loaded member provided that the calculated average stress does not exceed the allowable stresses given in Table 17, in which l is taken as $0 \cdot 85$ times the length of the strut, centre-to-centre of intersections at each end, and r is the minimum radius of gyration.

The angle struts with single-bolted or riveted connections shall be treated similarly, but the calculated stress shall not exceed 80 per cent of the values given in Table 17, and the full length l centre-to-centre of intersections shall be taken. In no case, however, shall the ratio of slenderness for such single angle struts exceed 180.

Example 1 A $70 \times 70 \times 10$ angle is used as a strut, the length of the strut between the intersections at each end being 2 metres. It is double-bolted at the ends. The least radius of gyration for the section $= 1 \cdot 35$ cm and its sectional area $= 13 \cdot 1$ cm². Calculate the safe axial load for the strut.

$$l = 0 \cdot 85 \times 200 = 170 \text{ cm}$$

$$\frac{l}{r} = \frac{170 \text{ cm}}{1 \cdot 35 \text{ cm}} = 126$$

From the table in fig. 11.16(a),

$$p_c = 55 \text{ N/mm}^2$$
$$\therefore \quad \text{Safe axial load} = (13 \cdot 1 \times 100 \times 55) \text{ N}$$
$$= 72\,050 \text{ N} = 72 \text{ kN}$$

Example 2 A strut having the same properties and dimensions as that in example 1 is single-bolted at the ends. Calculate the safe axial load for the strut.

$$l = 1 \cdot 0 \times 200 = 200 \text{ cm}$$
$$\frac{l}{r} = \frac{200 \text{ cm}}{1 \cdot 35 \text{ cm}} = 148$$

From the table in fig. 11.16(a)

$$p_c = 41 \text{ N/mm}^2$$
$$\text{Calculated stress} = \frac{80}{100} \times 41 = 32 \cdot 8 \text{ N/mm}^2$$
$$\therefore \quad \text{Safe axial load} = (13 \cdot 1 \times 100 \times 32 \cdot 8) \text{ N}$$
$$= 42\,968 \text{ N} = 43 \text{ kN}$$

It will be seen, however, that this value exceeds the value of the rivet or bolt, and consequently in almost every case the strength of a single-bolted single angle strut will be governed by the strength of the connection.

Case 2

Clause 30c(ii). For double angle discontinuous struts, back-to-back connected to both sides of a gusset or section by not less than two bolts or rivets in line along the angles at each end, or by the equivalent in welding, the load may be regarded as applied axially. The effective length *l* shall be taken as between 0·7 and 0·85 times the distance between intersections, depending on the degree of restraint, and the calculated average stress shall not exceed the values obtained from Table 17 for the ratio of slenderness based on the minimum radius of gyration about a rectangular axis of the strut. The angles shall be connected together in their length so as to satisfy the requirements of Clause 37.

Example A discontinuous compound strut consists of two equal angles back-to-back with a gusset plate in between. The strut is efficiently welded at the ends. The angles are $100 \times 100 \times 12$ and the length between intersections is 3 metres. The least radius of gyration for the section $= 3.02$ cm and the area of section $= 45.4$ cm^2. Calculate the safe axial load for the strut.

Effective length $= 0.7 \times 300 = 210$ cm

$$\frac{l}{r} = \frac{210 \text{ cm}}{3.02 \text{ cm}} = 69.5$$

From the table in fig. 11.16(a),
$$p_c = 116 \text{ N/mm}^2$$
$$\therefore \text{ Safe axial load} = (45.4 \times 100 \times 116) \text{ N}$$
$$= 526\,640 \text{ N} = 527 \text{ kN}$$

Case 3

Clause 30c(iii). Double angle discontinuous struts back-to-back, connected to one side of a gusset or section by one or more bolts or rivets in each angle, or by the equivalent in welding, shall be designed as for single angles in accordance with c(i) and the angles shall be connected together in their length so as to satisfy the requirements of Clause 37.

Example Two unequal angles having the long legs back-to-back, with a gusset on the backs of the angles (i.e. the shorter dimensions), are used as a compound strut. The length of the strut between the intersections at each end is 4 metres. The angles are $100 \times 75 \times 12$, and the strut is efficiently bolted at the ends. The area of the section is given in the appropriate section tables as 39.3 cm^2, and the least radius of gyration as 3.10 cm. Calculate the safe axial load for the strut.

$$\frac{l}{r} = \frac{0.85 \times 400 \text{ cm}}{3.10 \text{ cm}} = 109.7$$

From the table in fig. 11.16(a),
$$p_c = 69 \text{ N/mm}^2$$
$$\therefore \text{ Safe axial load} = (39.3 \times 100 \times 69) \text{ N}$$
$$= 271\,170 \text{ N} = 271 \text{ kN}$$

Clause 30c(iv) completes the provisions in respect of angle struts, and states:

The provisions in this clause are not intended to apply to continuous angle struts such as those forming the rafters of trusses, the flanges of trussed girders, or the legs of towers, which shall be designed in accordance with Clause 26 and Table 17.

Effective length of struts

The effective lengths of columns or struts are to be determined in accordance with the following table:

Type of strut	Effective length of strut = l
1. Effectively held in position and restrained in direction at both ends.	$0.7L$
2. Effectively held in position at both ends and restrained in direction at one end.	$0.85L$
3. Effectively held in position at both ends but not restrained in direction.	L
4. Effectively held in position and restrained in direction at one end, and at the other end partially restrained in direction but not held in position.	$1.5L$
5. Effectively held in position and restrained in direction at one end, but not held in position or restrained in direction at the other end.	$2.0L$

L = the length of the strut, centre-to-centre of intersections with supporting members.

Fig. 11.17 *Table of effective column lengths.*

Practical interpretation of restraints

The practical interpretation may be taken as given on page 238, with the appropriate factors, but with this important difference: a slab base, or a small bolted base to a column is not considered to give complete directional restraint. For example, the effective length of a column, with a four-way connection at one end and a slab base at the other, is $0.85L$.

Example 1 A solid steel circular column 100 mm diameter is 2·75 m high. Its ends are adequately restrained in position, but not in direction. Calculate the safe concentric load.

As this is a column of one storey, the effective length = the actual length (for the end fixture given).

$$\therefore\ l = 2750 \text{ mm}$$

$$\text{Radius of gyration} = r = \frac{D}{4} = \frac{100 \text{ mm}}{4} = 25 \text{ mm}$$

$$\therefore\ \frac{l}{r} = \frac{2750 \text{ mm}}{25 \text{ mm}} = 110$$

From the tables in fig. 11.16(a), $p_c = 69 \text{ N/mm}^2$
Area of column section $= 7854 \text{ mm}^2$
$\therefore\ $ Safe concentric load $= (7854 \times 69) \text{ N}$
$= 541\,900 \text{ N} = 542 \text{ kN}$

Example 2 A $203 \times 203 \times 71$ universal column is used as a continuous column through two floors, the connections to the column being two-way. Calculate the maximum load, regarded as concentric, which may be transmitted to the bottom length of the column, the distance between the floor levels being 3 m. Take effective length equal to $0.85 \times$ actual length in this problem.

Effective length $= 0.85 \times 300 \text{ cm} = 255 \text{ cm}$

From page 268, $\qquad r = 5.28 \text{ cm}$

$$\therefore\ \frac{l}{r} = \frac{255}{5.28} = 48.3$$

The corresponding $p_c = 134 \text{ N/mm}^2$
Area of section $= 91.1 \text{ cm}^2$
Safe concentric load for the bottom length of column
$= (134 \times 91.1 \times 100) \text{ N}$
$= 1\,220\,000 \text{ N} = 1220 \text{ kN}$

Eccentric loads

Under the requirements of Clause 34 of BS 449:1969, the effect of eccentric loading of columns has to be considered in the following manner. For the purpose of determining the stress in a stanchion or column section, the beam reactions or similar loads shall be assumed to be applied 100 mm from the face of the section or at the centre of the bearing, whichever dimension gives the greater eccentricity, and with the exception of the following two cases:
i) In the case of cap connections, the load shall be assumed to be applied at the face of the column shaft or stanchion section, or edge of packing if used, towards the span of the beam.

ii) In the case of a roof truss bearing on a cap, no eccentricity need be taken for simple bearings without connections capable of developing an appreciable moment.

Eccentric bending moments in continuous stanchions

BS 449:1969. *Clause* 34b.

b) In effectively jointed and continuous stanchions the bending moments due to eccentricities of loading at any one floor or horizontal frame level may be taken as being:

1. Ineffective at the floor or frame levels above and below that floor.
2. Divided equally between the stanchion lengths above and below that floor or frame level, provided that the moment of inertia of either stanchion section, divided by its *actual* length, does not exceed 1·5 times the corresponding value for the other length. In cases exceeding this ratio the bending moment shall be divided in proportion to the moments of inertia of the stanchion sections, divided by their respective actual lengths.

Combined stresses

Clause 14a states:

Members subject to both axial compression and bending stresses shall be so proportioned that the quantity

$$\frac{f_c}{p_c} + \frac{f_{bc}}{p_{bc}}$$

does not exceed unity at any point, where

f_c = the calculated average axial compressive stress.

p_c = the allowable compressive stress in axially loaded struts (see Table 17).

f_{bc} = the resultant compressive stress due to bending about both rectangular axes.

p_{bc} = the appropriate allowable compressive stress for members subject to bending (see Clause 19).

In cased struts for which allowance is made for the load carried by the concrete in accordance with Clause 30b, the ratio f_c/p_c in the above expression shall be replaced by the ratio of the calculated axial load on the strut to the allowable axial load determined from Clause 30b.

It is now necessary to digress a little in order to consider the implications of p_{bc}.

Clause 19a2. states that for parts in compression, the bending stress in the extreme fibres shall not exceed the lesser of the values given in Tables 2 and 3(a) (grade 43 steel).

These tables have already been reproduced on pages 27 and 28.

The allowable stresses set out in Table 3 depend on two factors, viz. l/r_y and D/T. In the case of rolled sections, the value of D/T in no case exceeds 44·7 and therefore, for all values of l/r_y up to 90, the maximum stress of 165 N/mm² is allowable. For universal column sections the value of D/T in only one case exceeds 20·3 and, therefore, the maximum stress is normally allowable for all values of l/r_y up to 100. Further, since the value of l in the expression l/r_y is the *effective* length, which is, in most cases, less than the *actual* length, it is unlikely that, in most cases of practical columns, the value of p_{bc} will be less than 165 N/mm².

Practical stress allowances

Example An intermediate length of a stanchion is 4·8 m long and is loaded as shown in fig. 11.18. The loads indicated are transmitted by floor beams and by the stanchion length above, which is 4·5 m long and consists of a $216 \times 206 \times 71$ kg UC. Test the suitability of the section, with Grade 43 steel.

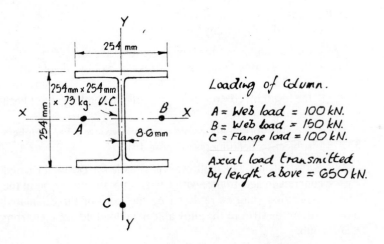

Fig. 11.18 *Example of continuing column.*

The overall depth of the 254×254 section is 254 mm, and therefore $e_{xx} = 127 \text{ mm} + 100 \text{ mm} = 227 \text{ mm}$.

The thickness of the web is 8·6 mm, and therefore

$$e_{yy} = 4·3 \text{ mm} + 100 \text{ mm} = 104·3 \text{ mm}$$

XX axis

Length of upper section $= 4·5$ m $= 450$ cm
Length of lower section $= 4·8$ m $= 480$ cm
I_{XX} of upper section $= 7647$ cm^4
I_{XX} of lower section $= 11\,360$ cm^4

$$\text{Stiffness of upper section} = \frac{7647}{450} = 17·0$$

$$\text{Stiffness of lower section} = \frac{11\,360}{480} = 23·7$$

YY axis

I_{YY} of upper section $= 2536$ cm^4
I_{YY} of lower section $= 3873$ cm^4

$$\text{Stiffness of upper section} = \frac{2536}{450} = 5·64$$

$$\text{Stiffness of lower section} = \frac{3873}{480} = 8·07$$

As in neither case does the stiffness ratio exceed 1·5, the bending moments may be assumed to be divided equally between the two lengths.

Total vertical load on lower length

$$= 650 + 100 + 150 + 100 = 1000 \text{ kN}$$

$$\therefore f_c = \frac{1000 \times 1000}{92·9 \times 100} = 107·6 \text{ N/mm}^2$$

Since the connections are three-way, the effective length

$$= 0·7 \times 480 = 336 \text{ cm}$$

$$\frac{l}{r} = \frac{336}{6·45} = 52·1 \qquad \therefore p_c = 132 \text{ N/mm}^2$$

f_{bc} = the sum of the bending stresses about both axes

$$f_{bc} = \frac{\text{B.M.}_{XX}}{Z_{XX}} + \frac{\text{B.M.}_{YY}}{Z_{YY}}$$

$$= \frac{100 \times 1000 \times 227}{2 \times 895 \times 1000} + \frac{(150-100) \times 1000 \times 104 \cdot 3}{2 \times 305 \times 1000}$$

(divided by 2, since the B.M.'s are divided equally between the two lengths)

$$= 12 \cdot 7 + 8 \cdot 55 = 21 \cdot 25 \text{ N/mm}^2$$

$$\therefore \quad \frac{f_c}{p_c} + \frac{f_{bc}}{165} = \frac{107 \cdot 6}{132} + \frac{21 \cdot 25}{165} = 0 \cdot 816 + 0 \cdot 129 = 0 \cdot 945$$

which is less than unity, and therefore the section is suitable.

Joints in column lengths

The requirements of BS 449:1969 Clause 32b are as follows:

    Where the ends of compression members are faced for bearing over the whole area, they shall be spliced to hold the connected members accurately in place and to resist any tension where bending is present.

    Where such members are not faced for complete bearing the joints shall be designed to transmit all the forces to which they are subjected.

    Wherever possible, splices shall be proportioned and arranged so that the centroidal axis of the splice coincides as nearly as possible with the centroidal axis of the members jointed, in order to avoid eccentricity; but where eccentricity is present in the joint, the resulting stresses shall be provided for.

Example Design a suitable joint for the two stanchion lengths of the previous example.

It is usual to make the splice points in stanchions at a short distance *above* the appropriate floor level, in order that the splice plates may be clear of the connections between the stanchion and the beams at that level. A common distance is 450 mm.

    The splice plates would normally be shop bolted to the lower section and site bolted to the upper section. Close-tolerance bolts may well be used.

$$\text{Load to be transmitted} = 650 \text{ kN}.$$

The appropriate bolt diameter for a 206 mm wide flange is 22 mm and the single shear value of one 22 mm bolt = 36·1 k N (see page 62).

A common rule for such splices is that the riveting or bolting should be capable of carrying 60% of the load at the splice.

$$60\% \text{ of } 650 \text{ kN} = 390 \text{ kN}$$

$$\text{Number of bolts required} = \frac{390}{36 \cdot 1} = 11 \text{ bolts.}$$

A convenient number will be 12, giving 6 on each flange. The usual 150 mm pitch will be reduced to 75 mm in this case, in order to reduce the length of the cover plates. 19 mm thick packings will be required, to make up the difference in overall depths of the sections.

The joint is shown in fig. 11.19.

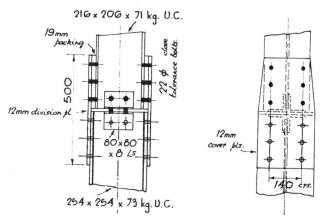

Fig. 11.19 *Column joint.*

Web angle cleats, with a division plate, are used when the joist section is not the same above and below the joint

Web splice plates are employed when the joint is subject to bending, and these, in conjunction with the flange plates, must be capable of resisting the applied moment.

Wind forces on stanchions

The following extracts and interpretations from BS 449 and from CP 3, Chapter V: Part 2:1972 dealing with the subject of wind force are given here for reference.

Chapter 5: Part 2 gives methods for calculating the wind loads that should be taken into account when designing buildings, structures, and components thereof. It sets out the procedures for determining the relevant wind pressures to be applied to:
  (1) the structure as a whole,
  (2) individual structural elements such as roofs and walls, and
  (3) individual cladding units and their fixings.
The steelwork designer will normally be principally concerned with (1) and (2). The procedure is more fully dealt with in the design example in Appendix 1 of this book.

BS 449. *Clause* 10*a*(*i*) *and* (*ii*). When considering the effect of wind pressure on buildings, due allowance shall be made for the resistance and stiffening effects of floors, roofs and walls.

When the floors, roof and walls are incapable of transmitting the horizontal forces to the foundation, the necessary steel framework shall be provided to transmit the forces to the foundations. . . .

Clause 13   Unless otherwise stated, the allowable stresses specified in the relevant Clauses of Chapter 4 of this standard may be exceeded by 25 per cent in cases where an increase in stress is solely due to wind forces, provided that the steel section shall be not less than that needed if the wind stresses were neglected.

This provision does not apply to grillage beams.

The following points should be noted:

i) Wind loads should be taken into account for both the structure as a whole, and individual structural elements.

ii) The wind pressure varies in accordance with the height and shape of the structure, and is influenced by the topography of the area, the ground roughness, and the designed life of the building.

iii) The stresses, as computed for all loadings, excluding wind pressure, may exceed by 25% those specified, . . . in cases where such increase is solely due to stresses induced by the wind. No increase shall be permitted to the permissible stress specified for grillage beams.

General assumptions

In the treatment of wind pressure given, the following assumptions will be made:

i) The effect of the wind is to bend the stanchions and beams as shown in fig. 11.20, giving points of contraflexure, or points where no B.M.

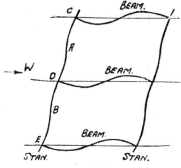

Fig. 11.20.

occurs in the stanchions, at A and B, which are in the centres of the floor heights, the maximum B.M. in the stanchion occurring at the floor levels.

ii) The wind force on any floor height, or panel, say CD, is assumed to act at the floor levels C and D, giving forces at these points which are equal, and, therefore, each equal to half the wind load on the floor height under consideration.

iii) As the points of contraflexure occur at the centre of the floor height, these positions are assumed to give the positions of maximum shear force, or reaction to the wind force, and the total load above any floor level, say C, is assumed to be resisted by a horizontal force at the point of contraflexure below that floor level—in this case at A. The bending moments at C and D due to this force at A are given by force at A multiplied by the distance AC or AD (which are equal to half the floor height).

iv) If there are a number of stanchions in the width of the building, each of these stanchions is assumed to take its share of the total wind force on them, provided that they are all equally capable of resisting this force, i.e. that they all present the same axis to the windward side. If, out of several stanchions, one or more present the 'weak' axis to the windward side, they may be neglected, and the wind force divided equally between those stanchions which present their 'strong' axis. This assumption only holds good if the beams connecting the stanchions are strong enough, both in themselves and in their connections to the stanchions, to transmit the moments that occur at the floor levels. In this connection it is essential to bear in mind that the moment to be transmitted by any beam is equal to the sum of the moments above and below the point of connection to the stanchion.

260 Structural Steelwork

(N.B. Strictly, the respective capability of the stanchions to resist the wind force will be in direct ratio with the *I* of the stanchion about the axis at right angles to the direction of the wind force, but for practical purposes the assumption made above may be used, where the stanchions are reasonably similar.)

v) The effect of the wind pressure on the vertical loads in the stanchions is as follows.

The *stress* in the stanchions will vary in proportion to their distances from the centre of gravity of the stanchions as a whole, i.e. the farther a stanchion is from the centre of gravity, the more heavily will it be stressed. The vertical load in each stanchion is, of course, the area of the stanchion multiplied by the appropriate stress.

The vertical loads in the stanchions to the leeward of the centre of gravity will be increased, while the loads in the stanchions to windward will be reduced. These reductions are called *uplifts*. The total uplift must, of course, be equal to the total increase in downward load.

The downward load on the base of any leeward stanchion must be added to the base load when designing the foundation, and the dead load on any stanchion must be greater than the maximum uplift.

Example in the worked example, a building one bay wide only is considered (fig. 11.21 on pages 262 and 263).*

The figures, showing the horizontal forces at the floor levels, are the actual forces due to wind; the figures at the points of contraflexure are the horizontal resistances supplied by the stanchions at these points; those at the points of connection of the beams to the stanchions are the moments in kNm in the various members of the structure, and the other figures at these points give the downward loads or the uplifts, respectively, at these points.

Maximum stress in stanchion

The maximum stress in any length of the stanchion is determined by the sum of the direct stresses and the bending stresses. The direct stresses are those due to loading from the beams and also due to the increase in vertical load owing to wind pressure, while the bending stresses are those caused by eccentric loading from beams and by wind pressure. The bending stress due to wind pressure is obtained by dividing the bending moment in the particular length of stanchion by the appropriate section modulus.

*The wind pressure has been taken as $50 N/m^2$.

The beams connecting the stanchions must be capable of resisting—in addition to the B.M. due to dead and superimposed loads—the B.M. due to wind loading, which, as previously explained, is the sum of the B.M.s in the lengths of stanchions respectively above and below the floor level under consideration.

(N.B. The maximum B.M. due to dead and superimposed loads will occur generally near the centre of the beam, while the maximum B.M. due to wind occurs at the ends, and if these two quantities are simply added together a total B.M. will be obtained which is greatly in excess of the actual maximum B.M. The actual maximum B.M. should be obtained by superimposing the two B.M. diagrams one on the other.)

Connection of beam to stanchion

The connection of any given beam to a stanchion must be capable of transmitting the total B.M. taken by the beam. In order to design such connections, the resistance moment of a connection is considered to be that of a couple. In the case of a cleated connection, the forces in the couple are provided by the strength of the rivets or bolts in the top and bottom cleats, and the lever arm of the couple is taken as the depth of the beam. The web connections, or side cleats, do not contribute much to the resisting of the bending moment and may, especially in the case of the more shallow sections, be neglected. The number of rivets or bolts connecting the cleats to the stanchion should be equal to the number of rivets or bolts connecting the cleats to the beam.

In the case of end-plate connections, the resistance moment is also considered to be that of a couple, the forces being the strength of the top or bottom bolts and the opposite beam flange, with the lever arm to suit.

Exercises 11

(Note: Grade 43 steel, unless otherwise stated.)

1 An angle strut in a truss has to carry an axial load of 50 kN. The length of the strut is 3 m and its ends are fixed. Using Rankine's formula and the usual constants (with a factor of safety of 4) show that a $90 \times 90 \times 12$ angle would be suitable.

Properties: Least radius of gyration = 1·75 cm, area of section = 20·3 cm^2.

2 Calculate the value of least r in each of the following cases:
  (a) rectangular section 75 mm × 200 mm;
  (b) solid circular section 150 mm diameter;

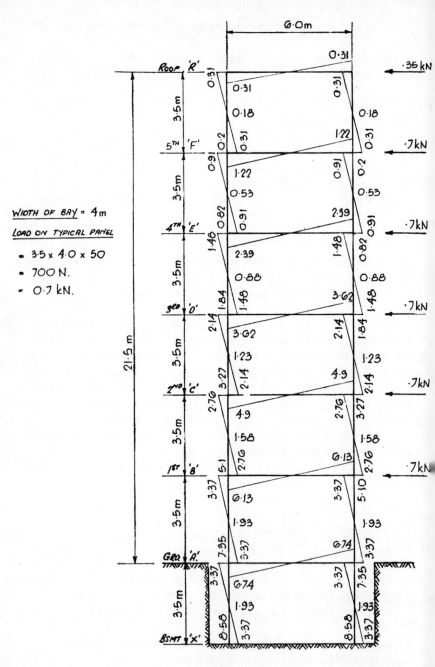

Fig. 11.21.

B.M. in stanchions	B.M. in beams	Vertical loads in stanchions
	0·31 kNm	
$0{\cdot}175 \times 1{\cdot}75 = 0{\cdot}31$ kNm		
	1·22 kNm	$\dfrac{0{\cdot}7 \times 1{\cdot}75}{6{\cdot}0} = 0{\cdot}20$ kN
$\dfrac{(0{\cdot}35+0{\cdot}7)}{2} \times 1{\cdot}75 = 0{\cdot}91$ kNm		
	2·39 kNm	$\dfrac{1{\cdot}4 \times 3{\cdot}50}{6{\cdot}0} = 0{\cdot}82$ kN
$\dfrac{(0{\cdot}35+0{\cdot}7+0{\cdot}7)}{2} \times 1{\cdot}75 = 1{\cdot}48$ kNm		
	3·62 kNm	$\dfrac{2{\cdot}1 \times 5{\cdot}25}{6{\cdot}0} = 1{\cdot}84$ kN
$1{\cdot}225 \times 1{\cdot}75 = 2{\cdot}14$ kNm		
	4·90 kNm	$\dfrac{2{\cdot}8 \times 7{\cdot}0}{6{\cdot}0} = 3{\cdot}27$ kN
$1{\cdot}575 \times 1{\cdot}75 = 2{\cdot}76$ kNm		
	6·13 kNm	$\dfrac{3{\cdot}5 \times 8{\cdot}75}{6{\cdot}0} = 5{\cdot}10$ kN
$1{\cdot}925 \times 1{\cdot}75 = 3{\cdot}37$ kNm		
	6·74 kN m	$\dfrac{4{\cdot}2 \times 10{\cdot}5}{6{\cdot}0} = 7{\cdot}35$ kN
$1{\cdot}925 \times 1{\cdot}75 = 3{\cdot}37$ kNm		
		$\dfrac{4{\cdot}2 \times 12{\cdot}25}{6{\cdot}0} = 8{\cdot}58$ kN

(c) $203 \times 203 \times 46$ kg UC; $Z_{YY} = 151 \cdot 5$ cm^3; area $= 58 \cdot 8$ cm^2;

(d) 253×253 compound stanchion, composed of one $203 \times 203 \times 46$ kg UC, with 250×25 plates on each flange.

3 Using the method of BS 449, calculate the safe concentric load for a $305 \times 305 \times 97$ kg uncased universal column for an effective height of $8 \cdot 0$ m. Look up the necessary section properties.

4 An uncased compound column, 253×250, is composed of one universal beam section, $203 \times 133 \times 25$ kg and plates on each flange to form 250×25. I_{YY} for the steel joist is 280 cm^4 and the area $= 32 \cdot 3$ cm^2. Calculate the least radius of gyration for the column section. Obtain the safe axial load for the column if the effective height $= 5 \cdot 0$ m.

5 An uncased compound strut consists of two equal angles, back to back, with a gusset between the angles. The angle size is $120 \times 120 \times 15$ and the effective height $= 4 \cdot 5$ m. The least radius of gyration of each angle is $2 \cdot 33$ cm and the area of each angle $= 33 \cdot 9$ cm^2.

Calculate the safe concentric load, for Grade 50 steel.

6 A cased stanchion has a core composed of one universal column, $254 \times 254 \times 73$ kg. Assuming the minimum concrete cover, obtain (i) the value of least r; (ii) the l/r value for an effective height of $4 \cdot 5$ m; (iii) the safe concentric load for this effective height, using the BS 449 method. Area of section $= 92 \cdot 9$ cm^2.

7 A discontinuous single-angle strut is connected to the leg of a rolled section by two bolts in line at each end of the angle. The distance, centre-to-centre, of the intersections is $2 \cdot 5$ m. The angle section is $90 \times 90 \times 12$ (least $r = 1 \cdot 75$ cm). Area of section $= 20 \cdot 3$ cm^2. Obtain the effective length of the strut by the usual method, and calculate the safe axial load for the strut, in grade 50 steel.

8 The middle length of a column continuous through three floors of a building consists of a $254 \times 254 \times 73$ kg UC. The effective length of the column length considered $= 3 \cdot 1$ m. The loads carried are:

(i) two web loads on the XX axis, one of 150 kN and the other of 230 kN;

(ii) a flange load on the YY axis of value 200 kN;

(iii) 300 kN transmitted concentrically from the upper column length.

NOTE: B.M. on the middle column length may be shared equally with the upper column length.

p_{bc} may be taken as 165 N/mm^2. Show that the problem represents a condition of loading permissible by BS 449, using the minimum eccentricities set out in Clause 34, for grade 43 steel.

UNIVERSAL COLUMNS

SAFE LOADS FOR GRADE 43 STEEL

BASED ON BS 449 1969

Serial Size in mm	Mass per metre in kg	SAFE CONCENTRIC LOADS IN KILONEWTONS FOR EFFECTIVE LENGTHS IN METRES										
		2.0	2.5	3.0	3.5	4.0	5.0	6.0	8.0	10.0	12.0	14.0	16.0
356 × 406	634	**11313**	**11313**	**11313**	**11313**	**11313**	11002	10517	9090	7287	5643	4382	3462
	551	**9826**	**9826**	**9826**	**9826**	**9826**	9527	9087	7801	6206	4782	3705	2923
	467	**8337**	**8337**	**8337**	**8337**	**8337**	8057	7668	6534	5158	3955	3056	2408
	393	**7012**	**7012**	**7012**	**7012**	7001	6755	6414	5427	4253	3247	2503	1970
	340	**6057**	**6057**	**6057**	**6057**	6040	5821	5518	4644	3620	2755	2121	1668
	287	5382	5309	5237	5174	5102	4911	4647	3889	3015	2288	1759	1381
	235	4406	4346	4286	4235	4174	4013	3790	3153	2432	1840	1412	1108
Column Core	477	**8501**	**8501**	**8501**	**8501**	8494	8201	7795	6615	5200	3978	3070	2417
356 × 368	202	3778	3723	3671	3621	3560	3398	3173	2555	1917	1430	1090	852
	177	3305	3257	3211	3168	3114	2970	2770	2224	1665	1241	945	739
	153	2857	2815	2776	2737	2690	2563	2388	1909	1426	1061	807	631
	129	2413	2377	2344	2311	2270	2161	2010	1601	1192	885	673	526
305 × 305	283	**5046**	**5046**	**5046**	4964	4840	4505	4048	2981	2116	1539	1158	
	240	4431	4356	4288	4200	4091	3796	3397	2480	1753	1272	957	
	198	3654	3593	3534	3459	3366	3113	2772	2006	1412	1023	768	
	158	2911	2862	2813	2751	2673	2462	2180	1562	1095	792	594	
	137	2524	2481	2438	2383	2314	2126	1877	1338	935	676	507	
	118	2164	2127	2090	2041	1981	1816	1599	1134	791	571		
	97	1780	1749	1718	1677	1626	1487	1304	920	640	461		

The above safe loads are tabulated for ratios of slenderness up to but not exceeding 180. Safe loads printed in ordinary type are calculated for the 'effective lengths' of stanchions in accordance with Table 17 a of BS 449 : 1969. Safe loads printed in bold type are based on the limiting stress stress of 140 N/mm^2 for material over 40mm thick.
1 kilonewton may be taken as 0.102 metre tonne (megagramme) force, but see page 102.

Table reproduced by permission and courtesy of the British Constructional Steelwork Association.

BASED ON BS 449 1969

UNIVERSAL COLUMNS

DIMENSIONS AND PROPERTIES

Serial Size	Depth of Section D	Width of Section B	Area of Section	Moment of Inertia		Radius of Gyration		Elastic Modulus		Ratio D/T
				Axis y–y	Axis x–x	Axis y–y	Axis x–x	Axis y–y	Axis x–x	
mm	mm	mm	cm²	cm⁴	cm⁴	cm	cm	cm³	cm³	
356 × 406	474.7	424.1	808.1	98211	275140	**11.0**	18.5	4632	11592	6.2
	455.7	418.5	701.8	82665	227023	**10.9**	18.0	3951	9964	6.8
	436.6	412.4	595.5	67905	183118	**10.7**	17.5	3293	8388	7.5
	419.1	407.0	500.9	55410	146765	**10.5**	17.1	2723	7004	8.5
	406.4	403.0	432.7	46816	122474	**10.4**	16.8	2324	6027	9.5
	393.7	399.0	366.0	38714	99994	**10.3**	16.5	1940	5080	10.8
	381.0	395.0	299.8	31008	79110	**10.2**	16.2	1570	4153	12.6
Column Core	427.0	424.4	607.2	68057	172391	**10.6**	16.8	3207	8075	8.0
356 × 368	374.7	374.4	257.9	23632	66307	**9.57**	16.0	1262	3540	13.9
	368.3	372.1	225.7	20470	57153	**9.52**	15.9	1100	3104	15.5
	362.0	370.2	195.2	17470	48525	**9.46**	15.8	943.8	2681	17.5
	355.6	368.3	164.9	14555	40246	**9.39**	15.6	790.4	2264	20.3
305 × 305	365.3	321.8	360.4	24545	78777	**8.25**	14.8	1525	4314	8.3
	352.6	317.9	305.6	20239	64177	**8.14**	14.5	1273	3641	9.4
	339.9	314.1	252.3	16230	50832	**8.02**	14.2	1034	2991	10.8
	327.2	310.6	201.2	12524	38740	**7.89**	13.9	806.3	2368	13.1
	320.5	308.7	174.6	10672	32838	**7.82**	13.7	691.4	2049	14.8
	314.5	306.8	149.8	9006	27601	**7.75**	13.6	587.0	1755	16.8
	307.8	304.8	123.3	7268	22202	**7.68**	13.4	476.9	1442	20.0

Each mass per metre is for the shaft only, mass of base, etc., to be added.
For explanation of tables, see notes commencing page 102.

Tables reproduced by permission and courtesy of

UNIVERSAL COLUMNS

BASED ON BS 449 1969

SAFE LOADS FOR GRADE 43 STEEL

Serial Size in mm	Mass per metre in kg	SAFE CONCENTRIC LOADS IN KILONEWTONS FOR EFFECTIVE LENGTHS IN METRES															
		1.0	1.5	2.0	2.5	3.0	3.5	4.0	4.5	5.0	6.0	7.0	8.0	9.0	10.0	11.0
254 × 254	167	3164	3100	3036	2980	2905	2809	2686	2537	2363	1977	1610	1307	1070	886	744
	132	2497	2445	2394	2348	2286	2206	2104	1981	1838	1526	1236	1000	817	676	567
	107	2033	1991	1949	1910	1858	1790	1704	1600	1480	1222	986	796	650	537	450
	89	1695	1659	1624	1592	1547	1490	1417	1328	1227	1010	813	656	535	442	370
	73	1381	1351	1323	1295	1259	1210	1149	1075	991	813	653	525	428	354	296
203 × 203	86	1622	1580	1541	1488	1416	1321	1205	1077	950	728	562	443	356		
	71	1341	1306	1273	1229	1168	1087	990	883	777	594	458	360	290		
	60	1116	1086	1058	1020	966	896	812	721	632	481	370	291	234		
	52	977	951	927	892	845	783	708	628	550	418	321	252	203		
	46	865	842	820	789	745	689	622	550	481	365	280	220	176		
152 × 152	37	685	661	626	574	505	429	357	298	249	180					
	30	552	533	504	460	403	341	283	235	197	142					
	23	429	413	388	351	303	253	208	172	144	103					

JOIST STANCHIONS

BASED ON BS 449 1969

SAFE LOADS FOR GRADE 43 STEEL

Nominal Size D×B mm	Mass per metre in kg	SAFE CONCENTRIC LOADS IN KILONEWTONS FOR EFFECTIVE LENGTHS IN METRES															
		1.0	1.2	1.4	1.6	1.8	2.0	2.2	2.4	2.6	2.8	3.0	3.2	3.4	3.6	4.0
203 × 102	25	441	422	398	368	333	297	263	232	204	180	160	143	128	115	94.6
178 × 102	22	375	359	339	314	284	254	225	198	175	154	137	122	109	98.6	80.9
152 × 89	17	290	272	250	224	197	171	148	129	112	98.4	86.8	77.1	68.9		
127 × 76	13	216	197	173	148	126	107	90.9	78.0	67.5	58.9	51.7				
102 × 64	10	142	122	101	82.6	68.1	56.6	47.6	40.6							
76 × 51	7	80.0	62.7	49.0	38.9	31.4	25.8									

The above safe loads are tabulated for ratios of slenderness up to but not exceeding 180.
Safe loads are calculated for the 'effective lengths' of stanchions in accordance with Table 17 *a* of BS 449 : 1969.
1 kilonewton may be taken as 0.102 metric tonne (megagramme) force, but see page 102.

the British Constructional Steelwork Association.

BASED ON BS 449 1969

UNIVERSAL COLUMNS

DIMENSIONS AND PROPERTIES

Serial Size	Depth of Section D	Width of Section B	Area of Section	Moment of Inertia		Radius of Gyration		Elastic Modulus		Ratio D/T
				Axis y–y	Axis x–x	Axis y–y	Axis x–x	Axis y–y	Axis x–x	
mm	mm	mm	cm²	cm⁴	cm⁴	cm	cm	cm³	cm³	
254 × 254	289.1	264.5	212.4	9796	29914	**6.79**	11.9	740.6	2070	9.1
	276.4	261.0	167.7	7444	22416	**6.66**	11.6	570.4	1622	11.0
	266.7	258.3	136.6	5901	17510	**6.57**	11.3	456.9	1313	13.0
	260.4	255.9	114.0	4849	14307	**6.52**	11.2	378.9	1099	15.1
	254.0	254.0	92.9	3873	11360	**6.46**	11.1	305.0	894.5	17.9
203 × 203	222.3	208.8	110.1	3119	9462	**5.32**	9.27	298.7	851.5	10.8
	215.9	206.2	91.1	2536	7647	**5.28**	9.16	246.0	708.4	12.5
	209.6	205.2	75.8	2041	6088	**5.19**	8.96	199.0	581.1	14.8
	206.2	203.9	66.4	1770	5263	**5.16**	8.90	173.6	510.4	16.5
	203.2	203.2	58.8	1539	4564	**5.11**	8.81	151.5	449.2	18.5
152 × 152	161.8	154.4	47.4	709	2218	**3.87**	6.84	91.78	274.2	14.1
	157.5	152.9	38.2	558	1742	**3.82**	6.75	73.06	221.2	16.8
	152.4	152.4	29.8	403	1263	**3.68**	6.51	52.95	165.7	22.4

BASED ON BS 449 1969

JOIST STANCHIONS

DIMENSIONS AND PROPERTIES

Nominal Size	Depth of Section D	Width of Section B	Area of Section	Moment of Inertia		Radius of Gyration		Elastic Modulus		Ratio D/T
				Axis y–y	Axis x–x	Axis y–y	Axis x–x	Axis y–y	Axis x–x	
mm	mm	mm	cm²	cm⁴	cm⁴	cm	cm	cm³	cm³	
203 × 102	203.2	101.6	32.26	162.6	2294	**2.25**	8.43	32.02	225.8	19.5
178 × 102	177.8	101.6	27.44	139.2	1519	**2.25**	7.44	27.41	170.9	19.8
152 × 89	152.4	88.9	21.77	85.98	881.1	**1.99**	6.36	19.34	115.6	18.4
127 × 76	127.0	76.2	17.02	50.18	475.9	**1.72**	5.29	13.17	74.94	16.7
102 × 64	101.6	63.5	12.29	25.30	217.6	**1.43**	4.21	7.97	42.84	15.4
76 × 51	76.2	50.8	8.49	11.11	82.58	**1.14**	3.12	4.37	21.67	13.6

Each mass per metre is for the shaft only, mass of base, etc., to be added.
For explanation of tables see notes commencing page 102.

Tables reproduced by permission and courtesy of

STRUTS
Equal Angles
Two or more rivets or bolts in line, or welded, at ends
SAFE LOADS FOR GRADE 43 STEEL

BASED ON BS 449 1969

Size d × b × t mm	SAFE LOADS IN KILONEWTONS FOR LENGTH IN METRES BETWEEN INTERSECTIONS														
	1.0	1.2	1.4	1.6	1.8	2.0	2.5	3.0	4.0	5.0	6.0	7.0	8.0	9.0	10.0
200 × 200 × 24	1323	1307	1292	1278	1261	1242	1179	1090	856	631	467	354	276	*221*	*181*
200 × 200 × 20	1116	1102	1089	1078	1064	1048	995	921	726	536	397	301	235	*188*	*154*
200 × 200 × 18	1010	998	986	976	963	949	901	835	659	487	361	274	214	*171*	*140*
200 × 200 × 16	903	892	882	872	862	849	807	748	591	438	324	246	192	*154*	*126*
150 × 150 × 18	730	720	707	691	672	648	573	482	320	217	155	*116*	*89*		
150 × 150 × 15	616	607	596	583	567	548	485	408	272	185	132	*98*	*76*		
150 × 150 × 12	499	492	483	473	460	445	394	333	222	151	108	*81*	*62*		
150 × 150 × 10	420	414	406	398	387	374	332	281	188	128	92	*68*	*53*		
120 × 120 × 15	477	465	450	431	408	380	304	235	144	95	67				
120 × 120 × 12	387	378	366	351	332	310	249	193	118	78	*55*				
120 × 120 × 10	326	318	308	296	280	262	211	164	101	66	*47*				
120 × 120 × 8	264	258	250	240	227	213	172	133	82	54	*38*				
100 × 100 × 15	382	367	346	321	292	261	192	142	84	*55*					
100 × 100 × 12	311	299	283	262	239	214	157	116	69	*45*					
100 × 100 × 8	213	205	194	180	165	148	109	81	48	*31*					
90 × 90 × 12	272	257	238	215	190	167	118	86	50	*33*					
90 × 90 × 10	230	218	202	182	162	141	100	73	43	*28*					
90 × 90 × 8	187	177	164	149	132	116	82	60	35	*23*					
90 × 90 × 6	142	135	125	114	101	89	63	46	27	*18*					
80 × 80 × 10	196	181	163	142	122	104	72	52	*30*						
80 × 80 × 8	160	148	133	116	100	86	59	42	*24*						
80 × 80 × 6	122	113	102	89	77	66	46	33	*19*						
70 × 70 × 10	161	143	122	103	86	72	49	*34*							
70 × 70 × 8	131	117	100	84	71	59	40	*28*	*16*						
70 × 70 × 6	101	90	77	65	55	46	31	*22*	*13*						
60 × 60 × 10	123	103	84	68	56	46	*31*	*22*							
60 × 60 × 8	101	85	69	56	46	38	*25*	*18*							
60 × 60 × 6	78	65	53	43	36	29	*19*	*14*							
60 × 60 × 5	66	55	45	37	30	25	*17*	*12*							
50 × 50 × 8	69	54	42	33	27	22	*14*								
50 × 50 × 6	53	42	33	26	21	17	*11*								
50 × 50 × 5	45	36	28	22	18	15	*10*								

The above safe loads are tabulated for ratios of slenderness up to, but not exceeding 250.
Safe loads printed in italics are for ratios of slenderness exceeding 180 and apply to wind forces only.
Safe loads are calculated for the length of strut centre to centre of intersections in accordance with clause 30.c.(i) of BS 449 : 1969 and require not less than 2 bolts or rivets in line or their equivalent in welding.
These safe loads allow for normal eccentricity in the end connection.

the British Constructional Steelwork Association.

BASED ON BS 449 1969

STRUTS
Equal Angles
DIMENSIONS AND PROPERTIES

Composed of One Equal Angle	Mass per metre in kg	Distance in cm			Radius of Gyration		Elastic Modulus	Area in cm²
		nx or ny	nv	nu	Axis v—v cm	Axis u—u cm	x—x or y—y cm³	
200 × 200 × 24	71.1	14.2	8.26	14.1	**3.90**	7.64	235	90.6
200 × 200 × 20	59.9	14.3	8.04	14.1	**3.92**	7.70	199	76.3
200 × 200 × 18	54.2	14.4	7.93	14.1	**3.93**	7.73	181	69.1
200 × 200 × 16	48.5	14.5	7.81	14.1	**3.94**	7.76	162	61.8
150 × 150 × 18	40.1	10.6	6.17	10.6	**2.92**	5.71	98.7	51.0
150 × 150 × 15	33.8	10.8	6.01	10.6	**2.93**	5.76	83.5	43.0
150 × 150 × 12	27.3	10.9	5.83	10.6	**2.95**	5.80	67.7	34.8
150 × 150 × 10	23.0	11.0	5.71	10.6	**2.97**	5.82	56.9	29.3
120 × 120 × 15	26.6	8.49	4.97	8.49	**2.33**	4.56	52.4	33.9
120 × 120 × 12	21.6	8.60	4.80	8.49	**2.35**	4.60	42.7	27.5
120 × 120 × 10	18.2	8.69	4.69	8.49	**2.36**	4.63	36.0	23.2
120 × 120 × 8	14.7	8.77	4.56	8.49	**2.37**	4.65	29.1	18.7
100 × 100 × 15	21.9	6.98	4.27	7.07	**1.93**	3.75	35.6	27.9
100 × 100 × 12	17.8	7.10	4.11	7.07	**1.94**	3.80	29.1	22.7
100 × 100 × 8	12.2	7.26	3.87	7.07	**1.96**	3.85	19.9	15.5
90 × 90 × 12	15.9	6.34	3.76	6.36	**1.74**	3.40	23.3	20.3
90 × 90 × 10	13.4	6.42	3.65	6.36	**1.75**	3.43	19.8	17.1
90 × 90 × 8	10.9	6.50	3.53	6.36	**1.76**	3.45	16.1	13.9
90 × 90 × 6	8.30	6.59	3.40	6.36	**1.78**	3.47	12.2	10.6
80 × 80 × 10	11.9	5.66	3.30	5.66	**1.55**	3.03	15.4	15.1
80 × 80 × 8	9.63	5.74	3.19	5.66	**1.56**	3.06	12.6	12.3
80 × 80 × 6	7.34	5.83	3.07	5.66	**1.57**	3.08	9.57	9.35
70 × 70 × 10	10.3	4.91	2.96	4.95	**1.35**	2.63	11.7	13.1
70 × 70 × 8	8.36	4.99	2.85	4.95	**1.36**	2.66	9.52	10.6
70 × 70 × 6	6.38	5.07	2.73	4.95	**1.37**	2.68	7.27	8.13
60 × 60 × 10	8.69	4.15	2.61	4.24	**1.16**	2.23	8.41	11.1
60 × 60 × 8	7.09	4.23	2.50	4.24	**1.16**	2.26	6.89	9.03
60 × 60 × 6	5.42	4.31	2.39	4.24	**1.17**	2.29	5.29	6.91
60 × 60 × 5	4.57	4.36	2.32	4.24	**1.17**	2.30	4.45	5.82
50 × 50 × 8	5.82	3.48	2.16	3.54	**.96**	1.86	4.68	7.41
50 × 50 × 6	4.47	3.55	2.04	3.54	**.97**	1.89	3.61	5.69
50 × 50 × 5	3.77	3.60	1.99	3.54	**.97**	1.90	3.05	4.80

Each mass per metre is for shaft only. Mass of connections, etc., to be added.
For explanation of tables, see notes commencing page 6.
1 kilonewton may be taken as 0.102 metric tonne (megagramme) force, but see page 6.

Tables reproduced by permission and courtesy of

COMPOUND STRUTS

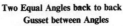
Two Equal Angles back to back
Gusset between Angles

BASED ON
BS 449
1969

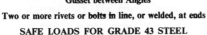
Two or more rivets or bolts in line, or welded, at ends

SAFE LOADS FOR GRADE 43 STEEL

Size d × b mm	SAFE LOADS IN KILONEWTONS FOR EFFECTIVE LENGTHS IN METRES														
	1.0	1.5	2.0	2.5	3.0	3.5	4.0	4.5	5.0	6.0	7.0	8.0	9.0	10.0	12.0
200 × 415	2686	2625	2568	2506	2421	2310	2170	2003	1820	1457	1152	920	745	614	*435*
200 × 415	2265	2214	2166	2114	2044	1953	1837	1698	1546	1241	983	786	637	525	*372*
200 × 415	2050	2004	1961	1915	1852	1770	1666	1542	1405	1129	896	716	581	478	*339*
200 × 415	1833	1792	1754	1713	1657	1584	1492	1382	1260	1015	805	644	522	431	*305*
150 × 312	1490	1447	1397	1323	1220	1092	952	818	699	516	391	305	*244*	*199*	
150 × 312	1257	1220	1179	1118	1033	927	810	696	596	440	334	261	*208*	*170*	
150 × 312	1018	989	956	907	840	755	661	569	487	361	274	214	*171*	*140*	
150 × 312	856	831	804	763	707	636	558	481	412	305	232	181	*145*	*118*	
120 × 252	975	938	879	790	678	564	463	382	318	228	*170*	*132*	*105*		
120 × 252	792	762	716	645	556	463	381	315	262	188	*141*	*109*	*87*		
120 × 252	667	642	603	545	470	393	324	267	223	160	*120*	*93*	*74*		
120 × 252	540	520	489	442	382	319	264	218	182	130	*98*	*76*	*60*		
100 × 210	790	743	662	552	439	346	276	223	184	*130*	*96*				
100 × 210	644	606	542	454	363	287	229	185	152	*108*	*80*				
100 × 210	440	415	373	314	252	200	160	129	107	*76*	*56*				
90 × 190	569	525	449	357	275	213	168	135	*111*	*78*					
90 × 190	481	444	382	304	235	182	144	116	*95*	*67*					
90 × 190	390	361	311	249	192	149	118	95	*78*	*55*					
90 × 190	297	275	238	191	148	115	91	73	*60*	*42*					
80 × 168	417	373	301	226	169	129	101	*81*	*66*	*46*					
80 × 168	339	304	247	186	139	106	83	*67*	*54*	*38*					
80 × 168	259	232	189	143	107	82	64	*51*	*42*	*30*					
70 × 148	353	297	221	157	115	86	67	*53*	*44*						
70 × 148	287	243	182	130	95	71	56	*44*	*36*						
70 × 148	220	187	140	101	74	55	43	*34*	*28*						
60 × 128	285	217	147	100	72	*54*	*41*								
60 × 128	233	179	122	83	60	*45*	*35*								
60 × 128	179	139	95	65	47	*35*	*27*	*21*							
60 × 128	151	117	80	55	40	*30*	*23*	*18*							
50 × 108	175	115	72	48	*34*	*25*									
50 × 108	136	90	57	38	*27*	*20*									
50 × 108	115	77	48	32	*23*	*17*									

The above safe loads are tabulated for ratios of slenderness up to, but not exceeding 250.
Safe loads printed in italics are for ratios of slenderness exceeding 180 and apply to wind forces only.
Safe loads are calculated for the length of strut, in accordance with clause 30.c.(ii) of BS 449 : 1969 and require not less than 2 bolts or rivets in line or their equivalent in welding.
These safe loads allow for normal eccentricity in the end connection.

the British Constructional Steelwork Association.

BASED ON
BS 449
1969

COMPOUND STRUTS
Two Equal Angles back to back
COMPOSITION AND PROPERTIES

Composed of Two Equal Angles	Mass per metre in kg	Space between Angles s mm	Distance nx cm	Area in cm²	Radius of Gyration		Elastic Modulus	
					Axis y—y cm	Axis x—x cm	Axis y—y cm³	Axis x—x cm³
200 × 200 × 24	142	15	14.2	181	8.95	6.06	700	470
200 × 200 × 20	120	15	14.3	153	8.87	6.11	579	398
200 × 200 × 18	108	15	14.4	138	8.83	6.13	520	361
200 × 200 × 16	97.0	15	14.5	124	8.79	6.16	460	323
150 × 150 × 18	80.2	12	10.6	102	6.73	4.54	296	197
150 × 150 × 15	67.6	12	10.8	86.0	6.66	4.57	245	167
150 × 150 × 12	54.6	12	10.9	69.7	6.59	4.60	194	135
150 × 150 × 10	46.0	12	11.0	58.5	6.54	4.62	161	114
120 × 120 × 15	53.2	12	8.49	67.9	5.48	3.62	162	105
120 × 120 × 12	43.2	12	8.60	55.1	5.41	3.65	128	85.5
120 × 120 × 10	36.4	12	8.69	46.4	5.37	3.67	106	72.1
120 × 120 × 8	29.4	12	8.77	37.5	5.32	3.69	84.1	58.2
100 × 100 × 15	43.8	10	6.98	55.8	4.61	2.98	113	71.2
100 × 100 × 12	35.6	10	7.10	45.4	4.55	3.02	89.5	58.2
100 × 100 × 8	24.4	10	7.26	31.0	4.45	3.06	58.6	39.9
90 × 90 × 12	31.8	10	6.34	40.6	4.16	2.70	73.8	46.7
90 × 90 × 10	26.8	10	6.42	34.3	4.11	2.72	60.9	39.5
90 × 90 × 8	21.8	10	6.50	27.8	4.06	2.74	48.2	32.1
90 × 90 × 6	16.6	10	6.59	21.1	4.00	2.76	35.7	24.4
80 × 80 × 10	23.8	8	5.66	30.2	3.64	2.41	47.8	30.9
80 × 80 × 8	19.3	8	5.74	24.5	3.60	2.43	37.8	25.2
80 × 80 × 6	14.7	8	5.83	18.7	3.54	2.44	28.0	19.1
70 × 70 × 10	20.6	8	4.91	26.2	3.25	2.09	37.4	23.3
70 × 70 × 8	16.7	8	4.99	21.3	3.21	2.11	29.6	19.0
70 × 70 × 6	12.8	8	5.07	16.3	3.16	2.13	21.9	14.5
60 × 60 × 10	17.4	8	4.15	22.1	2.86	1.78	28.4	16.8
60 × 60 × 8	14.2	8	4.23	18.1	2.82	1.80	22.4	13.8
60 × 60 × 6	10.8	8	4.31	13.8	2.77	1.82	16.5	10.6
60 × 60 × 5	9.14	8	4.36	11.6	2.74	1.82	13.6	8.89
50 × 50 × 8	11.6	8	3.48	14.8	2.43	1.48	16.2	9.37
50 × 50 × 6	8.94	8	3.55	11.4	2.38	1.50	11.9	7.22
50 × 50 × 5	7.54	8	3.60	9.61	2.35	1.51	9.85	6.10

Each mass per metre is for shaft only. Mass of connections, intermediate fastenings, etc., to be added.
For explanation of tables, see notes commencing page 6.
1 kilonewton may be taken as 0.102 metric tonne (megagramme) force, but see page 6.

Tables reproduced by permission and courtesy of

UNIVERSAL COLUMNS

SAFE LOADS FOR GRADE 50 STEEL

BASED ON BS 449 1969

Serial Size in mm	Mass per metre in kg	SAFE CONCENTRIC LOADS IN KILONEWTONS FOR EFFECTIVE LENGTHS IN METRES										
		2.0	2.5	3.0	3.5	4.0	5.0	6.0	8.0	10.0	12.0	14.0	16.0
356 × 406	634	**15427**	**15256**	**15085**	**14916**	**14724**	**14202**	**13443**	**11119**	8427	6287	4783	3735
	551	**13389**	**13238**	**13088**	**12938**	**12765**	**12291**	**11601**	9507	7150	5315	4037	3149
	467	12229	12086	11942	11799	11629	11154	10447	8326	6113	4490	3391	2637
	393	10278	10156	10033	9911	9761	9344	8721	6881	5019	3676	2773	2155
	340	8874	8767	8660	8552	8419	8046	7490	5866	4258	3113	2346	1823
	287	7502	7411	7319	7226	7110	6784	6298	4894	3536	2581	1943	1509
	235	6143	6067	5991	5913	5815	5539	5127	3953	2844	2071	1558	1210
Column Core	477	12465	12317	12169	12022	11844	11347	10608	8406	6148	4509	3403	2645
356 × 368	202	5268	5199	5130	5052	4952	4668	4245	3138	2209	1596	1196	926
	177	4609	4548	4488	4419	4330	4077	3702	2727	1916	1383	1036	803
	153	3985	3931	3879	3818	3740	3517	3186	2336	1638	1182	885	685
	129	3365	3320	3275	3223	3156	2963	2678	1955	1368	986	738	571
305 × 305	283	7299	7187	7064	6901	6686	6066	5208	3499	2373	1689	1258	
	240	6185	6089	5981	5837	5647	5098	4350	2900	1961	1395	1038	
	198	5101	5021	4928	4804	4641	4168	3534	2337	1577	1120	833	
	158	4064	3999	3922	3818	3681	3286	2765	1813	1220	866	644	
	137	3524	3467	3398	3306	3184	2832	2373	1549	1041	738	549	
	118	3022	2972	2912	2831	2723	2415	2016	1311	879	624		
	97	2485	2444	2393	2324	2233	1972	1639	1061	711	504		

The above safe loads are tabulated for ratios of slenderness up to but not exceeding 180.
Safe loads printed in ordinary type are calculated for the 'effective lengths of stanchions in accordance with Table 17 b of BS 449 : 1969. Safe loads printed in bold type are based on a yield stress of 325 N/mm^2, yield stress for sections over 65mm thick to be agreed with the manufacturer.
1 kilonewton may be taken as 0.102 metric tonne (megagramme) force, but see page 102.

the British Constructional Steelwork Association.

BASED ON BS 449 1969

UNIVERSAL COLUMNS

DIMENSIONS AND PROPERTIES

Serial Size	Depth of Section D	Width of Section B	Area of Section	Moment of Inertia		Radius of Gyration		Elastic Modulus		Ratio D/T
				Axis y–y	Axis x–x	Axis y–y	Axis x–x	Axis y–y	Axis x–x	
mm	mm	mm	cm^2	cm^4	cm^4	cm	cm	cm^3	cm^3	
356 × 406	474.7	424.1	808.1	98211	275140	**11.0**	18.5	4632	11592	6.2
	455.7	418.5	701.8	82665	227023	**10.9**	18.0	3951	9964	6.8
	436.6	412.4	595.5	67905	183118	**10.7**	17.5	3293	8388	7.5
	419.1	407.0	500.9	55410	146765	**10.5**	17.1	2723	7004	8.5
	406.4	403.0	432.7	46816	122474	**10.4**	16.8	2324	6027	9.5
	393.7	399.0	366.0	38714	99994	**10.3**	16.5	1940	5080	10.8
	381.0	395.0	299.8	31008	79110	**10.2**	16.2	1570	4153	12.6
Column Core	427.0	424.4	607.2	68057	172391	**10.6**	16.8	3207	8075	8.0
356 × 368	374.7	374.4	257.9	23632	66307	**9.57**	16.0	1262	3540	13.9
	368.3	372.1	225.7	20470	57153	**9.52**	15.9	1100	3104	15.5
	362.0	370.2	195.2	17470	48525	**9.46**	15.8	943.8	2681	17.5
	355.6	368.3	164.9	14555	40246	**9.39**	15.6	790.4	2264	20.3
305 × 305	365.3	321.8	360.4	24545	78777	**8.25**	14.8	1525	4314	8.3
	352.6	317.9	305.6	20239	64177	**8.14**	14.5	1273	3641	9.4
	339.9	314.1	252.3	16230	50832	**8.02**	14.2	1034	2991	10.8
	327.2	310.6	201.2	12524	38740	**7.89**	13.9	806.3	2368	13.1
	320.5	308.7	174.6	10672	32838	**7.82**	13.7	691.4	2049	14.8
	314.5	306.8	149.8	9006	27601	**7.75**	13.6	587.0	1755	16.8
	307.8	304.8	123.3	7268	22202	**7.68**	13.4	476.9	1442	20.0

Each mass per metre is for the shaft only mass of base, etc., to be added.
For explanation of tables, see notes commencing page 102.

Tables reproduced by permission and courtesy of

UNIVERSAL COLUMNS

BASED ON BS 449 1969

SAFE LOADS FOR GRADE 50 STEEL

Serial Size in mm	Mass per metre in kg	SAFE CONCENTRIC LOADS IN KILONEWTONS FOR EFFECTIVE LENGTHS IN METRES														
		1.0	1.5	2.0	2.5	3.0	3.5	4.0	4.5	5.0	6.0	7.0	8.0	9.0	10.0	11.0
254 × 254	167	4405	4325	4244	4156	4033	3863	3635	3351	3028	2385	1862	1472	1186	972	810
	132	3476	3411	3347	3274	3171	3029	2838	2603	2340	1830	1422	1123	903	740	617
	107	2831	2777	2724	2663	2576	2454	2293	2095	1877	1460	1132	892	717	588	489
	89	2360	2315	2271	2219	2144	2041	1903	1735	1551	1203	932	734	590	483	402
	73	1923	1886	1849	1805	1743	1656	1541	1401	1249	965	746	587	472	386	322
203 × 203	86	2260	2207	2149	2060	1928	1748	1534	1318	1125	826	623	485	387		
	71	1869	1825	1775	1700	1588	1436	1256	1077	918	673	507	394	315		
	60	1555	1518	1475	1409	1311	1178	1025	875	743	543	409	318	253		
	52	1362	1329	1292	1233	1146	1028	893	761	646	472	355	276	220		
	46	1206	1176	1142	1089	1009	903	781	665	563	411	309	240	191		
152 × 152	37	956	921	861	762	636	513	413	336	277	196					
	30	771	742	692	610	506	406	326	265	219	155					
	23	599	574	531	460	375	298	238	193	159	112					

JOIST STANCHIONS

BASED ON BS 449 1969

SAFE LOADS FOR GRADE 50 STEEL

Nominal Size D×B mm	Mass per metre in kg	SAFE CONCENTRIC LOADS IN KILONEWTONS FOR EFFECTIVE LENGTHS IN METRES													
		1.0	1.2	1.4	1.6	1.8	2.0	2.2	2.4	2.6	2.8	3.0	3.2	3.4	3.6	4.0
203 × 102	25	612	579	533	476	415	358	308	266	231	202	177	157	140	126	102
178 × 102	22	521	493	454	406	355	306	263	227	197	172	152	134	120	107	87.7
152 × 89	17	399	367	325	279	235	199	169	144	124	108	95.0	84.0	74.7		
127 × 76	13	293	256	214	176	145	120	101	86.0	73.9	64.1	56.1				
102 × 64	10	185	149	118	94.0	75.9	62.4	52.1	44.1							
76 × 51	7	96.9	72.2	54.9	42.8	34.2	28.0									

The above safe loads are tabulated for ratios of slenderness up to but not exceeding 180.
Safe loads are calculated for the 'effective lengths' of stanchions in accordance with Table 17 b of BS 449 : 1969.
1 kilonewton may be taken as 0.102 metric tonne (megagramme) force, but see page 102.

the British Constructional Steelwork Association.

BASED ON
BS 449
1969

UNIVERSAL COLUMNS

DIMENSIONS AND PROPERTIES

Serial Size mm	Depth of Section D mm	Width of Section B mm	Area of Section cm²	Moment of Inertia Axis y—y cm⁴	Moment of Inertia Axis x—x cm⁴	Radius of Gyration Axis y—y cm	Radius of Gyration Axis x—x cm	Elastic Modulus Axis y—y cm³	Elastic Modulus Axis x—x cm³	Ratio D/T
254 × 254	289.1	264.5	212.4	9796	29914	**6.79**	11.9	740.6	2070	9.1
	276.4	261.0	167.7	7444	22416	**6.66**	11.6	570.4	1622	11.0
	266.7	258.3	136.6	5901	17510	**6.57**	11.3	456.9	1313	13.0
	260.4	255.9	114.0	4849	14307	**6.52**	11.2	378.9	1099	15.1
	254.0	254.0	92.9	3873	11360	**6.46**	11.1	305.0	894.5	17.9
203 × 203	222.3	208.8	110.1	3119	9462	**5.32**	9.27	298.7	851.5	10.8
	215.9	206.2	91.1	2536	7647	**5.28**	9.16	246.0	708.4	12.5
	209.6	205.2	75.8	2041	6088	**5.19**	8.96	199.0	581.1	14.8
	206.2	203.9	66.4	1770	5263	**5.16**	8.90	173.6	510.4	16.5
	203.2	203.2	58.8	1539	4564	**5.11**	8.81	151.5	449.2	18.5
152 × 152	161.8	154.4	47.4	709	2218	**3.87**	6.84	91.78	274.2	14.1
	157.5	152.9	38.2	558	1742	**3.82**	6.75	73.06	221.2	16.8
	152.4	152.4	29.8	403	1263	**3.68**	6.51	52.95	165.7	22.4

BASED ON
BS 449
1969

JOIST STANCHIONS

DIMENSIONS AND PROPERTIES

Nominal Size mm	Depth of Section D mm	Width of Section B mm	Area of Section cm²	Moment of Inertia Axis y—y cm⁴	Moment of Inertia Axis x—x cm⁴	Radius of Gyration Axis y—y cm	Radius of Gyration Axis x—x cm	Elastic Modulus Axis y—y cm³	Elastic Modulus Axis x—x cm³	Ratio D/T
203 × 102	203.2	101.6	32.26	162.6	2294	**2.25**	8.43	32.02	225.8	19.5
178 × 102	177.8	101.6	27.44	139.2	1519	**2.25**	7.44	27.41	170.9	19.8
152 × 89	152.4	88.9	21.77	85.98	881.1	**1.99**	6.36	19.34	115.6	18.4
127 × 76	127.0	76.2	17.02	50.18	475.9	**1.72**	5.29	13.17	74.94	16.7
102 × 64	101.6	63.5	12.29	25.30	217.6	**1.43**	4.21	7.97	42.84	15.4
76 × 51	76.2	50.8	8.49	11.11	82.58	**1.14**	3.12	4.37	21.67	13.6

Each mass per metre is for the shaft only, mass of base, etc., to be added.
For explanation of tables, see notes commencing page 102.

Tables reproduced by permission and courtesy of

STRUTS
Equal Angles
Two or more rivets or bolts in line, or welded, at ends
SAFE LOADS FOR GRADE 50 STEEL

BASED ON BS 449 1969

Size d × b × t mm	SAFE LOADS IN KILONEWTONS FOR LENGTH IN METRES BETWEEN INTERSECTIONS													
	1.0	1.2	1.4	1.6	1.8	2.0	2.5	3.0	4.0	5.0	6.0	7.0	8.0	9.0	10.0
200 × 200 × 24	1846	1826	1806	1784	1758	1725	1612	1445	1038	721	518	388	300	*238*	*194*
200 × 200 × 20	1556	1539	1522	1504	1482	1455	1361	1222	880	613	441	330	255	*203*	*165*
200 × 200 × 18	1409	1394	1378	1362	1342	1318	1234	1109	800	557	401	300	232	*185*	*150*
200 × 200 × 16	1260	1246	1232	1218	1201	1179	1104	994	719	501	361	270	209	*166*	*135*
150 × 150 × 18	1021	1005	984	957	923	879	737	583	361	239	168	*125*	*96*		
150 × 150 × 15	861	848	830	808	779	743	624	495	307	203	143	*106*	*82*		
150 × 150 × 12	697	687	673	655	632	604	509	404	251	166	117	*87*	*67*		
150 × 150 × 10	586	577	566	551	532	508	430	342	213	141	100	*74*	*57*		
120 × 120 × 15	665	646	619	584	540	489	364	269	158	*103*	*72*				
120 × 120 × 12	540	525	504	476	440	399	298	220	130	*85*	*59*				
120 × 120 × 10	455	442	425	401	372	338	253	187	110	*72*	*50*				
120 × 120 × 8	368	358	344	326	302	275	206	153	90	*59*	*41*				
100 × 100 × 15	531	503	465	417	365	316	219	157	91	*59*					
100 × 100 × 12	432	410	380	341	299	259	180	129	75	*48*					
100 × 100 × 8	296	281	261	235	207	180	125	90	52	*34*					
90 × 90 × 12	376	349	312	270	230	195	132	94	*54*	*35*					
90 × 90 × 10	318	295	265	230	196	166	113	80	*46*	*30*					
90 × 90 × 8	258	240	216	188	160	136	92	66	*38*	*24*					
90 × 90 × 6	197	183	165	144	123	105	71	51	*29*	*19*					
80 × 80 × 10	268	240	206	172	143	119	79	56	*32*						
80 × 80 × 8	218	196	168	141	117	98	65	46	*26*						
80 × 80 × 6	167	150	129	108	90	76	50	36	*20*						
70 × 70 × 10	215	182	148	120	98	81	53	*37*							
70 × 70 × 8	176	149	121	98	80	66	44	*31*	*17*						
70 × 70 × 6	135	114	94	76	62	51	34	*24*	*13*						
60 × 60 × 10	158	125	97	77	62	51	*33*	*23*							
60 × 60 × 8	129	102	80	63	51	42	*27*	*19*							
60 × 60 × 6	100	79	62	49	39	32	*21*	*15*							
60 × 60 × 5	85	67	53	42	34	28	*18*	*13*							
50 × 50 × 8	83	62	47	37	29	24	*15*								
50 × 50 × 6	65	48	37	28	23	19	*12*								
50 × 50 × 5	55	41	31	24	19	16	*10*								

The above safe loads are tabulated for ratios of slenderness up to, but not exceeding 250.
Safe loads printed in italics are for ratios of slenderness exceeding 180 and apply to wind forces only.
Safe loads are calculated for the length of strut centre to centre of intersections in accordance with clause 30.c.(i) of BS 449 : 1969 and require not less than 2 bolts or rivets in line or their equivalent in welding. These safe loads allow for normal eccentricity in the end connection.

the British Constructional Steelwork Association.

BASED ON BS 449 1969

STRUTS

Equal Angles

DIMENSIONS AND PROPERTIES

Composed of One Equal Angle	Mass per metre in kg	Distance in cm			Radius of Gyration		Elastic Modulus	Area in cm²
		nx or ny	nv	nu	Axis y—y cm	Axis u—u cm	x—x or y—y cm³	
200 × 200 × 24	71.1	14.2	8.26	14.1	**3.90**	7.64	235	90.6
200 × 200 × 20	59.9	14.3	8.04	14.1	**3.92**	7.70	199	76.3
200 × 200 × 18	54.2	14.4	7.93	14.1	**3.93**	7.73	181	69.1
200 × 200 × 16	48.5	14.5	7.81	14.1	**3.94**	7.76	162	61.8
150 × 150 × 18	40.1	10.6	6.17	10.6	**2.92**	5.71	98.7	51.0
150 × 150 × 15	33.8	10.8	6.01	10.6	**2.93**	5.76	83.5	43.0
150 × 150 × 12	27.3	10.9	5.83	10.6	**2.95**	5.80	67.7	34.8
150 × 150 × 10	23.0	11.0	5.71	10.6	**2.97**	5.82	56.9	29.3
120 × 120 × 15	26.6	8.49	4.97	8.49	**2.33**	4.56	52.4	33.9
120 × 120 × 12	21.6	8.60	4.80	8.49	**2.35**	4.60	42.7	27.5
120 × 120 × 10	18.2	8.69	4.69	8.49	**2.36**	4.63	36.0	23.2
120 × 120 × 8	14.7	8.77	4.56	8.49	**2.37**	4.65	29.1	18.7
100 × 100 × 15	21.9	6.98	4.27	7.07	**1.93**	3.75	35.6	27.9
100 × 100 × 12	17.8	7.10	4.11	7.07	**1.94**	3.80	29.1	22.7
100 × 100 × 8	12.2	7.26	3.87	7.07	**1.96**	3.85	19.9	15.5
90 × 90 × 12	15.9	6.34	3.76	6.36	**1.74**	3.40	23.3	20.3
90 × 90 × 10	13.4	6.42	3.65	6.36	**1.75**	3.43	19.8	17.1
90 × 90 × 8	10.9	6.50	3.53	6.36	**1.76**	3.45	16.1	13.9
90 × 90 × 6	8.30	6.59	3.40	6.36	**1.78**	3.47	12.2	10.6
80 × 80 × 10	11.9	5.66	3.30	5.66	**1.55**	3.03	15.4	15.1
80 × 80 × 8	9.63	5.74	3.19	5.66	**1.56**	3.06	12.6	12.3
80 × 80 × 6	7.34	5.83	3.07	5.66	**1.57**	3.08	9.57	9.35
70 × 70 × 10	10.3	4.91	2.96	4.95	**1.35**	2.63	11.7	13.1
70 × 70 × 8	8.36	4.99	2.85	4.95	**1.36**	2.66	9.52	10.6
70 × 70 × 6	6.38	5.07	2.73	4.95	**1.37**	2.68	7.27	8.13
60 × 60 × 10	8.69	4.15	2.61	4.24	**1.16**	2.23	8.41	11.1
60 × 60 × 8	7.09	4.23	2.50	4.24	**1.16**	2.26	6.89	9.03
60 × 60 × 6	5.42	4.31	2.39	4.24	**1.17**	2.29	5.29	6.91
60 × 60 × 5	4.57	4.36	2.32	4.24	**1.17**	2.30	4.45	5.82
50 × 50 × 8	5.82	3.48	2.16	3.54	**.96**	1.86	4.68	7.41
50 × 50 × 6	4.47	3.55	2.04	3.54	**.97**	1.89	3.61	5.69
50 × 50 × 5	3.77	3.60	1.99	3.54	**.97**	1.90	3.05	4.80

Each mass per metre is for shaft only. Mass of connections, etc., to be added.
For explanation of tables, see notes commencing page 6.
1 kilonewton may be taken as 0.102 metric tonne (megagramme) force, but see page 6.

Tables reproduced by permission and courtesy of

COMPOUND STRUTS

Two Equal Angles back to back
Gusset between Angles
Two or more rivets or bolts in line, or welded, at ends
SAFE LOADS FOR GRADE 50 STEEL

BASED ON
BS 449
1969

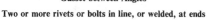

Size d × b mm	SAFE LOADS IN KILONEWTONS FOR EFFECTIVE LENGTHS IN METRES														
	1.0	1.5	2.0	2.5	3.0	3.5	4.0	4.5	5.0	6.0	7.0	8.0	9.0	10.0	12.0
200 × 415	3742	3665	3588	3487	3341	3135	2869	2560	2246	1701	1302	1020	817	668	*469*
200 × 415	3154	3090	3026	2943	2822	2653	2433	2176	1913	1452	1113	872	699	571	*401*
200 × 415	2856	2798	2740	2665	2557	2406	2209	1978	1740	1322	1014	795	637	521	*366*
200 × 415	2554	2502	2451	2385	2289	2155	1980	1775	1563	1189	912	715	573	469	*329*
150 × 312	2079	2021	1939	1807	1613	1378	1149	954	796	572	427	330	*263*	*214*	
150 × 312	1753	1705	1637	1528	1368	1172	979	814	680	488	365	282	*225*	*183*	
150 × 312	1420	1381	1327	1241	1113	957	801	666	557	400	299	232	*184*	*150*	
150 × 312	1194	1161	1116	1045	938	808	677	563	471	339	254	196	*156*	*127*	
120 × 252	1363	1305	1200	1033	836	662	527	427	351	248	*184*	*142*	*112*		
120 × 252	1107	1061	978	845	687	545	435	352	290	205	*152*	*117*	*93*		
120 × 252	932	894	825	715	582	463	370	299	246	174	*129*	*99*	*79*		
120 × 252	754	723	669	580	474	377	301	244	201	142	*105*	*81*	*64*		
100 × 210	1104	1024	872	677	510	390	305	244	200	*140*	*103*				
100 × 210	899	836	716	559	423	324	253	203	166	*116*	*86*				
100 × 210	615	573	494	389	295	226	177	142	116	*82*	*60*				
90 × 190	794	716	575	424	312	236	184	147	*120*	*84*					
90 × 190	671	607	489	363	267	202	157	126	*102*	*72*					
90 × 190	544	493	400	297	220	166	129	103	*84*	*59*					
90 × 190	414	376	306	228	169	128	99	79	*65*	*45*					
80 × 168	581	499	371	261	189	142	110	*87*	*71*	*50*					
80 × 168	472	408	304	215	156	117	91	*72*	*59*	*41*					
80 × 168	360	312	234	166	120	90	70	*56*	*45*	*32*					
70 × 148	487	384	260	177	126	94	*72*	*57*	*47*						
70 × 148	398	315	215	146	104	78	*60*	*48*	*39*						
70 × 148	304	243	166	113	81	60	*47*	*37*	*30*						
60 × 128	388	266	166	110	78	*58*	*44*								
60 × 128	318	220	139	92	65	*48*	*37*								
60 × 128	245	171	108	72	51	*38*	*29*	*23*							
60 × 128	207	145	92	61	43	*32*	*25*	*20*							
50 × 108	230	134	80	52	*37*	*27*									
50 × 108	179	105	63	41	*29*	*21*									
50 × 108	152	90	54	35	*25*	*18*									

The above safe loads are tabulated for ratios of slenderness up to, but not exceeding 250.
Safe loads printed in italics are for ratios of slenderness exceeding 180 and apply to wind forces only.

Safe loads are calculated for the 'effective length' of the strut in accordance with clause 30.c.(ii) of BS 449 : 1969 and require not less than 2 bolts or rivets in line or their equivalent in welding.

These safe loads allow for normal eccentricity in the end connection.

the British Constructional Steelwork Association.

BASED ON
BS 449
1969

COMPOUND STRUTS

Two Equal Angles back to back
COMPOSITION AND PROPERTIES

Composed of Two Equal Angles	Mass per metre in kg	Space between Angles mm	Distance nx cm	Area in cm²	Radius of Gyration		Elastic Modulus	
					Axis y—y cm	Axis x—x cm	Axis y—y cm³	Axis x—x cm³
200 × 200 × 24	142	15	14.2	181	8.95	**6.06**	700	470
200 × 200 × 20	120	15	14.3	153	8.87	**6.11**	579	398
200 × 200 × 18	108	15	14.4	138	8.83	**6.13**	520	361
200 × 200 × 16	97.0	15	14.5	124	8.79	**6.16**	460	323
150 × 150 × 18	80.2	12	10.6	102	6.73	**4.54**	296	197
150 × 150 × 15	67.6	12	10.8	86.0	6.66	**4.57**	245	167
150 × 150 × 12	54.6	12	10.9	69.7	6.59	**4.60**	194	135
150 × 150 × 10	46.0	12	11.0	58.5	6.54	**4.62**	161	114
120 × 120 × 15	53.2	12	8.49	67.9	5.48	**3.62**	162	105
120 × 120 × 12	43.2	12	8.60	55.1	5.41	**3.65**	128	85.5
120 × 120 × 10	36.4	12	8.69	46.4	5.37	**3.67**	106	72.1
120 × 120 × 8	29.4	12	8.77	37.5	5.32	**3.69**	84.1	58.2
100 × 100 × 15	43.8	10	6.98	55.8	4.61	**2.98**	113	71.2
100 × 100 × 12	35.6	10	7.10	45.4	4.55	**3.02**	89.5	58.2
100 × 100 × 8	24.4	10	7.26	31.0	4.45	**3.06**	58.6	39.9
90 × 90 × 12	31.8	10	6.34	40.6	4.16	**2.70**	73.8	46.7
90 × 90 × 10	26.8	10	6.42	34.3	4.11	**2.72**	60.9	39.5
90 × 90 × 8	21.8	10	6.50	27.8	4.06	**2.74**	48.2	32.1
90 × 90 × 6	16.6	10	6.59	21.1	4.00	**2.76**	35.7	24.4
80 × 80 × 10	23.8	8	5.66	30.2	3.64	**2.41**	47.8	30.9
80 × 80 × 8	19.3	8	5.74	24.5	3.60	**2.43**	37.8	25.2
80 × 80 × 6	14.7	8	5.83	18.7	3.54	**2.44**	28.0	19.1
70 × 70 × 10	20.6	8	4.91	26.2	3.25	**2.09**	37.4	23.3
70 × 70 × 8	16.7	8	4.99	21.3	3.21	**2.11**	29.6	19.0
70 × 70 × 6	12.8	8	5.07	16.3	3.16	**2.13**	21.9	14.5
60 × 60 × 10	17.4	8	4.15	22.1	2.86	**1.78**	28.4	16.8
60 × 60 × 8	14.2	8	4.23	18.1	2.82	**1.80**	22.4	13.8
60 × 60 × 6	10.8	8	4.31	13.8	2.77	**1.82**	16.5	10.6
60 × 60 × 5	9.14	8	4.36	11.6	2.74	**1.82**	13.6	8.89
50 × 50 × 8	11.6	8	3.48	14.8	2.43	**1.48**	16.2	9.37
50 × 50 × 6	8.94	8	3.55	11.4	2.38	**1.50**	11.9	7.22
50 × 50 × 5	7.54	8	3.60	9.61	2.35	**1.51**	9.85	6.10

Each mass per metre is for shaft only. Mass of connections, intermediate fastenings, etc., to be added.
For explanation of tables see notes commencing page 6.
1 kilonewton may be taken as 0.102 metric tonne (megagramme) force, but see page 6.

Table reproduced by permission and courtesy of the British Constructional Steelwork Association.

12

Stanchion and Column Bases

Column bases

The detail required at the foot of a column is governed by clauses contained in BS 449, Clause 38.

Stanchion and column bases.

Clause 38a. **Gusseted bases**. For stanchions with gusseted bases, the gusset plates, angle cleats, stiffeners, fastenings, etc., in combination with the bearing area of the shaft shall, where all the parts are fabricated flush for bearing, be sufficient to take the loads, bending moments and reactions to the base plate without exceeding the specified stresses.

Where the ends of the stanchion shaft and the gusset plates are not faced for complete bearing, the fastenings connecting them to the base plate shall be sufficient to transmit all the forces to which the base is subjected.

b. **Slab bases**. Stanchions with slab bases need not be provided with gussets, but fastenings shall be provided sufficient to retain the parts securely in place and to resist all moments and forces, other than direct compression, including those arising during transit, unloading and erection. When the slab alone will distribute the load uniformly, the minimum thickness of a rectangular slab shall be:

$$t = \sqrt{\left\{\frac{3w}{p_{bct}}\left(A^2 - \frac{B^2}{4}\right)\right\}}$$

where t = the slab thickness in mm.
  A = the greater projection of the plate beyond the stanchion in mm.
  B = the lesser projection of the plate beyond the stanchion in mm.

w = the pressure or loading on the underside of the base in N/mm².

p_{bct} = the permissible bending stress in the steel (185 N/mm² —see Table 2).

When the slab will not distribute the load uniformly or when the slab is not rectangular, special calculations shall be made to show that the stresses are within the specified limits.

For solid round steel columns in cases where the loading on the cap or under the base is uniformly distributed over the whole area including the column shaft, the minimum thickness, in mm, of a square cap or base shall be:

$$t = 10 \times \sqrt{\left(\frac{90W}{16p_{bct}} \cdot \frac{D}{D-d}\right)}$$

where t = the thickness of the plate in mm.

W = the total axial loading in kN.

D = the length of the side of cap or base in mm.

d = the diameter of the reduced end, if any, of the column in mm.

p_{bct} = the permissible bending stress in the steel (185 N/mm²).

When the load on the cap or under the base is not uniformly distributed or where the end of the column shaft is not machined with the cap or base, or where the cap or base is not square in plan, calculations shall be made based on the allowable stress of 185 N/mm².

The cap or base plate shall not be less than 1·5 ($d+75$) mm in length or diameter.

The area of the shoulder (the annular bearing area) shall be sufficient to limit the stress in bearing, for the whole of the load communicated to the slab, to the maximum values given in Clause 22 and resistance to any bending communicated to the shaft by the slab shall be taken as assisted by bearing pressures developed against the reduced end of the shaft in conjunction with the shoulder.

Bases for bearing upon concrete or masonry need not be machined on the underside provided the reduced end of the shaft terminates short of the surface of the slab, and in all cases the area of the reduced end shall be neglected in calculating the bearing pressure from the base.

Plate 5 Welded stanchion base.
Reproduced by courtesy of Conder Construction Ltd.

Plate 6 (a) *Section of fillet weld.* (b) *Section of butt weld.*
Reproduced by courtesy of the Welding Institute.

In cases where the cap or base is welded direct to the end of the column without boring and shouldering, the contact surfaces shall be machined to give a perfect bearing and the welding shall be sufficient to transmit the forces as required above for fastenings to slab bases. Plate 5, facing page 282, shows a typical welded stanchion base.

Theoretical considerations of bases

The modern treatment of slab bases assumes that the maximum bending moment occurs at an edge of the upper unit (see fig. 12.1). As the slab

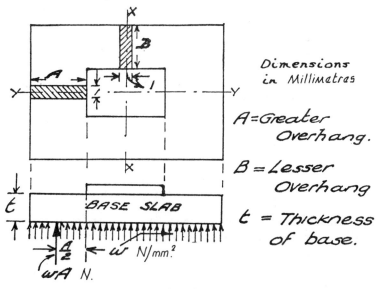

Fig. 12.1.

tends to bend simultaneously about the two principal axes of the slab, the stress caused by bending about one axis is affected by the stress due to bending about the other axis. The reader will recall the application of Poisson's ratio in this particular matter. Poisson's ratio is here taken as $\frac{1}{4}$.

Axis XX
Taking 1 mm strip of slab projection

$$\text{B.M. maximum} = w \times 1 \times A \times \frac{A}{2} = \frac{wA^2}{2} \text{ Nmm}$$

Axis YY
Taking 1 mm strip of slab projection,

$$\text{B.M. maximum} = w \times 1 \times B \times \frac{B}{2} = \frac{wB^2}{2} \text{ Nmm}$$

Taking the case of maximum tensile stress:
For overhang A, let F_1 = maximum tensile stress in N/mm².

$$\frac{wA^2}{2} = \frac{F_1 \times 1 \times t^2}{6} \qquad \therefore \quad F_1 = \frac{3wA^2}{t^2}$$

For overhang B, let F_2 = maximum tensile stress in N/mm².

$$\frac{wB^2}{2} = \frac{F_2 \times 1 \times t^2}{6} \qquad \therefore \quad F_2 = \frac{3wB^2}{t^2}$$

But the tensile stress $3wB^2/t^2$ in axial direction XX creates a compressive stress equal to $\frac{1}{4}(3wB^2/t^2)$ in axial direction YY, therefore the net tensile stress p_{bt} (axis YY)

$$= \frac{3wA^2}{t^2} - \frac{1}{4}\left(\frac{3wB^2}{t^2}\right) = \frac{3w}{t^2}\left(A^2 - \frac{B^2}{4}\right)$$

$$\therefore \quad t = \sqrt{\frac{3w}{p_{bt}}\left(A^2 - \frac{B^2}{4}\right)}$$

A similar expression would, of course, be obtained if we considered the case of p_{bc} being a compressive stress.

Permissible stress for slab bases

The permissible stress for slab bases shall not exceed 185 N/mm² in tension or compression for all steels (see Table 2, p. 27).

Example of slab base. Using the formula given in BS 449; calculate the necessary dimensions for a grade 43 steel slab base to carry an axial load of 3200 kN. The base rests on 1:2:4 concrete (safe bearing pressure = 4200 kN/m²). The column size is 340 mm × 314 mm.

Necessary area of base = 3200/4200 = 0·76 m²
 A rectangular base 900 mm × 900 mm gives 0·81 m²
 Using the BS formula,

$$t = \sqrt{\frac{3w}{p_{bct}}\left(A^2 - \frac{B^2}{4}\right)}$$

$$w = \frac{3200}{0.81} = 3.95 \text{ N/mm}^2$$

$$p_{bct} = 185 \text{ N/mm}^2$$

$$A = \text{greater overhang} = \frac{900-314}{2} = 293 \text{ mm}$$

$$B = \text{lesser overhang} = \frac{900-340}{2} = 280 \text{ mm}$$

$$\therefore \quad t = \sqrt{\frac{3 \times 3.95}{185}\left(293^2 - \frac{280^2}{4}\right)} = \text{say } 65 \text{ mm}$$

Example of gusseted base. Design a suitable gusseted base for a 254 × 254 × 73 kg universal column which transmits an axial load of 1100 kN. The base plate is to rest on 1:2:4 concrete.

The minimum area required in the base plate, taking 4200 kN/m² safe bearing stress for the concrete foundation

$$= \frac{1100}{4200} = 0.262 \text{ m}^2$$

If we assume 150 × 100 × 12 base angles for the column flanges (100 mm horizontal), and 12 mm gusset plates, the minimum width will be 254+(2 × 12)+(2 × 100) = 478 mm, say 500 mm.

$$\text{Length required} = \frac{0.262}{0.500} = 0.525 \text{ m say } (see \text{ fig. 12.2}).$$

$$\text{Average stress under base plate} = \frac{1100}{0.50 \times 0.525} \text{ kN/m}^2$$

$$= 4190 \text{ kN/m}^2 = 4.19 \text{ N/mm}^2$$

The overhang of the base plate beyond the vertical leg of the 150 × 100 angle = 99 mm. The total upward thrust on the strip of base plate 99 mm wide × 525 mm long, at 4·19 N/mm² = 217·8 kN.

The strip acts as a cantilever, as illustrated in fig. 12.3. The thickness of the cantilever is the combined angle and base thickness.

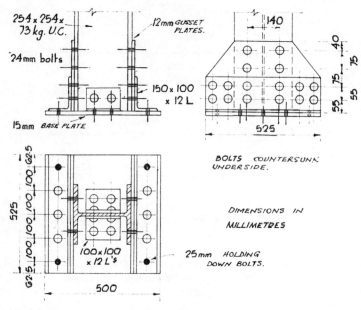

Fig. 12.2 Base of column.

Let t mm = the minimum combined thickness.

Maximum B.M. = $217·8 \times 10^3 \times 49·2$ N mm = 10716×10^3 N mm

$$M = \frac{f \times b \times d^2}{6}$$

If $f = 165$ N/mm², $b = 525$ mm, $d = t$ mm (the thickness required),

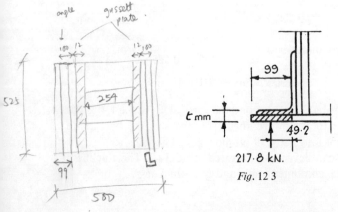

Fig. 12.3

$$10\,716 \times 10^3 = \frac{165 \times 525 \times t^2}{6}$$
$$t = 27 \cdot 2 \text{ mm}$$

As the angle is 12 mm thick, the base plate should be 18 mm.

The web cleats are often regarded as connecting cleats, assisting only in the distribution of the load uniformly to the base plate. The connection of the stanchion to the gussets may be calculated by assuming that the load under the portion of the baseplate outside the stanchion section is carried by the bolts. The length of the two outstands = $525 - 254 = 271$ mm. Therefore, in this case, slightly more than one-half of the load must be taken by the bolts. The bolts in the flange connection are in single shear and bearing (in the gusset plate or in the flange of the column).

In either case the single-shear value will be the lesser.

Assuming 24 mm dia. bolts, the S.S. value of one bolt = 36·2 kN. Number of bolts required = $1100 \times \dfrac{271}{525} \div 36.2 = 16$, i.e. 8 in each flange.

Steel grillages

Various methods are used in the design of grillage beams. The principles underlying the following treatment assume that the maximum bending moment in a given grillage tier occurs at the centre of its length.

Referring to fig. 12.4,

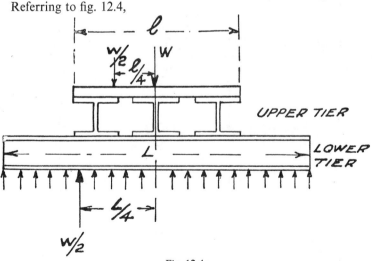

Fig. 12.4.

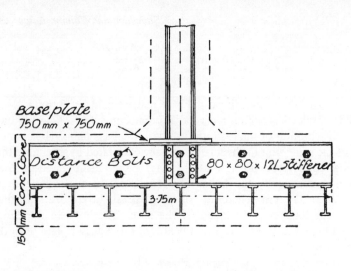

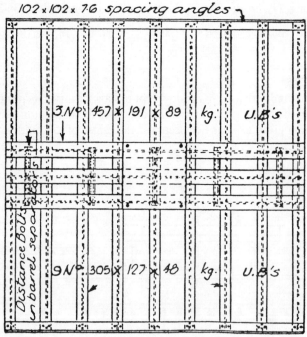

Fig. 12.5 *Steel grillage.*

B.M. at centre of length of lower tier

$$= \left(\frac{W}{2} \times \frac{L}{4}\right) - \left(\frac{W}{2} \times \frac{l}{4}\right)$$
$$= \frac{WL}{8} - \frac{Wl}{8} = \frac{W}{8}(L-l)$$

The accepted forms of the B.M. and S.F. diagrams are shown in fig. 12.7.

In the above treatment, W is taken to be the applied column load. In calculating the necessary plan dimensions of the grillage, an estimate must be made for the combined self-weight of the steel and concrete in the foundation.

Example A column transmits a load of 3100 kN to a steel grillage composed of two tiers of steel beams. The base plate of the column is 750 mm square. The subsoil safe bearing pressure is 250 kN/m². Select suitable UBs for the grillage, in grade 43 steel.

Allowing 400 kN for the combined weight of grillage beams and concrete filling, the total load carried by the ground = 3500 kN.

Necessary area at bottom of grillage

$$= \frac{3500}{250} = 14 \text{ m}^2$$

The grillage beams will be made 3·75 m long. The steel beams of a grillage are usually completely encased in concrete, and sufficient space must be left between the flange edges of adjacent beams to permit of the concrete completely filling in all internal spaces.

Clause 40 in BS 449 reads as follows:

> 40. Where grillage beams are enveloped in a solid block of dense concrete as specified in condition 2 in Clause 21, the permissible working stresses specified in this British Standard for uncased beams may be increased by $33\frac{1}{3}\%$ provided that
>
> (a) The beams are spaced apart so that the distance between the edges of adjacent flanges is not less than 75 mm.
>
> (b) The thickness of the concrete cover on the top of the upper flanges, at the ends, and at the outer edges of the sides of the outermost beams is not less than 100 mm.

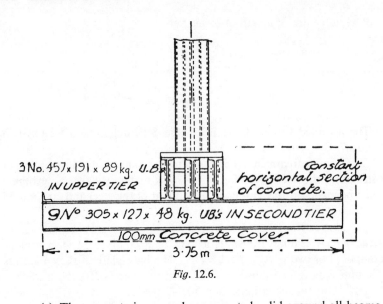

Fig. 12.6.

(c) The concrete is properly compacted solid around all beams. These increased stresses shall not apply to hollow compound girders.

For the first tier of beams l = length of base plate of column (or steel slab) and L = length of the beams in the first tier. In the case of the second tier, l = total width of beams in first tier and L = length of beams in second tier. For a square base plate (or steel slab) with a grillage, square in plan, the total maximum B.M. for the first and second tiers will be the same. Sometimes l is taken as the distance between the outer lines of rivets or bolts in the flange-to-base connections of the column. In the example, the load transmitted through the grillage = 3100 kN.

$$\text{B.M. maximum for each tier} = \frac{W}{8}(L-l)$$

$$= \frac{3100 \times 10^3}{8}(3750-750) \text{ Nmm}$$

$$= 1162 \cdot 5 \times 10^6 \text{ Nmm}$$

$$\text{Total section modulus required} = \frac{M}{f}$$

$$\frac{M}{f} = \frac{1162 \cdot 5 \times 10^6}{165 \times 4/3} = 5284 \times 10^3 \text{ mm}^3$$

$$= 5284 \text{ cm}^3$$

If there are 3 beams in the first tier, Z for each beam = 1761 cm^3. A $457 \times 191 \times 89$ kg UB has a Z value of 1767 cm^3, and hence will be suitable from the point of view of flexural stress.

The webs of grillage beams should always be tested for shear stress. In calculating the area of web of a rolled beam, the full depth of the beam is taken.

Total shear area provided by three $457 \times 191 \times 89$ kg UBs

$$= (3 \times 463 \cdot 6 \times 10 \cdot 6) \text{ mm}^2 = 14743 \text{ mm}^2$$

As shown in fig. 12.7, the maximum shear force is at the edge of the base plate. Its value may be obtained from $W/2$ (= 1550 kN) by proportion.

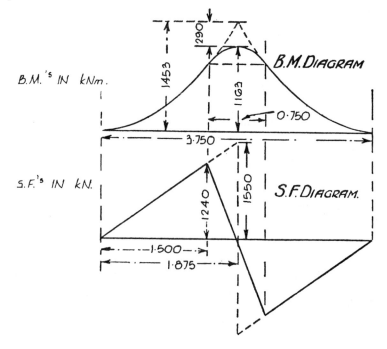

Fig. 12.7 B.M. and S.F. diagrams for grillage.

Thus maximum S.F. $= \left(\dfrac{1\cdot 5}{1\cdot 875} \times 1550\right)$ kN

$= 1240$ kN (see page 291)

Maximum shear stress in webs $= \dfrac{1240 \times 10^3}{14743}$ N/mm^2

$= 84$ N/mm^2

The beams in the upper tier should be provided with web stiffeners, machined to fit under the flanges, and placed, in the case of bolted construction, under the column flange gussets. Some form of distance pieces should also be provided to prevent movement during concreting. End angles as shown in the second tier in fig. 12.6 are commonly used for this purpose. For the second tier, nine beams have been selected.

In our example the B.M. maximum is the same for the second as for the first tier, so that each beam must have a section modulus of at least $5284/9 = 587$ cm^3. A $305 \times 127 \times 48$ kg UB has a Z value of $611\cdot 1$ cm^3, and hence will be a suitable selection.

The total shear area provided in second tier $= (9 \times 310\cdot 4 \times 8\cdot 9)$ cm^2 $= 24863$ cm^2.

Shear stress in webs $= \dfrac{1240 \times 10^3}{24863} = 49\cdot 8$ N/mm^2

The distance between the flange edges in the first tier

$= \dfrac{750 - (3 \times 192\cdot 0)}{2} = 87$ mm

There is ample room for concreting in the second tier.

Webs, as already indicated, must be stiffened if liable to buckling. The calculations involved in buckling considerations are illustrated in the following example of a compound grillage.

Design and detail of a combined grillage in grade 43 steel (fig. 12.8)

The two stanchions to be carried have base loads of 1800 kN and 1100 kN, respectively, and are at 6 m centres (fig. 12.11).

The first point to be remembered in dealing with combined grillages is that the centre of gravity of the foundation must be in the same vertical line as the centre of gravity of the loads.

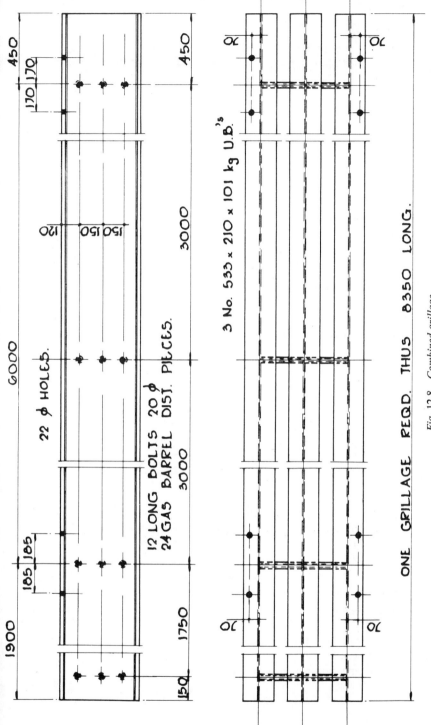

Fig. 12.8 Combined grillage.

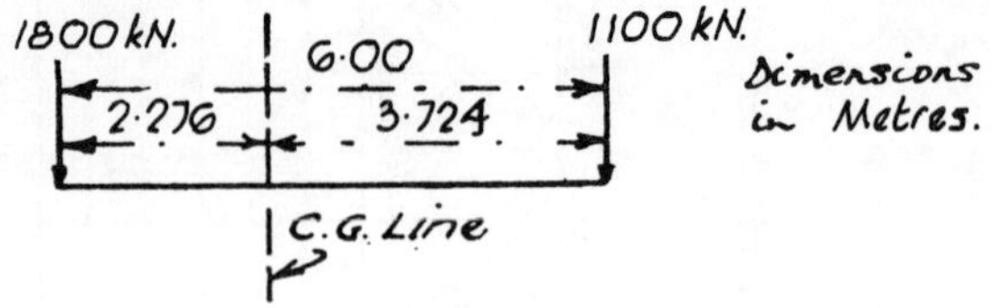

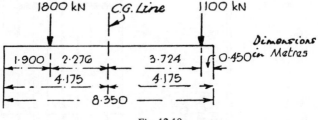

Fig. 12.9.

C.G. of loads from 1800 kN load $= \dfrac{1100 \times 6}{1100+1800} = 2\cdot 276$ m

Length of grillage

This is determined by the least projection required (from a consideration of the connection of a stanchion base to the grillage). This is usually about 450 mm, and, as the stanchion concerned is the one with the 1100 kN load, the length of the grillage

$$= 2 \times (3\cdot 724 + 0\cdot 45) = 2 \times 4\cdot 174 = 8\cdot 348 \text{ m}$$

say 8·35 m (fig. 12.10).

Fig. 12.10.

The grillage may be considered to be an inverted beam, loaded with a uniform load (the pressure under the grillage) and supported at two points (the stanchions), thus having a cantilevered portion at each end (fig. 12.11).

This load system will produce cantilever moments at each end, as well as a moment between the stanchions (the span moment) of the opposite character.

The B.M. and S.F. diagrams are shown inverted in order to correspond with the previous examples of overhanging beams considered in Chapter 5.

The maximum cantilever moment and the span moment must be

Stanchion and Column Bases

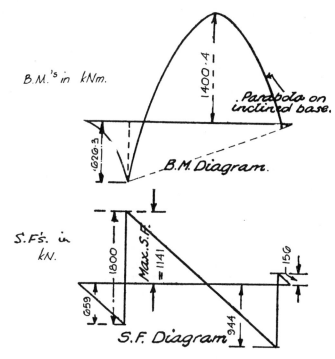

Fig. 12.11 *B.M. and S.F. diagrams for combined grillage.*

calculated and the grillage designed to withstand the greater of these.

$$\text{Load per metre of grillage} = \frac{2900}{8 \cdot 35} = 347 \text{ kN}$$

$$\text{Maximum cantilever moment} = \frac{347 \times 1 \cdot 900^2}{2} = 626 \cdot 3 \text{ kNm}$$

The maximum span moment occurs at the point of zero shear (see Chapter 9).

Point of zero shear from l.h. end of grillage

$$= \frac{1800}{347} = 5 \cdot 187 \text{ m}$$

$$\therefore \text{ Maximum span moment} = 1800(5 \cdot 187 - 1 \cdot 900) - 1800 \times \frac{5 \cdot 187}{2}$$

Maximum span moment = $1800 \times 3\cdot287 - 1800 \times 2\cdot509$
= $1800 \times 0\cdot778 = 1400\cdot4$ kNm

Assuming the grillage to be solidly encased in concrete, the allowable bending stress = 220 N/mm².

Section modulus required for grillage = $\dfrac{1400\cdot4 \times 10^6}{220}$

= 6366×10^3 mm³
= 6366 cm³

Use three $533 \times 210 \times 101$ kg UBs, each having a modulus of 2293 cm³, thus giving a total modulus of 6879 cm³.

Shear

From the shear diagram (fig. 12.11) we find that the maximum shear force in the grillage = 1141 kN. The web thickness of a $533 \times 210 \times 101$ kg UB is 10·9 mm, hence the total shear area for three such beams = $(3 \times 536\cdot7 \times 10\cdot9)$ mm². Using the $33\frac{1}{3}\%$ stress increase permitted, the working stress in shear = $100 + 33\frac{1}{3}\% = 133$ N/mm².

∴ Value in shear of the three beams

= $(3 \times 536\cdot7 \times 10\cdot9 \times 133)$ N = 2206 kN

Therefore the beams are adequate without any claim for concrete cover.

Web buckling

Bearing stiffeners should be provided to rolled beams at all points of concentrated load and points of support where the concentrated load or reaction exceeds the value $p_c tB$, where

- p_c = the axial stress for struts as given in the table in fig. 11.16 for a slenderness ratio of $d_3\sqrt{3}/t$;
- t = the web thickness;
- d_3 = the clear depth of the web between root fillets;
- B = the length of the stiff portion of the bearing, plus the additional length given by dispersion at 45° to the level of the neutral axis, plus the thickness of flange plates at the bearing, and the thickness of the seating angle (if any).

The stiff length of a bearing is that length which cannot deform appreciably in bending. In this case the length of the stiff portion of bearing may be taken as the length of the base plate of the stanchion, which will be assumed to be 500 mm.

The requirement above is strictly applicable to the bearing of a beam on a stiffened bracket. In the case of a grillage beam the analogy would be that of having two such brackets, one on either side of the centre line of the stanchion. The length l then becomes the stiff portion of the bearing (which may be taken as the appropriate dimension of the stanchion) plus the overall depth of the grillage beam.

$$\frac{d_3\sqrt{3}}{t} = \frac{472\cdot7 \times 1\cdot732}{10\cdot9} = 75\cdot1$$

$$\therefore \quad p_c = 109\cdot8 \text{ N/mm}^2$$
$$B = 536\cdot7 + 500 = 1036\cdot7 \text{ mm}$$
$$\therefore \quad p_c \times t \times B = (109\cdot8 \times 10\cdot9 \times 1036\cdot7) \text{ N} = 1241 \text{ kN}$$

The value may be increased by $33\frac{1}{3}\%$ for casing, giving a working value of 1655 kN. This is safe as there are three beams in the tier.

No bottom tier is necessary in this grillage as the area provided by the tier already designed makes the bearing pressure on the foundation quite a low figure.

Riveting and bolting at column ends

There are various methods employed, and assumptions made, in the design of the riveting or bolting at the base (or top) of a column. The problem turns upon the nature of the stress distribution beneath the base plate (or on top of the cap). An axial column load can only produce uniform stress over, say, a concrete foundation, if the base plate does not yield near the edges and, by such deflection, cause the more centrally placed portion—actually under the column—to take a higher stress. If a uniform stress under a base plate be assumed, the flexural stress in the portion of base projecting beyond the *vertical leg* of the column flange base angles should be checked.

Permissible bearing pressures on subsoils

As a guide to the permissible loads upon various subsoils, the following values may be taken.

	Load on ground kN/m²
Alluvial soil, made ground, very wet sand	50
Soft clay, wet or loose sand	100

Ordinary fairly dry clay, fairly dry fine sand,
sandy clay   200
Firm dry clay, firm ballast   300
Compact sand or gravel, London blue or similar
hard compact clay   425

Exercises 12

1 A load of 1500 kN is transmitted by a $305 \times 305 \times 97$ kg UC to a grade 43 steel slab base, 600 mm square, the base resting on concrete. Calculate a suitable thickness for the base. Take f as 185 N/mm².

2 A load of 1200 kN is to be carried by a two-tier steel grillage. The base slab of the column is 525 mm square. The safe bearing pressure of the foundation is 225 kN/m². Assuming the weight of grillage plus concrete filling to be 200 kN, work out suitable details for the grillage. $f = 220$ N/mm².

3 The footing of a wall is composed of a concrete slab 1·0 m thick. The projection of the footing is 0·5 m beyond the edge of the wall. The total loading is such as to produce a *net* upward pressure on the footing of 200 kN/m². Calculate the maximum tensile stress in the concrete.

(Find B.M. at edge of wall, taking 1 m *length* of footing, and treat as a cantilever with rectangular cross-section, 1 m wide × 1 m deep.)

4 A square base 1·3 m × 1·3 m, carries an eccentric load of 320 kN, the eccentricity being 300 mm with respect to one of the axes of symmetry. Calculate the maximum foundation pressure under the base.

(Maximum pressure $= (W/A)(1+6x/b)$, where $x =$ eccentricity and $b =$ breadth of base, parallel to direction of eccentricity.)

5 A concentric column load of 6000 kN is transmitted through a grade 43 steel slab base to a steel grillage. The column size is 356×406 and the slab is 600 mm × 600 mm. Calculate a practical thickness for the slab.

6 Two stanchions 3·800 m apart carry loads of 1400 kN and 2000 kN respectively. It is required to carry these two stanchions on a single grillage, top tier having three UBs in grade 43 steel. If the grillage is solidly embedded in concrete, determine the section required for each of the top-tier beams. Assume that the minimum overhang beyond a stanchion centre line is 450 mm. Investigate the grillage from the points of view of shear and web buckling, given that the dimension of both stanchions, in the direction of the grillage length, is 305 mm.

13

Encasement of Steelwork. Concrete Floors

The encasement of steelwork by concrete, or other fire-resisting material, is required by building regulations, except in certain permissible cases. The reader should consult such regulations, as the importance lies not wholly in the fire-protection point of view but also in the effect concrete embedment has on the working stresses permissible in steel in any given circumstance. The following excerpts, taken by permission from BS 449, illustrate the importance of the subject of the casing of steel by concrete.

Cased struts (see chapter 11)

Clause 30b has been reproduced in full on pages 246 and 247.

Grillage beams

Clause 76(c):

(c) *Encasing of steelwork in foundations and filling between grillage beams.* Grillage beams and all steel in foundations shall be solidly encased in dense concrete as specified in condition 2 of Clause 21, with a minimum cover of 100 mm.

Filler joists

Clauses 29(d) and 29(e):

(d) **Thickness of concrete.** Where the underside of the concrete is arched between the filler joists, the thickness at the crown shall be not less than 50 mm. The thickness of structural concrete over hollow blocks shall nowhere be less than 30 mm when the filler joists do not exceed 500 mm centre to centre and nowhere less than 50 mm when the spacing of the filler joists is greater than 500 mm centre to centre.

(e) **Stresses.** (i) *Bending stress.* The flexural stress in filler joists of grade 43 steel, other than those designed as part of a reinforced concrete slab, shall not exceed $(165+0\cdot6t)$ N/mm^2, where t is an allowance the value of which is equal to the thickness in mm of the structural concrete cover to the compression flanges of the filler joists; provided that the allowance t is applied only in cases where the filler joists are embedded at least flush with the underside of their bottom flanges in a solid concrete slab throughout and that any cover less than 25 mm and in excess of 75 mm shall be neglected.

In cases where the underside of the slab is flush with the bottom flanges of the filler joists, the allowance t shall not apply in respect of support moments; nevertheless if the top flange is covered the allowance may be made in calculating the resistance to the sagging moment.

(ii) *Shear and bearing stress.* The shear and bearing stresses at end supports of grade 43 steel shall be calculated as taken entirely on the filler joists, and the stresses shall not exceed those given for grade 43 steel in Clauses 23*b* and 22 respectively.

Compressive stress in beams

The determination of a permissible stress for the compression flange of an uncased beam has been dealt with in Chapter 11. The determination of the permissible stress in the compression flange of a cased beam is governed by Clause 21 of BS 449.

BENDING STRESSES (CASED BEAMS)

21. Beams and girders with equal flanges may be designed as cased beams when the following conditions are fulfilled:

1. The section is of single web and I-form or of double open channel form with the webs not less than 40 mm apart.

2. The beam is unpainted and is solidly encased in ordinary dense concrete, with 10 mm aggregate (unless solidity can be obtained with a larger aggregate), and of a works strength not less than 21 N/mm^2 at 28 days, when tested in accordance with BS 1881, 'Methods for testing concrete'.

3. The minimum width of solid casing is equal to $b+100$ mm where b is the overall width of the steel flange or flanges in millimetres.

4. The surface and edges of the flanges of the beam have a concrete cover of not less than 50 mm.

5. The casing is effectively reinforced with wire to BS 4449, 'Hot rolled steel bars for the reinforcement of concrete'. The wire shall be at least 5 mm diameter and the reinforcement shall be in the form of stirrups or binding at not more than 150 mm pitch,

and so arranged as to pass through the centre of the covering to the edges and soffit of the lower flange.

The compressive stress in bending shall not exceed the value of p_{bc} obtained from Clauses 19 and 20; for this purpose the radius of gyration r_y may be taken as $0.2\,(b+100\text{ mm})$, and D/T as for the uncased section.

The stress shall not, however, exceed $1\frac{1}{2}$ times that permitted for the uncased section.

See Clauses 29 and 40 for solidly encased filler joists and grillage beams respectively.

NOTE: This clause does not apply to beams and girders having a depth greater than 1000 mm or a width greater than 500 mm, or to box sections.

The remainder of this chapter is mainly devoted to the case of steel beams which are entirely embedded in concrete floors, and to concrete floors.

Floor loads

By kind permission of the BSI we can refer to typical loads. CP3, Chapter V: part 1 gives detailed particulars of dead loads, and superimposed floor loads. Superimposed loads are given in Chapter V in tabular form. The headings given in the table are in the following order:
1 Use to which building or structure is to be put.
2 Intensity of distributed load.
3 Concentrated load to be applied, unless otherwise stated, over any square with a 300 mm side.

As an example, we find under 'SHOP FLOORS for the display and sale of merchandise' the following figures:

Intensity of distributed load		4·0	kN/m²
	or	408	kgf/m²
	or	83·5	lbf/ft²
and concentrated load (as 3 above)		3·6	kN
	or	367	kg
	or	809	lbf.

Beams, ribs and joists spaced at not more than 1 m centres may be calculated for slab loadings.

302 *Structural Steelwork*

Chapter V gives a table of reductions permissible in superimposed floor loads when columns, piers, walls, their supports and foundations are being considered. The percentage reduction in load depends upon the number of floors carried by the member under consideration, the roof being considered as a floor. The percentage reduction of superimposed load on all floors carried by the member under consideration varies from 0 (when the member carries one floor) to 50 (when the member carries more than ten floors). For further details the reader should consult Chapter V or local regulations.

Fire-resisting floors

A solid unreinforced concrete floor, such as a filler-joist floor, possesses the necessary properties for a long period of fire resistance; unfortunately, such floors are heavy and costly to construct, and consequently are rarely used.

Reinforced-concrete floors may be of many different forms, either in-situ (cast in place in the structure) or precast, and the latter may also be pre-stressed. The cross-sectional shape of the slabs may vary enormously, and many proprietary systems of formwork for in-situ floors, and precast and/or prestressed units, are available.

It is not possible to review all of these systems here, but reference may be made to the technical literature of the various manufacturers. Also, Brochure M2 produced by the British Constructional Steelwork Association reviewed many of the proprietary floors then available in 1964. The properties of each are summarised, including the fire-resistance ratings, and fire protection generally is covered in BCSA Brochure FP1:1961.

Filler floor beams

NOTE: Whilst this form of construction is now very rarely used, this topic is included for use in assessing older buildings, where this may exist.

Solid unreinforced concrete floors are supported between main steel beams by having *filler joists* embedded in the concrete. Fig. 13.1 illustrates the construction of such a floor. The rules governing the design of filler joists are given in BS 449, Clause 29. This Clause is not reproduced here, and the reader is referred to the BS for the details. Two methods of calculation are in use for finding the strength of a filler joist concrete floor. One method regards the joists as simple beams,

Encasement of Steelwork. Concrete Floors

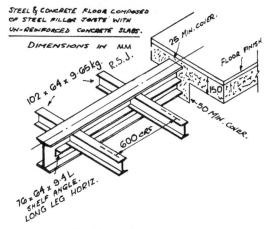

Fig. 13.1   *Filler joist floor.*

but a higher than normal working stress is used in the steel. The second method is to treat the problem as one in reinforced concrete, neglecting the strength of the concrete in tension.

Example Fig. 13.1 shows an existing solid concrete floor supported by $102 \times 64 \times 9.65$ kg joists (I value of 217.6 cm^4). Assuming a span of 3.0 m for the filler joists, calculate the safe uniformly distributed load for the floor (including the self-weight of the floor). Use a working stress of 180 N/mm^2 in the filler joists.*

Let W kN/m^2 = safe total U.D. load.

Area supported by one filler joist = $3.0 \times 0.6 = 1.8$ m^2

$\therefore$ Total load carried = $1.8\,W$ kN

B.M. maximum in joists = $\dfrac{1.8W \times 3.0}{8}$ kNm

$M = fI/y$

$\therefore \quad \dfrac{1.8W \times 3.0}{8} = \dfrac{180 \times 217.6}{10^2 \times 51}$

$W = 11.4$ kN/m^2

*180 N/mm$^2 = (165 + 0.6t) = (165 + 0.6 \times 25)$ N/mm^2. See BS 449, Clause 29e.

Readers acquainted with the theory of reinforced concrete will be able to follow a solution of the problem based upon a method referred to in regulations affecting filler floor beams, e.g. BS 449, Clause 29b.

In this method the embedded steel filler joist and the surrounding concrete are treated as a reinforced concrete section.

Fig. 13.2 shows the problem reduced to one in reinforced concrete.

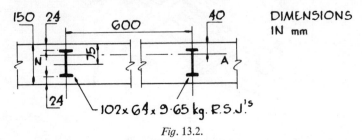

Fig. 13.2.

The solution will be given briefly, using the usual symbols, and taking $C = 7.0$ N/mm²

$$\text{Area of joist} = 12.3 \text{ cm}^2$$
$$I_{xx} = 217.6 \text{ cm}^4$$
$$d = 75 \text{ mm}$$
$$b = 600 \text{ mm}$$

$$r = \frac{A}{bd} = \frac{12.3}{600 \times 75} \times 10^2 = 0.0273$$

$$\frac{n}{d} = \sqrt{(2mr + m^2 r^2)} - mr$$

$$= \sqrt{(2 \times 15 \times 0.0273 + 15^2 \times 0.0273^2)} - 15 \times 0.0273$$
$$= 0.537$$
$$\therefore n = 0.537 \times 75 = 40 \text{ mm}$$

$$I_e = \frac{bn^3}{3} + mI_{xx} + mA(d-n)^2 \text{ expressed in cm}^4$$

$$= \frac{60 \times 4.0^3}{3} + (15 \times 217.6) + (15 \times 12.3 \times 3.5^2).$$

(NOTE: $(d-n) = 75 - 40 = 35$ mm)
$$\therefore I_e = 6804 \text{ cm}^4$$

$$\text{Moment of resistance} = \frac{C \times I_e}{n} = \frac{7 \cdot 0 \times 6804 \times 10^4}{40}$$

$$= 11 \cdot 91 \times 10^6 \text{ Nmm}$$

(Check on steel stress: d_s = depth to bottom fibre.)

$$11 \cdot 91 \times 10^6 = \frac{t}{m}\left(\frac{I_e}{d_s - n}\right)$$

$$t = \frac{15 \times 11 \cdot 91 \times 10^6 \times 86}{6804 \times 10^4} = 225 \text{ N/mm}^2.$$

This stress is greater than 180 N/mm². The moment of resistance of the floor is therefore governed by the steel and

$$= \frac{180 \times 6804 \times 10^4}{15 \times 86} = 949 \cdot 4 \times 10^4 \text{ Nmm}$$

As before, $\dfrac{1 \cdot 8W \times 3 \cdot 0 \times 10^3}{8} = 949 \cdot 4 \times 10^4$

$$W = 14 \cdot 1 \text{ kN/m}^2$$

This result shows that the method involving the composite reinforced-concrete theory is more economical than the approximate method.

Exercises 13

1 (a) State the conditions under which struts may be designed as cased struts.

(b) What is the minimum cover, and the requirements of the concrete mix for steelwork in foundations and filling between grillage beams?

2 (a) Explain the terms in the expressions l/r_y and D/T.

(b) What is the value of r_y in respect of a cased beam?

What do the symbols l and r mean in this formula (See Chapter 11).

3 A filler joist floor is composed of filler joists, $152 \times 89 \times 17 \cdot 09$ kg. They have 50 mm concrete cover on the top flange, and the concrete is flush with the bottom flange. The fillers are spaced at 450 mm centres. The area of one joist $= 21 \cdot 8$ cm² and the I_{xx} value is $881 \cdot 1$ cm⁴. The effective span of the joists $= 5 \cdot 0$ m. The limiting stress for the concrete $= 7 \cdot 0$ N/mm². Using the 'equivalent concrete' method, find:

(a) the depth of the neutral axis;

(b) the equivalent moment of inertia of one floor section 450 mm wide;

(c) the safe total uniformly distributed load on the floor (the maximum steel stress may be assumed to be permissible).

4 State the following requirements in order that beams may be designed as 'cased beams':
   (a) form of section;
   (b) nature of concrete encasement;
   (c) minimum width of solid casing;
   (d) minimum surface and edge cover of flanges;
   (e) type of reinforcement and casing.

5 A filler joist floor is composed of steel joists, $76 \times 51 \times 6\cdot67$ kg, at 600 mm centres. The concrete cover to the compression flange = 50 mm. The effective span of the joists = $3\cdot0$ m. The joists are completely embedded in concrete. Using the appropriate stress for this case, and calculating the moment of resistance from the properties of the steel joist only, obtain the total safe load per sq. m for the floor.

Z_{xx} for a $76 \times 51 \times 6\cdot67$ kg joist = $21\cdot67$ cm^3.

Do these fillers conform, in respect of spacing, to the requirements of BS 449, Clause 29c?

6 With respect to the superimposed loads set out in CP3: Chapter V: part 1:
   (a) in which two forms are the loads expressed?
   (b) what percentage reduction is permissible in the superimposed load carried by a member supporting 10 floors or more?

14

Introductory Principles of the Welding of Steelwork

Welding of structural steelwork

An alternative method to the connecting together of structural members by riveting or bolting is to effect the joint by means of a *weld*. British Standards 1856 ('General requirements for the metal-arc welding of mild steel'), 2462 ('General requirements for the arc welding of carbon manganese steels'), and 499 ('Glossary of terms relating to the welding and cutting of metals') deal with the subject of welding.

The subject-matter of this chapter is merely of an introductory character and an attempt to explain the underlying and basic principles of welding. BS 449:1969 must be consulted to obtain more detailed information of modern techniques. The permissible working stresses used in the numerical questions worked out in this chapter are based on the recommendations of this BS.

Methods of welding

There are two methods of welding,

(i) oxy-acetylene welding and (ii) metal-arc welding. In the former process a flame of high temperature is produced by burning a mixture of acetylene and oxygen. The gases are obtained from cylinders and mixed and ignited in a special form of blowpipe. The hot flame fuses the steel pieces at their point of contact and, by means of steel welding rods, a *fusion weld* is obtained. This method is not now extensively used.

In the metal-arc process, electricity is employed, and the metal, required for deposition at the weld, itself forms part of the electrical circuit. This method is the one which is almost always used now, and this chapter is devoted largely to it.

It should be noted that only metal-arc welding is permitted by BS 449 (Clause 63).

Metal-arc welding

The principle of metal-arc welding is illustrated in the simple diagram given in fig. 14.1.

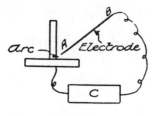

Fig. 14.1

The two given steel plates are to be joined by a weld, to form a T-section. C is the source of electrical energy, giving either alternating or direct current—both being suitable for the welding process. AB is the *electrode*, which is a metallic conductor used up in the production of the weld. By rubbing end A on the lower steel plate the circuit is closed; the end may then be slightly withdrawn from the plate, an electric arc being maintained by the ionised gases produced in the volatilisation of the material forming the electrode.

Electrodes may be bare mild-steel wire, or may be steel covered with substances possessing certain necessary physical and chemical properties.

The temperature of the steel at the weld in the welding operation is such that iron oxide tends to form, by the combination of iron with the oxygen of the air. To prevent this chemical combination, the electrode covering is made of material which forms a *slag*. The slag covers the welding metal immediately after deposition has taken place, and protects it from the air. The slag is removed, subsequently, by gently tapping with a hammer, followed by wire brushing. In addition to the property of suitable slag formation, the covering should be composed of substances which possess a higher affinity for oxygen than has iron. Certain firms specialise in the manufacture of electrodes, using their own patented coverings.

Forms of weld

The two types of welds are (i) *butt welds* (fig. 14.2) and (ii) *fillet welds* (fig. 14.3). There are many forms of butt welds. Details of some are given later. See Plate 6, facing page 283, for photographs of typical welds.

Introductory Principles of the Welding of Steelwork 309

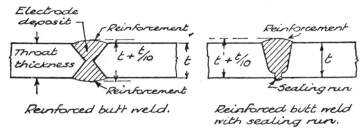

Fig. 14.2   *Typical butt welds.*

Fillet welds are distinguished as *end fillets*, *side fillets*, *diagonal fillets*, and *tee-fillets*.

Allowable stresses in welds

BS 449, *Clause* 53a—*General*

When electrodes complying with Sections 1 and 2 of BS 639 are used for the welding of grade 43 steel, or with Sections 1 and 4 of BS 639 are used for the welding of grade 50 steel, or of grade 55 steel and the yield stress of an all-weld tensile test specimen is not less than 430 N/mm^2 when tested in accordance with Appendix D of BS 649, the following shall apply:

i) Butt welds. Butt welds shall be treated as parent metal with a thickness equal to the throat thickness (or a reduced throat thickness as specified in Clause 54 for certain butt welds) and the stresses shall not exceed those allowed for the parent metal.

ii) Fillet welds. The allowable stress in fillet welds, based on a thickness equal to the throat thickness, shall be 115 N/mm^2 for grade 43 steel or 160 N/mm^2 for grade 50 steel or 195 N/mm^2 for grade 55 steel.

iii) When electrodes appropriate to a lower grade of steel are used for welding together parts of material of a higher grade of steel, the allowable stresses for the lower grade of steel shall apply.

iv) When a weld is subject to a combination of stresses, the stresses shall be combined as required in Subclause 14c and d, the value of the equivalent stress f_e being not greater than that permitted for the parent metal.

Clause 53b deals with the welding of tubular members.

310 *Structural Steelwork*

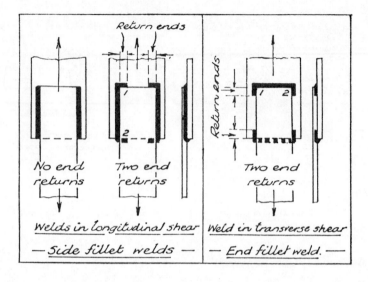

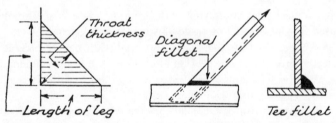

The throat thickness is to be taken as 0.7 × the length of leg (shorter leg if legs are unequal). Area of weld for stress calculation equals 'effective length × throat thickness'

Fig. 14.3 *Forms of fillet welds.*

Description of butt welds

The butt welds shown in the following diagrams are included to show the type of preparation given to surfaces to be butt-welded. They are reproduced by kind permission of the BSI and have been extracted from BS 1856:

(b) open-square butt weld (without backing);
(d) single-V butt weld (without backing);
(f) double-V butt weld;
(h) single-U butt weld;
(j) double-U butt weld;
(k) single-J butt weld;
(l) double-J butt weld;
(m) single-bevel butt weld;
(n) double-bevel butt weld.

It should be noted that, at the time of writing, BS 1856 is not published in a metric version. Therefore, whilst the format is reproduced here, the dimensions have been directly converted to metric units.

Note on butt welds

Incomplete-penetration weld

Single V, U, or J, etc. welds should be sealed. It may be impossible to deposit the weld metal through the full thickness and to seal; in this case we have an *incomplete-penetration weld.*

The size of a butt weld shall be specified by the *throat thickness* of the weld. The throat thickness of a complete-penetration butt weld shall be taken as the thickness of the thinner part joined. In the case of incomplete-penetration the throat thickness is the minimum depth of the weld metal common to the parts joined, the reinforcement being omitted. For certain incomplete-penetration butt welds a reduced throat thickness must be used (see BS 449).

Effective length

The *effective length* is taken as the length of the butt weld which has the specified weld size.

Sealing run

As mentioned above and illustrated in fig. 14.2, a *sealing run* is commonly deposited at the back of certain butt joints. It may be that, due to contact with another steel face or for some other reason, the deposition of weld metal can be effected only from one side. To assume a throat thickness equal to that of the full plate, complete fusion must be ensured with the contacting plate or a suitable backing strip of material.

312 *Structural Steelwork*

(b) Open-square butt weld (without backing)

Weld detail		Welding position	Thickness T	Gap G
Welded from both sides		Flat	mm 3–6	$T/2$
		Horizontal-vertical or vertical	3–5	$T/2$

(d) Single-V butt weld (without backing)

Weld detail		Welding position	Thickness T	Flat position only		
				Gap G	Angle α	Root face R
Welded from both sides or one side only (see Clause 8 and Appendix A)		All positions	mm Over 5	mm 2	60°	mm 2

(f) Double-V butt weld

Weld detail		Welding position	Thickness T	Flat position only		
				Gap G	Angle α	Root face R
Welded from both sides		All positions	mm Over 12	mm 2	60°	mm 2

The dimensions of the weld preparation may have to be modified for welding in positions other than flat, in which case they should be the subject of agreement between the contracting parties.

(h) Single-U butt weld

Weld detail		Welding position	Thickness T	Flat position only		
				Angle* α	Radius* r	Root face R
Welded from both sides		All positions	mm Over 20	10°–30°	mm 10–3	mm 5

(j) Double-U butt weld

Weld detail		Welding position	Thickness T	Flat position only		
				Angle* α	Radius* r	Root face R
Welded from both sides		All positions	mm Over 35	10°–30°	mm 10–3	mm 5

(k) Single-J butt weld

Weld detail		Welding position	Thickness T	Flat position only			
				Land L	Angle* α	Radius* r	Root face R
Welded from both sides		All positions	mm Over 20	mm 3	20°–40°	mm 10–5	mm 5

*The use of the min. angle should be associated with the max. radius, conversely, the max. angle should be associated with the min. radius.

The dimensions of the weld preparation may have to be modified for welding in positions other than flat, in which case they should be the subject of agreement between the contracting parties.

(l) Double-J butt weld

Weld detail		Welding position	Thickness T	Flat position only			
				Land L	Angle* α	Radius* r	Root face R
			mm	mm		mm	mm
Welded from both sides		All positions	Over 35	3	20°–40°	10–5	5

(m) Single-bevel butt weld

Weld detail		Welding position	Thickness T	Flat position only		
				Gap G	Angle α	Root face R
			mm	mm		mm
Welded from both sides		All positions	Over 5	3	45°–50°	2

(n) Double-bevel butt weld

Weld detail		Welding position	Thickness T	Flat position only		
				Gap G	Angle α	Root face R
			mm	mm		mm
Welded from both sides		All positions	Over 12	3	45°–50°	2

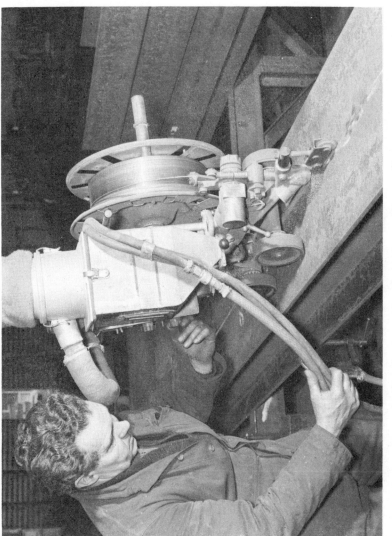

Plate 7 Automatic submerged-arc welding machine.
Reproduced by courtesy of Conder Construction Ltd.

*Plate 8 Examples of welded construction.
(a) Tubular grid structure.*
Reproduced by courtesy of British Steel Corporation.

(b) Portal framed building.

Plate 9   Detail of welded plate construction.
Reproduced by permission and courtesy of Messrs. Yorke Rosenberg Mardall.

The reader is referred to BS 1856 for further cases of butt-welded joints. In the case of butt welding of unequal cross-section parts, the wider or thicker part is reduced at the butt to the dimensions of the smaller part under certain conditions laid down.

Note on fillet welds

A normal fillet weld is one made between two parts approximately at right angles to each other, the weld metal being approximately a triangle in section. The surfaces of the parts joined form the two sides of the weld, the penetration beyond the root being less than 2·25 mm. In a *deep-penetration fillet weld* the depth of penetration beyond the root is 2·25 mm or more. The effect of deep penetration is to increase the throat thickness. Such welds will not be further referred to.

Throat thickness

BS 1856 gives a table of throat thicknesses for various angles between the fusion faces. For 90° angle, the throat thickness is 0·7 × fillet size (see fig. 14.3). The *fillet size* is taken as the minimum leg length of the fillet.

Effective length

The *effective length* of a fillet weld shall be taken as that length only which continues to have the specified weld size, and throat thickness corresponding. It was usual practice in the case of fillet welds with open ends, i.e. not specially end-treated, to obtain the effective length, for calculation purposes, by taking twice the weld size from the overall length. It is not considered that details of preparation of surfaces for welding come within the scope of this book.

End returns

Fillet welds terminating at the ends or sides of parts or members shall, wherever practicable, be returned continuously around the corners for a distance not less than twice the size of the weld. This provision shall, in particular, apply to side and top fillet welds in tension which connect brackets, beam seatings, and similar parts.

BS 1856 does not allow any increase in *effective length* of a fillet weld because the ends are returned. However, the ends are then not *open* and the actual length of the weld may be considered as having the specified weld size.

In the case of butt-welded joints, BS 1856 requires the welding to be such that the full throat thickness is maintained to the ends of joints by 'cross run' or other methods approved by the engineer.

Example 1 Calculate the throat thickness to be taken in the calculation of weld strength in each of the following cases: (a) 6 mm fillet weld, (b) a fillet weld having one leg dimension 10 mm and the other 12 mm, (c) a butt weld joining two 12 mm plates, (d) a butt weld joining two plates, one being 12 mm thick and the other 15 mm.

(a) Throat thickness = 0·7 × specified size of leg = 0·7 × 6 mm
= 4·2 mm.
(b) Throat thickness = 0·7 × length of shorter leg = 0·7 × 10 mm
= 7 mm.
(c) Throat dimension = plate thickness = 12 mm.
(d) Throat dimension = thickness of thinner plate = 12 mm.

Example 2 Obtain the throat area to be taken in the case of a 10 mm end fillet of 150 mm overall length, (a) if no return side fillets are used, (b) if return fillets 25 mm long are employed at both ends of the weld.
(a) In this case we have to take $(150 - 2 \times 10)$ as the effective fillet length = 130 mm.
The throat dimension = 0·7 × fillet size = 0·7 × 10 mm = 7 mm.
Effective throat area = (130×7) mm^2 = 910 mm^2.
(b) The full 150 mm may be taken in this case,
∴ throat area = 150×7 = 1050 mm^2.

Example 3 Calculate the safe load for the butt-welded grade 43 steel ties shown in fig. 14.4. Each bar is 100 mm × 10 mm thick.

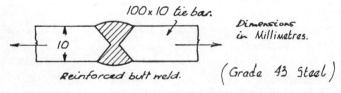

Fig. 14.4 *Butt weld.*

The throat thickness is 10 mm and the throat area = 100×10 = 1000 mm^2.
At 155 N/mm^2 the safe load = 155×1000 N = 155 kN.

Example 4 Find the safe value of P, in the example shown in fig. 14.5, from the point of view of the side fillet welds, for grade 43 steel.

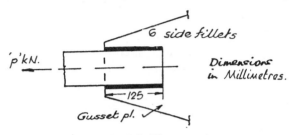

Fig. 14.5 *Side fillet connection.*

Effective length of each side fillet
 = (125 mm − 2 × fillet size) = 125 − (2 × 6) = 112 mm
Throat thickness = 0·7 × 6 mm = 4·2 mm
∴ Total throat area = (2 × 112 × 4·2) mm² = 941 mm²
∴ Safe load at 115 N/mm² = 115 × 941 N = 108·2 kN*
[For the effect of returning the ends see example 6.]

Example 5 Fig. 14.6 shows a connection with two end welds both having 25 mm return side fillets. Calculate the strength of the welded joint.
Effective length of each fillet = actual length, as the ends are returned as side fillets.

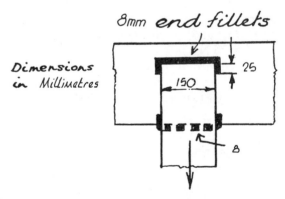

Fig. 14.6 *End fillet connection.*

*For end and side fillets the working stress is taken as 115 N/mm² for grade 43 steel.

318 *Structural Steelwork*

∴ Effective length = 150 mm for each fillet.
Throat thickness = 0·7 × 8 mm = 5·6 mm
∴ Throat area = 2 × 150 × 5·6 = 1680 mm^2
Safe load at 115 N/mm^2 = 193·2 kN

Example 6 In fig. 14.7 a welded bracket connection is shown. Calculate the safe load for the bracket (assuming no bending moment on the welds). Find also the safe load if the side fillets shown had returned ends at the top and bottom of the bracket. The steel is grade 43.

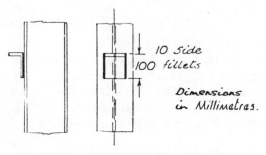

Fig. 14.7 *Simple bracket connection.*

Assuming no returned ends, the effective length of each side fillet = $(100 - (2 \times 10)) = 80$ mm.
∴ Total length of side fillet = 2 × 80 = 160 mm
Throat thickness = 0·7 × 10 = 7·0 mm
∴ Effective throat area = 160 × 7·0 = 1120 mm^2
∴ Safe load at 115 N/mm^2 = 115 × 1120 N = 128·8 kN

If the ends are returned at top and bottom, no deduction need be made in the overall fillet length as the end *crater* of the welds will be in the returned portions (which are omitted from the calculations).

∴ Effective length of side fillet = 2 × 100 = 200 mm
∴ Safe load = (200 × 7·0 × 115) N
  = 161 kN

This result may be checked by inspection of the safe-load tables for simple brackets given in fig. 14.8. Looking across horizontally from the fillet size, 10 mm, until the column headed '100' is reached we find 161 kN.

SIMPLE BRACKETS
Connected by
SIDE FILLET WELDS

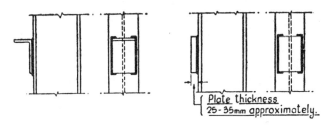

TYPICAL BRACKETS

LOAD IN KN FOR VARIOUS SIZES OF WELDS
CARRYING VERTICAL SHEAR ONLY

Size of fillet welds (mm)	Depth of bracket (mm)							Return end factor (kN)
	75	85	100	115	125	150	175	
4	48·3	54·7	64·4	74·1	80·5	96·6	112·7	2·6
6	72·5	82·1	96·6	111·2	120·8	144·9	169·1	5·8
8	96·6	109·4	128·8	148·2	161·0	193·2	225·4	10·3
10	120·8	136·9	161·0	185·2	201·3	241·6	281·8	16·1
15	181·2	205·4	241·5	277·2	301·9	362·3	422·6	36·3
20	241·6	273·8	322·0	370·4	402·6	483·2	563·6	64·4

The above values are for brackets having returned end welds as illustrated in the diagrams. Where the welds cannot be returned at the top, deduct the return end factor value from the tabulated value. If returned ends are omitted at top and bottom, deduct twice the end factor value.

NOTE: In using plate brackets, special attention must be given to the buckling strength of the web of the supported beam.

Fig. 14.8.

For no returned ends we have to deduct twice the end factor value
i.e. $2 \times 16 \cdot 1 = 32 \cdot 2$ kN.

∴ Safe load when the returned ends are omitted, top and bottom
$= (161 \cdot 0 - 32 \cdot 2)$ kN $= 128 \cdot 8$ kN, as before.

Example 7 A simple bracket connection of the form shown in fig. 14.8 has side fillet welds and a top fillet weld so that the sides and top form one continuous weld. The side fillets are returned at the bottom of the bracket. Calculate the vertical shear load this bracket can safely take, given the following particulars:

depth of angle bracket = 150 mm
breadth of angle bracket = 125 mm
size of both fillet welds = 6 mm

Length of top end fillet = 125 mm. This is the effective length as there will be no end craters to allow for.

∴ Effective area of throat = $(125 \times 0.7 \times 6)$ mm^2 = 525 mm^2
Weld strength at 115 N/mm^2 = 115×525 = 60·4 kN
As the side fillets are returned at the ends the effective length of each = 150 mm.
Total length of side fillet = 300 mm
Effective area of throat = $(300 \times 0.7 \times 6)$ = 1260 mm^2
Weld strength at 115 N/mm^2 = 115×1260 = 144·9 kN
∴ Safe shear load for bracket = (60.4×144.9) = 205·3 kN

The tables, given in figs 14.8 and 14.9 are based on the following permissible stresses: end and side fillet welds—115 N/mm^2 in grade 43 steel.

Welded structures

The examples given illustrate the method of computation of the strength of a simple weld. The theory involved in the design of welded structures, such as a welded steel frame, will be generally concerned with the mechanics of rigid structures. For the nature of the detailed calculations which welded structures entail, the reader is recommended to consult the 'Handbook for Welded Structural Steelwork', published by the Institute of Welding, 54 Princes Gate, SW7. The simple bracket tables given in fig. 14.8 are taken from the handbook by kind permission of the Institute. Those specially interested in the welding processes should write to the Secretary for particulars of the activities of the Institute.

Safe-load tables

It is convenient to evaluate and tabulate the strength in kN *per linear mm of weld* for the various types of welds and for the usual sizes. We

Introductory Principles of the Welding of Steelwork 321

have only then to multiply the 'strength per mm' by the 'effective weld length in mm' to obtain the total strength of the weld.

Assuming, for example, a 6 mm end fillet, the throat area per linear mm of weld $= 0.7 \times 6$ mm $\times 1$ mm $= 4.2$ mm^2. At 115 N/mm^2 the strength per mm $= 115 \times 4.2 = 483$ kN.

Similarly for a 6 mm side fillet, the strength per linear inch of weld would be $115 \times 4.2 = 483$ kN.

The table given in fig. 14.9 is built up by calculations similar to the foregoing. A few weld sizes only are shown.

Practical considerations in welding

The production of a successful weld is an operation demanding skilled and trained workmanship. Such training is commonly provided by welding firms themselves, and also in technical institutions.

SAFE LOADS PER MM RUN OF FILLET WELDS

Size of Fillet	End and Side Fillets		
	Working Stresses: 115 N/mm^2		
mm	Calculations	N	kN
4	0.7 × 4 × 115	322	0.32
6	0.7 × 6 × 115	483	0.48
8	0.7 × 8 × 115	644	0.64
10	0.7 × 10 × 115	805	0.81
12	0.7 × 12 × 115	966	0.97
15	0.7 × 15 × 115	1208	1.21
20	0.7 × 20 × 115	1610	1.61
25	0.7 × 25 × 115	2013	2.01

Fig. 14.9.

BS 449 includes a list of tests for use in the approval of welders.

Welding operations have to be carried out in horizontal, vertical, and overhead positions, each possessing a special technique to ensure a successful weld. It is essential, of course, that the electrode should deposit high-grade material to build up the weld, but many other considerations are involved. The electrical current used is a material factor. Welds are usually made in more than one *run*, and the current required may differ for different runs. Anything, such as plate thickness, which affects the heat conditions of the welding operation, has to be taken into account in fixing the welding current. Important data relating to the physical and metallurgical aspects of welding will be

STRENGTH OF FILLET WELDS
FOR GRADE 43 STEEL
PERMISSIBLE LOADS IN KILONEWTONS PER mm RUN
WITH ELECTRODES TO BS 639, SECTIONS 1 AND 2.

Leg length in mm	Throat thickness in mm	Load at 115 N/mm²	Leg length in mm	Throat thickness in mm	Load at 115 N/mm²
3	2.1	0.24	12	8.4	0.97
4	2.8	0.32	15	10.5	1.21
5	3.5	0.40	18	12.6	1.45
6	4.2	0.48	20	14.0	1.61
8	5.6	0.64	22	15.4	1.77
10	7.0	0.80	25	17.5	2.01

STRENGTH OF FULL PENETRATION BUTT WELDS
FOR GRADE 43 STEEL
PERMISSIBLE LOADS IN KILONEWTONS PER mm RUN
WITH ELECTRODES TO BS 639, SECTIONS 1 and 2

Thickness in mm	Shear at 100 N/mm²	Tension or Compression at 155 N/mm²	Thickness in mm	Shear at 100 N/mm²	Tension or Compression at 155 N/mm²
6	0.60	0.93	22	2.20	3.41
8	0.80	1.24	25	2.50	3.88
10	1.00	1.55	28	2.80	4.34
12	1.20	1.86	30	3.00	4.65
15	1.50	2.33	35	3.50	5.43
18	1.80	2.79	40	4.00	6.20
20	2.00	3.10			

For incomplete penetration welds, the above loads are to be multiplied by a factor not exceeding 0.625.
For explanations of tables, see Notes on page 135.
1 kilonewton may be taken as 0.102 metric tonne (megagramme) force, but see page 102.

Tables reproduced by permission and courtesy of

STRENGTH OF FILLET WELDS
FOR GRADE 50 STEEL

PERMISSIBLE LOADS IN KILONEWTONS PER mm RUN
WITH ELECTRODES TO BS 639, SECTIONS 1 and 4

Leg length in mm	Throat thickness in mm	Load at 160 N/mm²	Leg length in mm	Throat thickness in mm	Load at 160 N/mm²
3	2.1	0.34	12	8.4	1.34
4	2.8	0.45	15	10.5	1.68
5	3.5	0.56	18	12.6	2.02
6	4.2	0.67	20	14.0	2.24
8	5.6	0.90	22	15.4	2.46
10	7.0	1.12	25	17.5	2.80

STRENGTH OF FULL PENETRATION BUTT WELDS
FOR GRADE 50 STEEL

PERMISSIBLE LOADS IN KILONEWTONS PER mm RUN
WITH ELECTRODES TO BS 639, SECTIONS 1 and 4

Thickness in mm	Shear at 140 N/mm²	Tension or Compression at 215 N/mm²	Thickness in mm	Shear at 140 N/mm²	Tension or Compression at 215 N/mm²
6	0.84	1.29	22	3.08	4.73
8	1.12	1.72	25	3.50	5.38
10	1.40	2.15	28	3.92	6.02
12	1.68	2.58	30	4.20	6.45
15	2.10	3.23	35	4.90	7.53
18	2.52	3.87	40	5.60	8.60
20	2.80	4.30			

For incomplete penetration welds, the above loads are to be multiplied by a factor not exceeding 0.625.
For explanation of tables, see Notes on page 135.
1 kilonewton may be taken as 0.102 metric tonne (megagramme) force, but see page 102.

the British Constructional Steelwork Association.

found in the handbooks of firms which specialise in welding and in the reports issued as a result of practical research.

Certain data in this chapter has been kindly supplied by the Institute of Welding. British Standards may be obtained from the British Standards Institution, 2 Park Street, London, W1A 2BS.

Exercises 14

1 Distinguish between a *butt weld* and a *fillet weld*. Give simple diagrams illustrating the various forms of fillet welds.

Explain the meaning of the terms '*throat thickness*' and '*effective weld length*'.

2 Write down the working stresses given in BS 449 for fillet welds. Under what circumstances may the tensile and compressive stresses respectively in butt welds be taken at 155 N/mm^2? State the importance of *returned ends* in fillet welds.

3 Find (a) the throat thickness for a 10 mm fillet weld, (b) the throat area for a 12 mm fillet weld which has an effective length of 200 mm, (c) the effective length of a 12 mm fillet weld, 150 mm long, which has no returned ends.

4 Calculate the safe load per mm for (a) a 6 mm side fillet weld, (b) a 8 mm end fillet weld, (c) a 15 mm butt weld, reinforced and sealed, in grade 50 steel.

5 A flat tie-bar laps on to a gusset plate to which it is connected by two side fillets each 215 mm long. Assuming 10 mm fillets, and grade 43 steel, find the safe axial load for the tie-bar from the point of view of its welded end connection. Each fillet has two returned ends.

6 Two lengths of a grade 43 steel tie member, each 150 mm wide × 12 mm thick, are joined by a double-V butt weld which is suitably reinforced. Calculate the safe axial load for the welded member.

7 A simply supported beam AB is supported at A and B by similar welded-angle-bracket connections. The angles are 125 mm deep and have 6 mm side fillets which are returned at the bottom ends only. Assuming the welds to be subjected to simple vertical shear, calculate the safe total uniformly distributed load for the beam AB, from the point of view of its end supports, in grade 43 steel.

8 The flange of a grade 43 steel plate girder is joined to the web by two 8 mm fillet welds. Calculate the maximum permissible horizontal shear, in kN per mm run of girder, at the level of the junction of flange and web.

NOTE: In all examples use:

	Grade 43 steel	Grade 50 steel
for butt welds	155 N/mm^2	215 N/mm^2
for fillet welds	115 N/mm^2	160 N/mm^2

15

Plate and Lattice Girders. Theory and Practical Design

M.R. of plate-girder section

Sometimes in steel-beam design, the large span and heavy loads carried demand deeper sections than can be supplied by compounding standard sections together. In these cases, plates, possibly with angles, may be connected to obtain a section which will satisfy the requirements of both strength and stiffness (see plate 9). The relative dimensions of plate girder details are governed by theoretical considerations, and by regulations which are the result of practical experience.

The true M.R. of a plate girder section may be derived from its moment of inertia (M.R. $= fI/y$) or from its section modulus (M.R. $= fZ$). The building up of a girder section to give the required modulus is not a difficult matter, especially if section books be employed to give the I-value of flange plates at a stated distance apart. In any case, a trial section can easily be tested for suitability.

Another method commonly employed is to use a formula for the M.R. which involves the 'flange area' and 'depth of the girder' (*flange-area* method). The formula is approximate, but sufficiently accurate for practical design.

In fig. 15.1, F represents the resultant compressive and tensile forces acting, respectively, in the top and bottom flanges of the girder. D is the distance between the lines of action of these forces, i.e. the 'arm' of the couple resisting bending. The M.R. of the girder section is clearly very nearly given by the product $F \times D$; an error on the safe side arises by neglecting the relatively small contribution of the web. If $f =$ the average stress in the flange, and $A =$ the flange area, $F = fA$. The expression for M.R. is therefore fAD.

Writing M for moment of resistance (and for the applied bending

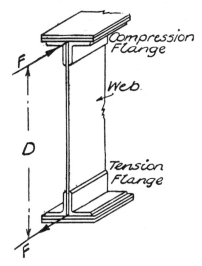

Fig. 15.1 Couple resisting bending in plate girder section.

moment) we have the general expression

$$M = fAD.$$

A slight error—on the wrong side—occurs if f is regarded as the maximum skin stress in the steel.

Flange area

The constitution of the flange area is variously taken by different designers. Three methods are common where connecting angles are used. The flange area is assumed to be made up of:

i) horizontal legs of angles only, plus the flange plates;
ii) whole angle area, plus the flange plates;
iii) whole angle area and flange plates, plus a portion of the web.

Connection holes are deducted in both tension and compression flanges, in bolted construction. Rivet holes would be allowed for in tension flanges, but gross areas would be taken in compression flanges, where riveted construction is used.

Depth of girder

D is the distance between the centres of gravity of the respective flanges. This has to be estimated from the chosen web depth and is frequently taken as the web depth itself. It should not be taken greater than the web depth.

Example A plate girder of 13 m effective span is to carry the concentrated load system given in fig. 15.2. Calculate suitable flange details for the girder, assuming the compression flange to be embedded in a concrete floor, and assuming holes for bolted construction. The overall depth of the girder—measured over the flange plates—is not to exceed 900 mm (limitation due to headroom requirement). The maximum bending stress must not exceed 155 N/mm².

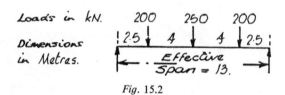

Fig. 15.2

B.M. maximum for concentrated loads

$$R_A = R_B = \frac{200+250+200}{2} \text{ kN} = 325 \text{ kN}$$

B.M. max. = $(325 \times 6.5) - (200 \times 4)$ kNm = 1312·5 kNm
B.M. maximum due to self-weight of girder, etc.
Assuming girder + casing to weigh 120 kN,

$$\text{B.M. maximum} = \frac{120 \times 13}{8} = 195 \text{ kNm}$$

Total B.M. maximum = 1312·5 + 195 = 1507·5 kNm
Flange area required

$$M = fAD$$

$f = 155$ N/mm², $D = 800$ mm (assuming a tentative value for web depth).

$$A = \frac{M}{fD} = \frac{1507 \cdot 5 \times 10^3}{155 \times 800} \text{ mm}^2$$

$$= 12\,157 \text{ mm}^2$$

Compression flange As the girder is shallow, the horizontal legs only of the main angles will be taken in the flange area.

Assuming $150 \times 150 \times 15$ angles, the horizontal legs give a gross area of $(2 \times 150 \times 15)$ mm^2 = 4500 mm^2.

∴ Area to be supplied by the flange plates

$$= (12157 - 4500) \text{ mm}^2 = 7657 \text{ mm}^2$$

If the flange plates be taken 350 mm wide, the necessary total thickness of plates $= \dfrac{7657}{350} = 22$ mm.

Tension flange Taking two holes 22 mm diameter (for 20 mm bolts) from the gross angle area, the net area supplied by angles $= (4500 - 2 \times 15 \times 22)$ mm^2 = 3840 mm^2.

The plates will have to supply a net area of $(12157 - 3840)$ mm^2 = 8317 mm^2.

The effective solid width of the flange plates

$$= (350 - 2 \times 22) \text{ mm} = 306 \text{ mm}$$

∴ Necessary total plate thickness $= \dfrac{8317}{306} = 27 \cdot 2$ mm.

(When the same working stress is used for tension and compression flanges in *riveted* construction, it is good practice to design the tension flange, and then make the gross area of compression flange equal to it, i.e. have similar detail for both flanges.)

Design and detail of a plate girder in accordance with BS 449

In designing the girder, the following points will be taken into consideration:

1 The depth of the girder should be within the limits $L/10 - L/14$, where L is the length of the girder. The depth of the girder may be less than the minimum given by this rule, provided that the deflection is calculated and proved to be not excessive. Shallow girders lead to high web stress and connection difficulties.

2 It is common practice to camber plate girders. If a girder be made without camber, the deflection of the girder when fully loaded is marked, and in fact appears to the eye to be greater than actually is the case. The camber is therefore made equal to the maximum calculated deflection, so that under full load the girder appears substantially horizontal.

3 The width of the girder should be within the limits $L/40 - L/50$. The maximum width given by this rule may be exceeded—if made necessary by a consideration of the unsupported length of flange plates (see Note 5)—provided the flange plates are adequately stiffened.

4 At least $33\frac{1}{3}\%$ of the gross flange area should be provided by the flange angles.

5 The maximum stresses to be used are as follows:

a) *For parts in tension* (on the effective section) 155 N/mm² for material up to and including 40 mm thickness *or* 140 N/mm² for material over 40 mm thickness.

b) *For parts in compression* Clause 20.2 of BS 449 deals with this matter. It is a rather long and somewhat complicated sub-clause and will not be repeated here in full. The reader is referred to clause 20.2 and also to the BCSA publication No. 12, 1959, which deals fully with this point. For the purpose of this book a simplified method of procedure will be adopted. This simplification takes into consideration the following points:

i) Only sectional shapes having I_y smaller than I_x are dealt with (Clause 20.2).
ii) The value of the compression stress depends on l/r_y and D/T, and depends on the value of the critical stress C_s. It will be found that the maximum allowable stress can *always* be used if the value of l/r_y does not exceed 40, and by selecting a section which conforms to this requirement, a great deal of time will be saved in the design stage.
iii) The value of l depends on the nature of the end restraints to the girder (Clause 26a) or, where there are *effective* lateral restraints at points along the span, the maximum distance between such points (Clause 26b).

For a girder supporting a uniformly distributed floor load, the compression flange is considered to be adequately restrained laterally.

c) *For parts in shear* (on the gross section of the web) For unstiffened webs, where d/t does not exceed 95, 80, or 70, for grades 43, 50, or 55 respectively, Table 11 in BS 449 sets out the allowable average shear stress. If the web is stiffened, the thickness may be reduced and the allowable shear stress will vary in accordance with Table 12 a, b, and c in BS 449. The gross section of the web = depth of web plate multiplied by its thickness.

Particulars of girder

We will consider the design of a welded plate girder for the balcony of a small cinema, in grade 43 steel. The girder is 18 m span between the centres of bearings, and the loads come on to it from the main balcony rakers. These are at 3 m centres, and provide a maximum reaction of 180 kN each. The rakers pass over the top of the girder.

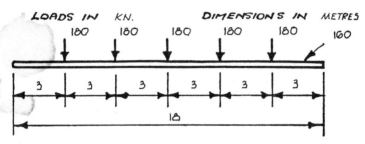

Fig. 15.3.

Own weight of girder (no casing), say, 160 kN.
Total load on girder = 1060 kN
∴ each reaction = 1060/2 = 530 kN
Maximum B.M. = $(530 \times 9 - 180 \times 3 - 180 \times 6 - 80 \times 4 \cdot 5)$ kN m
= 2790 kN m

NOTE: The weight of a girder is given approximately by $WL/200$, where W = equivalent distributed load on girder in kN; L = length of girder in metres;

or by $\dfrac{\text{B.M. (kN m)}}{25}$.

In this case $\dfrac{\text{B.M.}}{25} = \dfrac{2790}{25} = 111 \cdot 6$ kN

The assumed weight of 160 kN is, therefore, on the right side.

Web plate

Taking $L/14$ for depth of girder,

$$\text{depth} = \frac{18 \times 10^3}{14} = 1286 \text{ mm}$$

Assume web 1250 mm deep.

[handwritten note at top: d_2 = twice the dist. from comp'r flange to neutral axis]

332 Structural Steelwork

Maximum shear = 530 kN.
Assume thickness of web = 10 mm

$$\frac{d}{t} = \frac{1250}{10} = 125$$

Reference to Table 12a of BS 449 will show that, with this value of d/t the stiffener spacing may be up to $0 \cdot 8d$ before the maximum allowable stress of 100 N/mm² need be reduced. It will be convenient to space the stiffeners at 1·0 m centres and, since $0 \cdot 8d = 0 \cdot 8 \times 1 \cdot 250 = 1 \cdot 000$, the maximum allowable stress of 100 N/mm² can be used.

$$\therefore \quad \text{Allowable shear} = 1250 \times 10 \times 100 \times 10^{-3} = 1250 \text{ kN}.$$

BS 449, Clause 27f states that the web thickness shall not be less than 1/180 of the smaller clear panel dimension, and $d_2/200$, where $d_2 = 2 \times (1250/2) = 1250$ mm.

$$\therefore \quad \text{Minimum web thickness} = \frac{1000}{180} \text{ or } \frac{1250}{200} = 5 \cdot 55 \text{ mm or } 6 \cdot 25 \text{ mm}$$

Make web 1250 × 10 as calculated.

Flanges

In BS 449 it is required that plate girders shall be proportioned on the basis of the moment of inertia of the section. This may be done either on the gross I with the neutral axis taken at the centroid of the section or on the net I. In arriving at the maximum flexural stresses, the stresses calculated on the gross I shall be increased in the ratio of gross area to effective area of the flange section.

This means that the previous alternative, permitted in earlier editions of BS 449, of an approximate method is not now permissible, but it will be found that the approximate method used in the previous example will give a generally adequate result.

Approximate method

$M = fAD$

D is assumed to be 1280 mm.
$f = 155$ N/mm² on net flange area.

$$2790 \times 10^6 = 155 \times A \times 1280$$

$$A = \frac{2790 \times 10^6}{155 \times 1280} = 13960 \text{ mm}^2$$

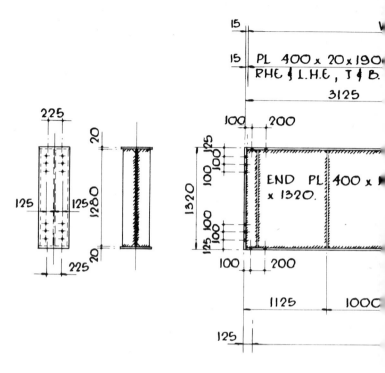

ALL WELDING TO BE 6mm CONT. FILLET. U.N.O.

FULL PENETRATION BUTT WELD.

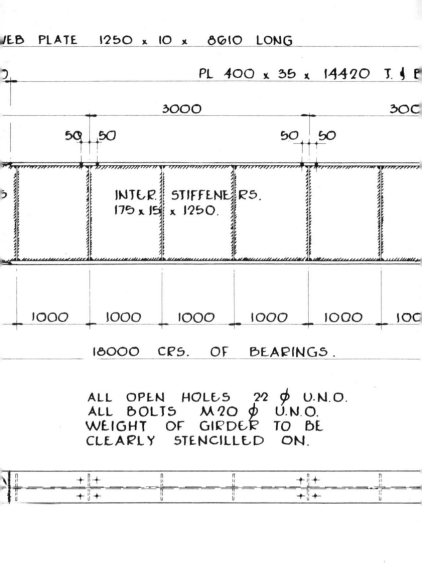

Chart 1 *Plate girder*.

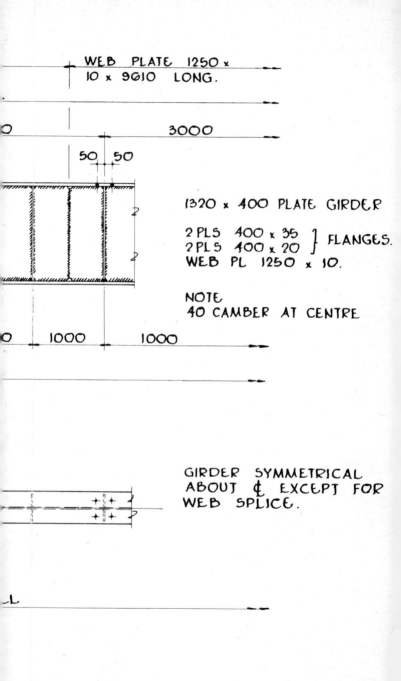

Make plates 400 mm wide = $L/45$ (see Note 3)

$$\therefore \text{ necessary thickness} = \frac{13960}{400} = 34 \cdot 9 \text{ mm}$$

Use a 35 mm plate.

Care should be taken in the selection of the quality of steel which is to be welded, in cases of thickness greater than 30 mm. It may be necessary to use grade 43c, which is not readily available.

Exact method

I_{XX} (gross):

$$I_{web} = \frac{bd^3}{12} = \frac{10 \times 1250^3}{12} = 1627 \times 10^6 \text{ mm}^4$$

$$I_{plates} = \frac{b}{12}(D^3 - d^3) = \frac{400}{12}(1320^3 - 1250^3)$$

$$= \frac{400}{12} \times 346 \cdot 843 \times 10^6 = 11\,562 \times 10^6 \text{ mm}^4$$

$$\therefore I_{XX} \text{ (gross)} = (1627 + 11\,562) \times 10^6 = 13\,189 \times 10^6 \text{ mm}^4$$

Tension flange:

$$\text{Gross flange area} = 400 \times 35 = 14\,000 \text{ mm}^2$$

Note that the effective area of the tension flange is the gross area with certain deductions as set down in Clause 17a of BS 449. It will be assumed, in this case, that there are no deductions.

Flexural stress on gross I

$$= \frac{2790 \times 10^6 \times 660}{13\,189 \times 10^6}$$

$= 139 \cdot 6 \text{ N/mm}^2$, which is less than 155 N/mm².

Compression flange:

It will be assumed that the girder is uncased, in which case the allowable stress is 155 N/mm² if the value of l/r_y is 40 or less.

The value of l is the distance between points of lateral restraint and equals 3·000 m in this case.

I_{YY} (net):

$$I_{web} = \frac{1250 \times 10^3}{12} = 0 \cdot 104 \times 10^6 \text{ mm}^4$$

$$I_{plates} = \frac{70 \times 400^3}{12} = 373 \times 10^6 \text{ mm}^4$$

$$\overline{373 \cdot 1 \times 10^6} \text{ mm}^4 \text{ gross}$$

Again, it will be assumed that there are no deductions for holes.
Net area of section $= 2 \times 400 \times 35 + 1250 \times 10$

$$= (28\,000 + 12\,500) \text{ mm}^2$$
$$= 40\,500 \text{ mm}^2.$$

$$r_y = \sqrt{\frac{373 \cdot 1 \times 10^6}{40\,500}} = \sqrt{9212 \cdot 3} = 96 \text{ mm}$$

$$\frac{l}{r_y} = \frac{3000}{96} = 31, \text{ which is less than 40.}$$

∴ The allowable stress is 155 N/mm².

It will be seen that the approximate section has provided an adequate solution.

Welding

Welds connecting the flanges to the webs of plate girders shall be proportioned to carry the resultant of the longitudinal and vertical shears. It is usual to assume that the horizontal shear is equal to the vertical shear.

$$\text{Horizontal shear/mm} = \frac{S}{D}.$$

Taking D as the depth between the weld lines in the top and bottom flanges,

$D = 1250$ mm

∴ At the ends of the girder, the horizontal shear/mm

$$= \frac{530}{1250} = 0.424 \text{ kN}$$

Therefore a continuous 3 mm fillet weld on both sides would be adequate, but 6 mm welds will be provided.

Stiffeners

Stiffeners are usually placed at the ends of plate girders and under point loads, and at equal distances between the point loads—the spacing being determined mainly by the distances apart of the loads. The stiffeners should not, however, be farther apart than the depth of the web plate.

It has been previously shown—when determining the allowable shear stress—that the web is effectively stiffened, so that it is now only necessary to show that the stiffeners provided are adequate. Bearing stiffeners should be symmetrical about the web where practicable, and should project as nearly as practicable to the outer edges of the flanges. They should be designed as a column, assuming the section of the column to consist of the pair of stiffeners, together with a length of web on each side of the centre line of the stiffeners, equal, where available, to twenty times the web thickness. The radius of gyration shall be taken about the axis parallel to the web of the girder, and the working stress shall be in accordance with the appropriate permissible value as a column, assuming an effective length of 0·7 of the length of the stiffener.

Clause 28a and b in BS 449 sets out the full requirements for stiffeners.

In the example, the criterion is at the ends of the girder, and here the length of web specified is not available on one side. The relationship of thickness to outstand is regulated by Clause 28b(iii).

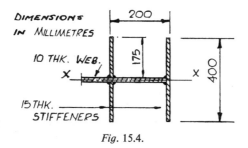

Fig. 15.4.

Assuming that the load is shared equally by the two pairs of stiffeners, it follows that the outer pair must be taken in conjunction with a 100 mm length of web, and it will thus be necessary to investigate the outer pair only.

$$I_{xx} = \frac{100 \times 10^3}{12} + 2\left(\frac{15 \times 175^3}{12} + 175 \times 15 \times 92 \cdot 5^2\right)$$

$$= 0 \cdot 008 \times 10^6 + 2(6 \cdot 70 \times 10^6 + 22 \cdot 46 \times 10^6)$$
$$= 58 \cdot 33 \times 10^6 \text{ mm}^4$$

Area $= 100 \times 10 + 2 \times 175 \times 15 = 6250 \text{ mm}^2$

$$r_{xx} = \sqrt{\frac{58\cdot 33 \times 10^6}{6250}} = 96\cdot 7 \text{ mm}$$

$$\frac{l}{r} = \frac{0\cdot 7 \times 1250}{96\cdot 7} = 9\cdot 04$$

$$\therefore \quad p_c = 151 \text{ N/mm}^2$$

$$\text{Actual stress} = \frac{265 \times 10^3}{6250} = 42\cdot 4 \text{ N/mm}^2$$

∴ Sections chosen are adequate. For practical reasons, the end stiffeners will be provided in the form of a single full-width end plate.

The area of each pair of stiffeners should be such that the bearing stress does not exceed 190 N/mm².

$$\text{Approximate area} = 4 \times 175 \times 15 = 10\,500 \text{ mm}^2$$

$$\text{Actual bearing stress} = \frac{530 \times 10^3}{10\,500} = 50\cdot 5 \text{ N/mm}^2.$$

The welds connecting the stiffeners to the web should be capable of taking the vertical shear.

$$\frac{530}{8 \times 1250} = 0\cdot 053 \text{ kN/mm},$$

therefore nominal 6 mm welds will be provided.

Curtailment of flange plates

This is best done by drawing the B.M. diagram and superimposing upon it the resistance moment diagrams of the various sections composing the girder. In welded girders, it is convenient to have only single plates in each flange, and any reduction in plate section is achieved by full-penetration butt welding of the ends of the abutting plates. In the case of riveted or bolted girders, with flange angles, it is usual practice to run the plate next to the angles for the full length, and also to place the thicker plates nearer the flange angles.

In this example, we have a 400×35 plate on each flange. We will determine at what point this could conveniently be reduced.

Construction of B.M. diagram.

$$\text{B.M. at centre} = 2790 \text{ kNm}$$

The forces in the members in the end bays are so much smaller than those in the remaining bays that a smaller section will be checked for the main compression boom at the panel points, 3·5 m from either end.

Design of end compression members

Maximum load (member 9Q) = 249 kN compression
   Length of member 3·7 m.
   Use two angles, 100×75×8.

$$\text{Gross area} = 26·9 \text{ cm}^2$$
$$\text{Least radius of gyration} = 3·14 \text{ cm}$$

$$\frac{\text{Effective length}}{\text{Least radius of gyration}} = \frac{0·7 \times 3700}{31·4} = 82·5$$

$$\text{Allowable stress} = 100·5 \text{ N/mm}^2$$

$$\text{Actual stress} = \frac{249 \times 10^3}{2690} = 92·5 \text{ N/mm}^2$$

Whilst this section is satisfactory, it is not considered worthwhile to change this panel only, but rather to continue the same section through, thus simplifying the details.

Actual bolting used

$$\begin{aligned}\text{Five 24 mm bolts in S.S.} &= 5 \times 36·2 = 181·0 \\ \text{Four 24 mm bolts in bearing} &= 4 \times 57·6 = \underline{230·4} \\ & \qquad\qquad\qquad\quad \underline{\underline{411·4}} \text{ kN}\end{aligned}$$

This bolting would be adequate to take the force in the member 8N which is jointed to the member 9Q, if the boom was broken at this point.

At the outer ends of the end members, the three bolts in bearing are adequate.

Where the section of the frame is unbroken over a panel point, the number of bolts which connect the gusset plate to the unbroken member need not be more than is sufficient to take the difference between the forces in the portions of the section, on either side of the panel point.

Main tension boom

Maximum force in boom (members OG and OK) = 486 kN tension.

$$\text{Net area required} = \frac{486 \times 10^3}{155} = 31·35 \text{ cm}^2$$

Use two angles, 150 × 75 × 10.
Gross area = 43·3 cm²
Net area = 43·3 − 2 × 2·6 × 10 (deduction for bolt holes)
         = 43·3 − 5·2
         = 38·1 cm²
∴ Section chosen is sufficient.

Bolting

As for compression boom, value = 519·6 kN.

As there is no force in the end members OA and OR, the main section could be reduced at the panel points, 3·5 m from each end, where two angles 80 × 80 × 8 would be adequate. For convenience this will not be done in this case.

Bolting required at end of members OC and OP

Force in OC and OP = 228 kN.
Bolting supplied:
    Five 24 mm bolts in bearing = 5 × 57·6 = 288·0 kN

Intermediate members

Compression members

Maximum load (member BC) = 162 kN
Length of member = 2·5 m
Use two angles 70 × 70 × 6
Gross sectional area = 16·3 cm²
Least radius of gyration = 2·13 cm

This is a double-angle, discontinuous strut, back-to-back connected to both sides of a gusset by not less than two bolts in line, and as such the allowable stresses are as given in fig. 11.16.

$$\frac{\text{Effective length}}{\text{Least radius of gyration}} = \frac{0·7 \times 2500}{21·3} = 82$$

Allowable stress = 101 N/mm²

$$\text{Actual stress} = \frac{162 \times 10^3}{1630}$$
$$= 99 \text{ N/mm}^2$$

Bolting
Member BC

$$\text{Number of 24 mm bolts in S.S.} = \frac{162}{36 \cdot 2} = 5 \text{ bolts (min.)}$$

Members DE and MN:

$$\text{Number of 24 mm bolts in S.S.} = \frac{115}{36 \cdot 2} = 4 \text{ bolts}$$

Members FG, HJ, KL, and PQ:

$$\text{Number of 24 mm bolts in S.S.} = \frac{69}{36 \cdot 2} = 2 \text{ bolts}$$

Members FG, HJ, and KL are double-angle, discontinuous struts back-to-back connected to both sides of a gusset with only one bolt in line, and as such are not specifically catered for by the regulations. As the maximum force in these members is 69 kN, the section comprising two angles $70 \times 70 \times 6$ is obviously adequate.

Tension members
Members QR and AB:

$$\text{Maximum force} = 279 \text{ kN}$$

$$\text{Area required} = \frac{279 \times 10^3}{155} = 1800 \text{ mm}^2 = 18 \text{ cm}^2$$

Use two angles $80 \times 80 \times 8$.

$$\begin{aligned}
\text{Gross area} &= 24 \cdot 5 \text{ cm}^2 \\
\text{Net area of section} &= 24 \cdot 5 - 2 \times 26 \times 8 \times 10^{-2} \\
&= 24 \cdot 5 - 4 \cdot 2 \\
&= 20 \cdot 3 \text{ cm}^2
\end{aligned}$$

Bolting used

$$\text{Five 24 mm bolts in bearing} = 5 \times 57 \cdot 6 = 288 \text{ kN}$$

Remaining tension members
Maximum load (member CD) = 198 kN

$$\begin{aligned}
\text{Area required} &= \frac{198 \times 10^3}{155} = 1277 \text{ mm}^2 \\
&= 12 \cdot 8 \text{ cm}^2
\end{aligned}$$

Use two angles $70 \times 70 \times 8$

$$\text{Gross area} = 21 \cdot 3 \text{ cm}^2$$
$$\text{Net area} = 21 \cdot 3 - 2 \times 26 \times 8 \times 10^{-2}$$
$$= 21 \cdot 3 - 4 \cdot 2$$
$$= 17 \cdot 1 \text{ cm}^2$$

Make the four central diagonal ties two angles $70 \times 70 \times 6$.

Bolting

Members CD and NP:

$$\text{Number of 24 mm bolts in bearing} = \frac{198}{57 \cdot 6} = 4 \text{ bolts}$$

Members EF and LM:

$$\text{Number of 24 mm bolts in bearing} = \frac{118}{57 \cdot 6} = 3 \text{ bolts}$$

Members GH and JK:

$$\text{Number of 24 mm bolts in bearing} = \frac{43}{57 \cdot 6} = 1 \text{ bolt (min.)}$$

Method of sections

The method of sections is an analytical method of determining the forces in the members of a frame. It is sometimes used instead of drawing a stress diagram. With complicated loading on frames, however, the process becomes cumbersome, and it will be found that the stress diagram method is at once the speedier and more simple.

Typical calculations for obtaining the forces by the method of sections are given below.

Force in 5H (see fig. 15.6)

Take moments of external forces about the point X and divide by the perpendicular distance between X and the member 5H.

Force in member 5H

$$= \frac{162 \cdot 2 \times 14 \cdot 0 - 1 \cdot 2 \times 14 \cdot 0 - 46 \times 10 \cdot 5 - 46 \times 7 \cdot 0 - 46 \times 3 \cdot 5}{2 \cdot 5}$$

$$= \frac{2271 - 983}{2 \cdot 5} = \frac{1288}{2 \cdot 5} = 515 \text{ kN}$$

This compares with 520 kN from the stress diagram.

Force in OG

Take moments about the point Y.
Force in member OG

$$= \frac{162\cdot 2 \times 10\cdot 5 - 1\cdot 2 \times 10\cdot 5 - 46 \times 7\cdot 0 - 46 \times 3\cdot 5}{2\cdot 5}$$

$$= \frac{1703 - 496}{2\cdot 5} = \frac{1207}{2\cdot 5} = 483 \text{ kN}$$

This compares with 486 kN from the stress diagram.

Force in BC

The force in a vertical member is equal to the shear force at the point under consideration.

Shear force at $Z = 162\cdot 2 - 1\cdot 2 = 161$ kN
This compares with 162 kN from the stress diagram.

Force in AB

Considering the point Z, it follows that the vertical component of the force in AB must equal the vertical component (i.e. the force, since the member is vertical) of the force in BC.

$$\therefore \text{ Force in AB} = \text{force in BC} \times \frac{\text{length of AB}}{\text{length of BC}} = 161 \times \frac{4\cdot 32}{2\cdot 5}$$

$$= 278 \text{ kN}$$

This compares with 279 kN from the stress diagram.

Appendix 1

Design of a Steel Frame for a Small Warehouse Building, with Typical Details

Introduction

It is intended in this chapter to demonstrate how the calculations for a structural steel frame may be set out and arranged in a form convenient for practical application. Points of theory not already explained in previous chapters are dealt with as they arise. In order to avoid repetition, calculations are not included for all the structure, but only the roof trusses and associated steelwork, the fourth floor, and typical stanchions and foundations.

The building, as will be seen by reference to the plans, has four floors and a pitched roof, and is one actually erected for a commercial firm, in South East England.*

The superimposed load to be allowed for on all floors is 1650 kgf per square metre.

In the design of the steel frame BS 449 will be referred to, and the allowable pressures on soil, and concrete (1:2:4) are taken at 300 kN/m^2 and 4200 kN/m^2, respectively.

On the data sheet it will be noticed that the various floor and wall loads are each converted from kgf/m^2 to kN/m^2—a step which simplifies and quickens the ensuing calculations.

The following abbreviations are made use of in the calculations:

E.D.L. = equivalent distributed load,

O.W. and C. = own weight and casing,

R/L and R/R (in tables) = left- and right-end reactions, respectively.

Contracted forms for units, e.g. cm^3 and cm^4 for Z and I respectively are used as is usual in design calculations. The symbols N/mm^2 and

---

*In this volume the building has been re-worked to agree with BS 449, with metric units.

Plate 10 *Lattice girder construction.*
Reproduced by courtesy of The Brixton Estate Ltd.

m², or mm² are employed to indicate respectively a 'stress' and an 'area'.

To describe superimposed load per unit of surface area, the unit would be, e.g., kgf/m² converted to kN/m².

The weights of the lift motors and machinery are taken as centrally applied point loads on beams R51, R53, and R91 (see roof-steel plan), the actual joists required for these loads not being shown.

The ground floor is assumed to be laid on a hardcore filling, which is placed on the soil, and therefore is not a 'suspended' floor, i.e. its weight is not taken by the frame, and consequently does not enter into the calculations.

Data sheets

Flat roof over lifts

$$\begin{aligned} 125 \text{ mm reinforced concrete} &= 293 \\ \text{asphalt} &= \underline{44} \\ &\,337 \\ \text{Superimposed load} &= \underline{146} \\ &\,483 \text{ kgf/m}^2 = 4{\cdot}8 \text{ kN/m}^2 \end{aligned}$$

Floor to lift-motor room

$$125 \text{ mm reinforced concrete} = 293 \text{ kgf/m}^2 = 2{\cdot}9 \text{ kN/m}^2$$

Floors

$$\begin{aligned} \text{slab construction} &= 365 \\ \text{finish} &= \underline{95} \\ &\,460 \\ \text{Superimposed load} &= \underline{1650} \\ &\,2110 \text{ kgf/m}^2 = 21{\cdot}0 \text{ kN/m}^2 \end{aligned}$$

Stairs and landings

$$\begin{aligned} \text{Construction, say,} &= 488 \\ \text{Superimposed load} &= \underline{488} \\ &\,976 \text{ kgf/m}^2 = 9{\cdot}7 \text{ kN/m}^2 \end{aligned}$$

Walls

$$\text{Wall at } 440 \text{ kgf/m}^2 = 4{\cdot}4 \text{ kN/m}^2.$$

Lifts

100 kN each

Calculations

Vertical load (kgf/m²)

	Slope	Flat
Insulated decking and Finish	40	
Purlins	10	
	50	= 58
Truss	12	
	70	= 0·69 kN/m²
Snow	75	
	145 kgf/m²	= 1·42 kN/m²

Wind loads

Pitch of roof = 30°. (For single-storey pitched-frame construction, a very much shallower pitch is likely to be adopted.)

The pressures given in CP3, Chapter V, for this slope are determined on the following basis:

The basic wind speed V appropriate to the location is determined. In this case, the value of V, determined from fig. 1 in CP3, Chapter V, is 40 metres per second.

The basic wind speed is now multiplied by three factors S_1, S_2, and S_3, influenced by the topography of the area, the ground roughness, and the designed life of the building.

Factor S_1 is normally 1·0, except for particularly exposed or very sheltered situations, and is applicable in this case.

Factor S_2 relates to the environment in which the building is situated, in this case the outskirts of a suburban town. Therefore S_2 is taken as in category 3, and for a building with neither height nor horizontal dimensions exceeding 50 m. Table 3 gives S_2 for a building height of 19 m as being 0·90.

Factor S_3 relates to the statistical concept of the life of the building, and in this case is taken as the normal figure of 1·0.

The design wind speed is therefore:

$$V_s = V \times S_1 \times S_2 \times S_3$$
$$= 40 \times 1·0 \times 0·90 \times 1·0$$
$$= 36 \text{ m/s}$$

$q = 0.613 V_s^2$
$= 0.613 \times 36^2$
$= 794 \text{ N/m}^2$

The dynamic pressure, q, relating to this design wind speed is given in Table 4 of the Code as 794 N/m^2.

The pressure coefficients are set out in Table 11 of the Code, relative to the height:width ratio of the building, and the roof pitch. In this case, $\dfrac{h}{w} = \dfrac{16\cdot000}{18\cdot000} = 0\cdot889$, i.e. less than unity.

The roof pitch is 30°, and therefore the pressure coefficients C_{pe} for the roof slopes are $-0\cdot2$, $-0\cdot6$, $-0\cdot2$, and $-0\cdot5$ respectively, for wind parallel to the trusses or $-0\cdot8$ overall for wind normal to the trusses.

The internal pressure coefficient, assuming that the four faces of the building are of approximately equal permeability, is given in Appendix E of the Code as being $C_{pi} = -0\cdot3$.

The wind loading on the four respective roof slopes is then given by:

$$F = (C_{pe} - C_{pi})\,qA$$

where A is the surface area of the structural element.

In the worst case for uplift (wind normal to trusses),

$$F = [-0\cdot8 - (-0\cdot3)]\,qA$$
$$= -0\cdot5\,qA$$

or 396 N/m^2 uplift pressure; or (for wind parallel to trusses),

$$F = [-0\cdot2 - (-0\cdot3)]\,qA$$
$$= +0\cdot1\,qA$$

or 79 N/m^2 downward pressure

As the dead loading = 0·69 kN/m^2, as compared with 0·40 kN/m^2 uplift from wind, it is apparent that reversal of forces will not occur. With regard to the downward pressure, the wind pressure of 0·08 kN/m^2 is only a small percentage of the vertical dead and/or superimposed load, and, bearing in mind the 25% increase in permissible stresses when considering wind effects, this is clearly not a critical condition.

It must be obvious that, unless the roof is constructed of the lightest possible material and the building situated in a very exposed position, reversals of force will not occur in roofs of pitches of $22\frac{1}{2}°$ to 30°.

With regard to frictional drag, clause 7.4 indicates the dimensional ratios of a building which must exist before this need be considered, and this building is clearly not required to be so considered.

The regulations say that the structure shall be capable of sustaining the most adverse combination of loads to which the building may reasonably be subjected.

350 *Structural Steelwork*

In this particular case, since the wind pressure reduces the total load on the structure, there are two possible combinations to consider:
a) dead load plus superimposed load without the wind load—this will give the maximum forces in the members;
b) dead load plus wind load without superimposed load—this would give the worst possible reversals of force, if any.

The force diagrams for both combinations will be drawn so that the reader may see for himself the forces involved.

Case (*a*)

$$\text{Total vertical load} = 9 \cdot 144 \times 3 \cdot 05 \times 1 \cdot 42 \text{ kN}$$
$$= 39 \cdot 6 \text{ kN } (= 6 \cdot 6 \text{ kN per panel point})$$

Case (*b*)

$$\text{Total vertical load} = 9 \cdot 144 \times 3 \cdot 05 \times 0 \cdot 69 \text{ kN}$$
$$= 19 \cdot 2 \text{ kN } (= 3 \cdot 4 \text{ kN per panel point})$$

Wind loads

For maximum uplift we will consider wind normal to the trusses, when all slopes will be similar:

$$\text{Load} = 5 \cdot 27 \times 3 \cdot 05 \times 0 \cdot 79 \times 0 \cdot 8 \text{ kN}$$
$$= 10 \cdot 2 \text{ kN } (= 3 \cdot 4 \text{ kN per panel point})$$
$$\text{Horizontal component} = 5 \cdot 1 \text{ kN (i.e. } 10 \cdot 2 \cos 60°)$$
$$\text{Vertical component} = 8 \cdot 1 \text{ kN}$$

Reactions

Horizontal:
   Total horizontal force = 0 (as the loads are symmetrical)
$$\therefore R_L = R_R = 0$$
Vertical:

$$\text{Total vertical force} = 19 \cdot 2 - (8 \cdot 1 \times 2) = 3 \cdot 0 \text{ kN}$$
$$\therefore R_L = R_R = 1 \cdot 5 \text{ kN upward}$$

For maximum out-of-balance forces, we will also consider the windward truss under conditions of lateral wind:

(a) Windward slope = $5 \cdot 27 \times 3 \cdot 05 \times 0 \cdot 79 \times 0 \cdot 2$ kN
$$= 2 \cdot 6 \text{ kN } (= 0 \cdot 85 \text{ kN per panel point})$$
   Horizontal component = $1 \cdot 3$ kN (i.e. $2 \cdot 6 \cos 60°$)
   Vertical component = $2 \cdot 0$ kN

Appendix 1 351

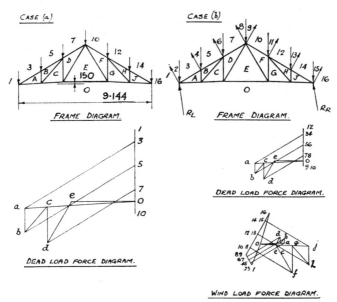

Fig. A.1.

(b)   Second slope = $5\cdot27 \times 3\cdot05 \times 0\cdot79 \times 0\cdot6$ kN
                                 = $7\cdot6$ kN (= $2\cdot5$ kN per panel point)
      Horizontal component = $3\cdot9$ kN
      Vertical component = $6\cdot0$ kN

Reactions (including wind load) of windward truss
Horizontal:
      Total horizontal force = $-1\cdot3 + 3\cdot9 = 2\cdot6$ kN
                $\therefore R_L = R_R = 1\cdot3$ kN

Vertical:
      Total vertical force = $19\cdot2 - 2\cdot0 - 6\cdot0 = 11\cdot2$ kN

$$R_R = \frac{19\cdot2}{2} - \frac{2\cdot0}{4} - \frac{6\cdot0 \times 3}{4} + \frac{2\cdot6 \times 1\cdot32}{9\cdot144}$$

            = $9\cdot6 - 0\cdot5 - 4\cdot5 + 0\cdot38$ = $4\cdot98$ kN upward
     $R_L = 11\cdot2 - 4\cdot98$ = $6\cdot22$ kN upward

(NOTE: Height of truss = $4\cdot57$ m $\times \tan 30° = 2\cdot65$ m. The components enable the line 16—0 to be drawn in the wind-force diagram.)

352 *Structural Steelwork*

The usual assumption is made (for trusses supported by steel columns) that the horizontal reactions at each end are equal; i.e. in this case, each equals 2·6 kN/2 = 1·3 kN. In finding R_R (vertical) the correct proportion of each vertical load is taken and, in addition, the value of R_R required to resist the moment about R_L of the horizontal force of 2·6 kN acting at 1·32 m ($\frac{1}{2}$ truss height) above, is taken into account.

It will be seen from the table of forces that, as premised above, the maximum force in every member is given by case (a), while in case (b) no reversal of forces occurs.

Member	Case (a)		Case (b)			
	Vert. load only		Vert. load only		Wind left	
	C	T	C	T	C	T
3A, 14J	36		17			4
5B, 12H	36		17			4
7D, 10F	29		13			4
OA, OJ		31		14	3	
OC, OG		25		12	2	
OE		18		9	2	
AB, HJ	7		3			1
BC, GH		9		5	1	
CD, FG	10		5			1
DE, EF		13		6	1	

Forces in kN
Table of forces.

Design of trusses

Compression boom

$$\text{Maximum load} = 36 \text{ kN, compression}$$
$$\text{Length} = 1·753 \text{ m}$$

Use one angle 70 × 70 × 6.

$$\text{Gross area} = 8·13 \text{ cm}^2$$
$$\text{Least } r = 1·37 \text{ cm}$$

Taking effective strut length as 0·7 × actual length,

$$\frac{L}{r} = \frac{0·7 \times 1753}{13·7} = 90$$

OR
NG & SHOES.

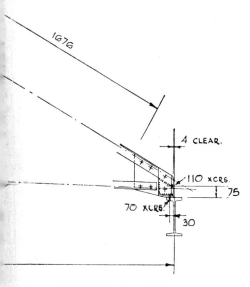

MATERIAL IN SHOE.
% 305 x 127 x 48 kg. U.B. x 275 LONG.
2 Ls 100 x 75 x 8 x 175 LONG SHAPED.
1 PACK 100 x 6 x 100 LONG SHAPED.

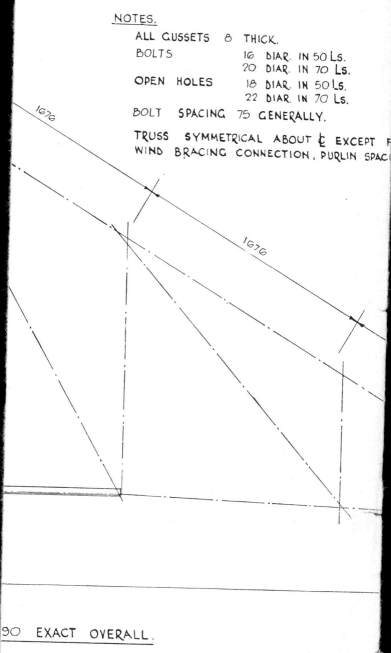

NOTES.

ALL GUSSETS 8 THICK.
BOLTS   16 DIAR. IN 50 Ls.
           20 DIAR. IN 70 Ls.
OPEN HOLES   18 DIAR. IN 50 Ls.
           22 DIAR. IN 70 Ls.
BOLT SPACING 75 GENERALLY.

TRUSS SYMMETRICAL ABOUT ℄ EXCEPT F
WIND BRACING CONNECTION, PURLIN SPAC

90 EXACT OVERALL.

10. 9·190. EXACT OVERALL.

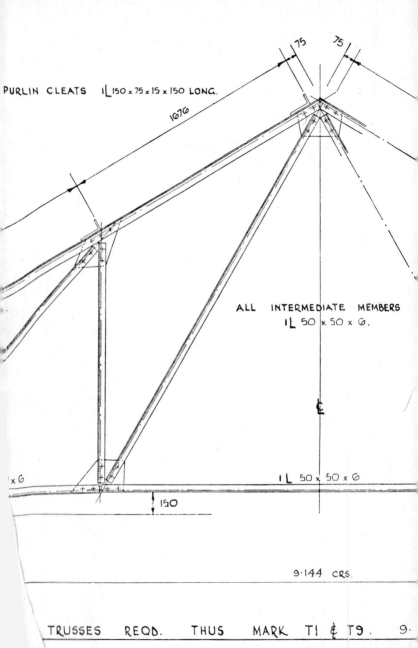

Chart 3 Roof trusses.

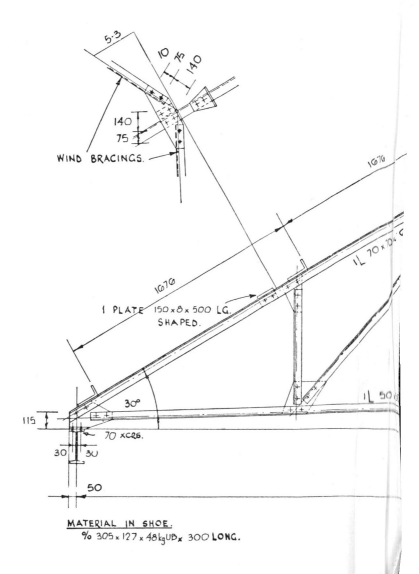

Allowable stress = 91 N/mm²

$$\text{Actual stress} = \frac{36 \times 10^3}{813} = 44 \cdot 3 \text{ N/mm}^2$$

Use two 20 mm bolts.

Tension boom

Maximum load = 31 kN, tension

$$\text{Area required} = \frac{31 \times 10^3}{155} = 200 \text{ mm}^2$$
$$= 2 \text{ cm}^2$$

Use one angle $50 \times 50 \times 6$.
The net area to be used is governed by Clause 42a (i) of BS 449.
a_1 (the *net* sectional area of the connected leg)

$$= 50 \times 6 - 18 \times 6 = 300 - 108 = 192 \text{ mm}^2$$

a_2 (the sectional area of the unconnected leg)

$$= 44 \times 6 = 264 \text{ mm}^2$$

$$\text{Net area} = 192 + \frac{264 \times 3 \times 192}{3 \times 192 + 264}$$

$$= 192 + 181 = 373 \text{ mm}^2$$

Use two 16 mm bolts.

Intermediate members

Maximum compressive force = 10 kN
Maximum tensile force = 13 kN

Use one angle $50 \times 50 \times 6$ throughout with two 16 mm bolts.

Wind bracing

Horizontal load per truss = 2·6 kN.

Each pair of braces is assumed to take all the horizontal load out of the truss and transfer it to the stanchions,

∴ horizontal load in each wind brace = 1·3 kN.

The actual load in the wind brace (which is in the plane of the rafter)

is given by dividing the actual length of the wind brace by its horizontal flat projection and multiplying by the horizontal load.

Horizontal flat projection = 1·52 m

$$\text{Actual length} = \sqrt{3\cdot05^2 + 1\cdot52^2 + \left(\frac{2\cdot65}{3}\right)^2}$$
$$= \sqrt{(9\cdot30 + 2\cdot31 + 0\cdot78)}$$
$$= 3\cdot520 \text{ m}$$

$$\therefore \text{ Load in wind brace} = 1\cdot3 \times \frac{3\cdot52}{1\cdot52} = 3\cdot0 \text{ kN}$$

Use one angle 50 × 50 × 6 with two 16 mm bolts.

Purlins (Reference should be made to Clause 45 of BS 449.)

In the case of sloping roofs not exceeding 30° pitch, the requirements ... as regards limiting deflections of beams and lateral instability may be waived in the design of simple angle purlins ... provided that where L is the distance in millimetres centre-to-centre of the steel principals or other supports, the following requirements are fulfilled:

The leg or the depth of the purlin in the plane appropriate to the incidence of the maximum load or maximum component of the load is not less than $L/45$, and the numerical value of the section modulus of the purlin in cm units is not less than $(WL/1\cdot8) \times 10^{-3}$ about the axis parallel to the roof slope, provided that the other leg or width of the purlin is not less than $L/60$. The loading W is the total distributed load in kN on the purlin arising from dead weight and snow, but excluding wind, both assumed as acting normal to the roof.

In the example, $L = 3050$ mm.

∴ The depth of the purlin should be at least 3050/45 = 67 mm and the width of the purlin should be at least 3050/60 = 51 mm.

The load per m² acting normal to the roof is 1·22 kN/m² made up of 0·49 kN dead weight and 0·73 kN for snow (the latter is not strictly normal to the roof slope).

Purlin spacing = 1·676 m.

$$\therefore W = 3\cdot050 \times 1\cdot676 \times 1\cdot22 \text{ kN}$$
$$= 6\cdot2 \text{ kN}$$

$$\therefore \text{ Modulus required} = \frac{6\cdot2 \times 3050}{1\cdot8} \times 10^{-3} = 10\cdot51 \text{ cm}^3.$$

REMAINING MEMBERS
ALL 1L 50×50×6

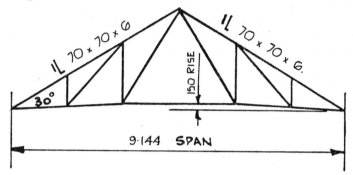

7 OFF THUS MARKED T1, T2, T4, T6, T8, T9 AND T10

REMAINING MEMBERS
ALL 1L 50×50×6

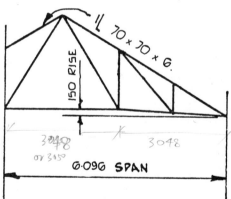

3 OFF THUS MARKED T3, T5 AND T7

Fig. A.2 Truss scantlings.

356 Structural Steelwork

Use one angle $80 \times 60 \times 7$ (modulus = $10 \cdot 7 \text{ cm}^3$), which is the lightest angle that satisfies all the conditions.

Design of roof steel (see chart 4, facing page 368)

The load values used in the calculations are obtained from the data sheet. The left- and right-end reactions are tabulated for the purpose of the stanchion calculations later. The reader should identify the particular beam section being considered, in the steelwork plans (Chart 4), by means of the reference numbers in the last column of the calculation sheet. The figures beneath the chosen section, thus (264·9) indicate the modulus actually provided, in cm^3.

Appendix 1 357

Calculation sheet

Motor roof steel

Loads and calculations	R/L kN	R/R kN	Section	Beam ref.
Span 3·048 m Ties			150 × 89 × 17·09 I	51, 91

Span 6·096 m
  Roof 6·096 × 1·524 × 4·8 = 44·6
  O.W. & C.   = 11·4
                          ─────
                          56·0 kN
 	
$Z \text{ reqd} = \dfrac{56 \cdot 0 \times 6096}{8 \times 165} = 258 \cdot 6 \text{ cm}^3$
 	
$\left[Z = \dfrac{W \times \text{span in m}}{8 \times 165} \right]$
 	
Since the superimposed load is less than ⅓rd of the total load, deflection will not be specially investigated.
	28·0	28·0	254 × 102 × 25 kg
			UB
			(264·9)
				52, 54
				(See
				Ch. 8,
				page
				178.)

Span 6·096 m
  Wall 6·096 × 0·600 × 4·4
                  = 16·1   18·1
  O.W.   = 2·0   13·2
                  ─────   ─────
                   18·1 kN 31·3
 	
Pt load at centre from T5
$= \dfrac{19 \cdot 8 \times 6 \cdot 096}{9 \cdot 144} = 13 \cdot 2 \text{ kN}^*$
∴ E.D.L. = 18·1 + 26·4 = 44·5 kN
∴ $Z \text{ reqd} = \dfrac{44 \cdot 5 \times 6096}{8 \times 165} = 205 \cdot 5 \text{ cm}^3$
	15·7	15·7	254 × 102 × 22 kg
			UB
			(225·4)
				54 ×
				low
				level

*13·2 kN central load = 26·4 kN U.D. load.   E.D.L. = equivalent U.D. load.

Roof steel

Loads and calculations	R/L kN	R/R kN	Section	Beam ref.
Span 6·096 m Wall $6·096 \times 1·54$ (average height) $\times 4·4$ $\qquad = 41·3$ Roof $6·096 \times 1·52 \times 1·42 \quad = 13·2$ O.W. & C. $\qquad\qquad\qquad = \underline{11·5}$ $\qquad\qquad\qquad\qquad\qquad \overline{66·0}$ $Z \text{ reqd} = \dfrac{66 \times 6096}{8 \times 165} = 304·8 \text{ cm}^3$ NOTE: Roof load $= 1·42 \text{ kN/m}^2$ (see page 348).	33·0	33·0	$254 \times 102 \times 28$ kg UB (307·6)	11, 31, 131, 151
Span 6·096 m O.W. & C. $\qquad = 11·5 \qquad 11·5$ Point load at centre $\Big\} = 39·6 \qquad 39·6$ from trusses $\qquad\qquad\qquad\qquad\qquad \overline{51·1}$ E.D.L. $= 11·5 + 79·2 = 90·7$ kN $Z \text{ reqd} = \dfrac{90·7 \times 6096}{8 \times 165} = 418·9 \text{ cm}^3$	25·6	25·6	$254 \times 146 \times 37$ kg UB (433·1)	22, 102
Span 6·096 m Wall $= 6·096 \times 0·86 \times 4·4$ $\qquad\qquad = 23·1$ Roofs as 11 etc. $= 13·2 \qquad 47·8$ O.W. & C. $\qquad = \underline{11·5} \qquad \underline{25·6}$ $\qquad\qquad\qquad \overline{47·8} \qquad \overline{73·4}$ Point load at centre $= 25·6$ kN E.D.L. $= 47·8 + 51·2 = 99·0$ kN $Z \text{ reqd} = \dfrac{99·0 \times 6096}{8 \times 165} = 457·2 \text{ cm}^3$	36·7	36·7	$305 \times 127 \times 37$ kg UB (470·3)	21, 141

Appendix 1   359

Roof steel (*continued*)

Load and calculations	R/L kN	R/R kN	Section	Beam ref.

Span 6·096 m
  O.W. & C.   = 11·5   11·5
  Point load at centre = 33·0   33·0
                                              44·5
E.D.L. = 11·5 + 66·0 = 77·5 kN
$Z \text{ reqd} = \dfrac{77\cdot5 \times 6096}{8 \times 165} = 357\cdot9 \text{ cm}^3$
	22·3	22·3	305 × 102 × 33 kg
			UB
			(414·6)
				62

Span 6·096 m
  O.W. & C.   = 11·5   11·5
  Point load at centre = 19·8   19·8
                                              31·3
E.D.L. = 11·5 + 39·6 = 51·1 kN
$Z \text{ reqd} = \dfrac{51\cdot1 \times 6096}{8 \times 165} = 236\cdot0 \text{ cm}^3$
	15·7	15·7	254 × 102 × 25 kg
			UB
			(264·9)
				12, 42,
				82, 92,
				122

Span 3·048 m
  Roof 3·048 × 1·52 × 1·42
                      = 6·6
  Wall 3·048 × 2·44 × 4·4
                      = 32·7   41·3
  O.W.   = 2·0   50·0
                        41·3   91·3
Point load at centre
from lift } = 50 kN
E.D.L. = 41·3 + 100·0 = 141·3 kN
$Z \text{ reqd} = \dfrac{141\cdot3 \times 3048}{8 \times 165} = 326\cdot3 \text{ cm}^3$
	45·7	45·7	254 × 146 × 43 kg
			UB
			(504·3)
			
			It is usual to make beams carrying lift gear larger than necessary, to allow for dynamic effects.
				51, 91

Span 3·048 m
  Wall as 51   = 32·7   34·7
  O.W.   = 2·0   100·0
                     34·7   134·7
Point load at centre = 100·0 kN
E.D.L. = 34·7 + 200·0 = 234·7 kN
$Z \text{ reqd} = \dfrac{234\cdot7 \times 3048}{8 \times 165} = 541\cdot9 \text{ cm}^3$
	67·4	67·4	305 × 165 × 40 kg
			UB
			(559·6)
				53

Roof steel (continued)

Load and calculations			R/L kN	R/R kN	Section	Beam ref.
Span 6·096 m Floor $\quad 6{\cdot}096 \times 1{\cdot}52 \times 2{\cdot}9 \quad = 26{\cdot}9$ Wall $\quad 6{\cdot}096 \times 2{\cdot}44 \times 4{\cdot}4 \quad = 65{\cdot}4 \qquad 96{\cdot}3$ O.W. $\qquad\qquad\qquad = \underline{4{\cdot}0} \qquad \underline{67{\cdot}4}$ $\qquad\qquad\qquad\qquad\qquad 96{\cdot}3 \qquad 163{\cdot}7$ Point load at centre $= 67{\cdot}4$ kN E.D.L. $= 96{\cdot}3 + 134{\cdot}8 = 231{\cdot}1$ kN $Z \text{ reqd} = \dfrac{231{\cdot}1 \times 6096}{8 \times 165} = 1067{\cdot}3 \text{ cm}^3$			81·8	81·8	381 × 152 × 67 kg UB (1095·0)	52
Span 6·096 m Floor as 52 $\qquad\quad = 26{\cdot}9$ Wall $\quad 6{\cdot}096 \times 1{\cdot}75 \times 4{\cdot}4 \quad = 46{\cdot}9 \qquad 77{\cdot}8$ O.W. $\qquad\qquad\qquad = \underline{4{\cdot}0} \qquad \underline{67{\cdot}4}$ $\qquad\qquad\qquad\qquad\qquad 77{\cdot}8 \qquad 145{\cdot}2$ Point load at centre $= 67{\cdot}4$ kN E.D.L. $= 77{\cdot}8 + 134{\cdot}8 = 212{\cdot}6$ kN $Z \text{ reqd} = \dfrac{212{\cdot}6 \times 6096}{8 \times 165} = 981{\cdot}8 \text{ cm}^3$			72·6	72·6	381 × 152 × 67 kg UB (1095·0)	54

Design of floor steel (see chart 4)

The floors are designed in order from the fourth floor downwards. The second floor steel is the same as the third floor steel.

The reader should carefully read the notes at the bottom of Chart 4. The identification letters for the various floors must be noted. Stanchions marked with a cross, thus '6 ×', are supported by beams (the particular stanchion resting on beam D61), as will be seen by reference to stanchion schedule and the note on fourth-floor steel plan. It is essential that the architects' and steel plans be carefully considered before working through the calculation sheets.

Fourth-floor steel

Loads and calculations	R/L kN	R/R kN	Section	Beam ref.
Span 6·096 m 　Wall $6·096 \times 2·75 = 16·76$ 　less window $= 5·20$ 　　　　　　　　$\overline{4·4 \times 11·26} = 49·5$ 　Floor $= 6·096 \times 1·52 \times 21·0 = 194·6$ 　O.W. $= 6·096 \times 0·82 = \underline{5·0}$ 　　　　　　　　　　　　　　$249·1$ 　Z reqd $= \dfrac{249·1 \times 6096}{8 \times 165} = 1150 \text{ cm}^3$	125	125	$406 \times 152 \times 67$ kg UB (1155)	11, 21, 31, 141, 151
Span 6·096 m 　Floor $= 6·096 \times 3·04 \times 21·0 = 389·2$ 　O.W. & C. $= \underline{30·0}$ 　　　　　　　　　　　　　　$419·2$ 　Z reqd $= \dfrac{419·2 \times 6096}{8 \times 165} = 1935·9 \text{ cm}^3$	209·6	209·6	$533 \times 210 \times 92$ kg UB (2072)	13, 23, 33, 63, 71, 73, 103, 111, 113
Span 3·048 m 　Floor $3·048 \times 3·048 \times 21·0 = 195·1$ 　O.W. & C. $= \underline{7·0}$ 　　　　　　　　　　　　　　$202·1$ 　Z reqd $= \dfrac{202·1 \times 3048}{8 \times 165} = 466·7 \text{ cm}^3$	101	101	$305 \times 127 \times 37$ kg UB (470·3)	55, 97

Fourth-floor steel

Loads and calculations	R/L kN	R/R kN	Section	Beam ref.
Span 6·096 m Wall as 11 etc. $\quad = 49·5 \quad 57·5$ O.W. $\qquad\qquad = \underline{8·0} \quad \underline{209·6}$ $\qquad\qquad\qquad\quad\;\; \overline{\underline{57·5}} \quad \overline{\underline{267·1}}$ Point load at centre $= 209·6$ kN E.D.L. $= 57·5 + 419·2 = 476·7$ kN Z reqd $= \dfrac{476·7 \times 6096}{8 \times 165} = 2201·5$ cm³	133·5	133·5	533 × 210 × 101 kg UB (2293)	12, 42, 82, 122
Span 6·096 m O.W. & C. $\qquad\quad = 34·0 \quad 34·0$ Point load at centre $= 310·6 \quad 310·6$ $\qquad\qquad\qquad\qquad\qquad\;\; \overline{\underline{344·6}}$ E.D.L. $= 34 + 621·2 = 655·2$ kN Z reqd $= \dfrac{655·2 \times 6096}{8 \times 165} = 3025·8$ cm³	172·3	172·3	610 × 229 × 125 kg UB (3217)	62, 102
Span 6·096 m O.W. & C. $\qquad\quad = 40·0 \quad 40·0$ Point load at centre $= 419·2 \quad 419·2$ $\qquad\qquad\qquad\qquad\qquad\;\; \overline{\underline{459·2}}$ E.D.L. $= 40·0 + 838·4 = 878·4$ kN Z reqd $= \dfrac{878·4 \times 6096}{8 \times 165} = 4056·6$ cm³	229·6	229·6	610 × 305 × 149 kg UB (4079)	22, 32, 72, 112

Fourth-floor steel (*continued*)

Loads and calculations	R/L kN	R/R kN	Section	Beam ref.
Span 3·048 m Wall $\quad= 36·0$ O.W. $\quad= \underline{2·0}$ $\qquad\qquad\quad\overline{\overline{38·0}}$ $Z \text{ reqd} = \dfrac{38·0 \times 3048}{8 \times 165} = 87·7 \text{ cm}^3$	19·0	19·0	$152 \times 89 \times 17$ kg I (115·6)	53
Span 6·096 m Wall $\quad= 72·0$ $\qquad\qquad\qquad\qquad\quad 76·0$ O.W. $\quad= \underline{4·0} \quad \underline{19·0}$ $\qquad\qquad\quad\overline{\overline{76·0}} \quad \overline{\overline{95·0}}$ Point load at centre $= 19·0$ kN E.D.L. $= 76·0 + 38·0 = 114·0$ kN $Z \text{ reqd} = \dfrac{114·0 \times 6096}{8 \times 165} = 526·5 \text{ cm}^3$	47·5	47·5	$305 \times 165 \times 40$ kg UB (559·6)	52
Span 6·096 m *Fig.* A.3. $\qquad\qquad\qquad\qquad\quad 4·75$ $\qquad\qquad\qquad\qquad\quad 4·75$ $\qquad\qquad\qquad\qquad\quad 9·5$ O.W. & C. $\quad = 26·0 \quad 26·0$ End piers $=$ $\quad 0·45 \times 2·4 \times 4·4 = 4·75$ ea. $120·0$ $\qquad\qquad\qquad\qquad\quad \overline{165·0}$ Central pier $= 0·9 \times 2·4 \times 4·4 = 9·5$ kN Point load at centre $\qquad = 120$ kN $M = 82·5 \times 3·048 - 4·75 \times 2·823 - 4·75$ $\quad \times 0·225 - 13 \times 1·524 = 251·5 - 13·4$ $\quad - 1·1 - 19·8 = 217·2$ kNm $Z \text{ reqd} = \dfrac{217·2 \times 10^3}{165} = 1316 \text{ cm}^3$	82·5	82·5	$457 \times 152 \times 74$ kg UB (1404)	54

Fig. A.3 shows a beam with loads 4.75, 11.9, 9.5, 2.6, 4.75 kN at spacings 0.450, 2.148, 0.900, 2.148, 0.450 m.

364 *Structural Steelwork*

Fourth-floor steel (*continued*)

Loads and calculations	R/L kN	R/R kN	Section	Beam ref.

Span 6·096 m
  Floor as 13, etc. = 389·2   429·2
  O.W. & C.   = 40·0   85·9
                     ─────   ─────
                     429·2   515·1
Point load at centre
from S6× (see page = 85·9 kN
374)
E.D.L. = 429·2 + 172 = 601·2 kN
Z reqd $= \dfrac{601 \cdot 2 \times 6096}{8 \times 165}$ = 2776·5 cm^3
	257·6	257·6	610 × 229 × 113 kg
			UB
			(2874)
				61, 101

Span 6·096 m
 	
Fig. A.4.
Floor (i)   32
  = 6·096 × 1·52 × 21·0 = 193·0   267
O.W. & C.   = 41·0   96
                                  234
                       ─────   ─────
                       234·0   629
Floor (ii)
  = 3·048 × 1·52 × 21·0 = 96
Wall
  = 3·048 × 2·4 × 4·4 = 32
Point load at centre
  = 184 from S5× (see page 374)
     83 from D54
    ───
    267
$M = 298 \times 3 \cdot 048 - 32 \times 1 \cdot 524 - 117$
$\times 1 \cdot 524 = 908 \cdot 3 - 48 \cdot 8 - 178 \cdot 3$
  = 681·2 kNm
Z reqd $= \dfrac{681 \cdot 2 \times 10^3}{165}$ = 4128 cm^3
	24
	133
	24
	117
	───
	298
		8
		133
		72
		117
		───
		330
			686 × 254 × 152 kg
			UB
			(4364)
				51

Diagram (Fig. A.4): Simply supported beam of two spans 3·048 m + 3·048 m. Loads on beam: 32 (left), point loads 267 and 96 near centre, 234 to the right.

Appendix 1   365

Fourth-floor steel (*continued*)

Loads and calculations	R/L kN	R/R kN	Section	Beam ref.
Span 2·743 m Stairs $2·743 \times 0·65 \times 9·7 = 17·3$ O.W. & C. $= \underline{3·0}$ $20·3$ $Z \text{ reqd} = \dfrac{20·3 \times 2743}{8 \times 165} = 42·2 \text{ cm}^3$	10·2	10·2	for convenience 152×89 I (115·6)	Raker 94
Span 3·048 m Landing $= 24·0 \quad\quad 10·2$ O.W. & C. $= \underline{4·0} \quad\quad \underline{28·0}$ $28·0 \quad\quad 38·2$ Point load at centre $= 10·2$ kN E.D.L. $= 28 + 20·4 = 48·4$ kN $Z \text{ reqd} = \dfrac{48·4 \times 3048}{8 \times 165} = 111·8 \text{ cm}^3$	19·1	19·1	152×89 I (115·6)	93

Span 3·048 m
Landing and O.W. & C. as 93
                                 $= 28·0 \quad\quad 28·0$
Point load at centre $= \underline{20·4} \quad\quad \underline{20·4}$
                                                      $48·4$
E.D.L. $= 28 + 40·8 = 68·8$ kN
$Z \text{ reqd} = \dfrac{68·8 \times 3048}{8 \times 165} = 158·9 \text{ cm}^3$
	24·2	24·2	178×102 I
			(170·9)
				95
				Low
				level

Span 6·096 m *Fig.* A.5 $\begin{array}{r}24·2\\19·1\\66·7\\\hline 110·0\end{array}$	$\begin{array}{r}17·6\\5·3\\33·4\\\hline 56·3\end{array}$	$\begin{array}{r}6·6\\13·8\\33·4\\\hline 53·8\end{array}$		92

Fig. A.5 shows a beam with loads 24·2, 19·1, and 66·7 at positions 1·073, 2·750, 1·073.

Fourth-floor steel (*continued*)

Loads and calculations	R/L kN	R/R kN	Section	Beam ref.
Wall = $6 \cdot 096 \times 2 \cdot 75$ = $16 \cdot 76$ less windows = $2 \cdot 5$ $ 4 \cdot 4 \times \overline{14 \cdot 26} = 62 \cdot 7$ O.W. = $4 \cdot 0$ $\overline{66 \cdot 7}$ Zero shear from l.h. end $= \dfrac{56 \cdot 3 - 24 \cdot 2}{66 \cdot 7} \times 6 \cdot 096 = 2 \cdot 93$ m $M = 56 \cdot 3 \times 2 \cdot 93 - 24 \cdot 2 \times 1 \cdot 26$ $ - \dfrac{66 \cdot 7 \times 2 \cdot 93^2}{6 \cdot 096 \times 2}$ $ = 165 \cdot 0 - 30 \cdot 5 - 47 \cdot 0 = 87 \cdot 5$ kNm Z reqd $= \dfrac{87 \cdot 5 \times 10^3}{165} = 530$ cm^3			$305 \times 165 \times 40$ kg UB (559·6)	92 (*cont'd*)
Span 6·096 m *Loading diagram: point loads 24·2, 101, 19·1, and 68·3 on a beam with spacings 1·673, 1·375, 1·375, 1·673* Fig. A.6. $ 24 \cdot 2$ $ 101 \cdot 0$ $ 19 \cdot 1$ Wall = $6 \cdot 096 \times 2 \cdot 75$ = $16 \cdot 7 68 \cdot 3$ less = $2 \cdot 6 \overline{212 \cdot 6}$ $ \overline{14 \cdot 1} \times 4 \cdot 4$ = $62 \cdot 3$ O.W. = $6 \cdot 0$ $ \overline{68 \cdot 3}$ $M = 107 \cdot 6 \times 3 \cdot 048 - 24 \cdot 2 \times 1 \cdot 375$ $ - 34 \cdot 2 \times 1 \cdot 542 = 328 \cdot 0 - 33 \cdot 3 - 52 \cdot 7$ $ = 242 \cdot 0$ kNm Z reqd $= \dfrac{242 \cdot 0 \times 10^3}{165} = 1466 \cdot 7$ cm^3	17·6 50·5 5·3 34·2 107·6	6·6 50·5 13·8 34·2 105·1	$457 \times 152 \times 82$ kg UB (1555)	96

Fourth-floor steel (*continued*)

Loads and calculations	R/L kN	R/R kN	Section	Beam ref.

Span 6·096 m

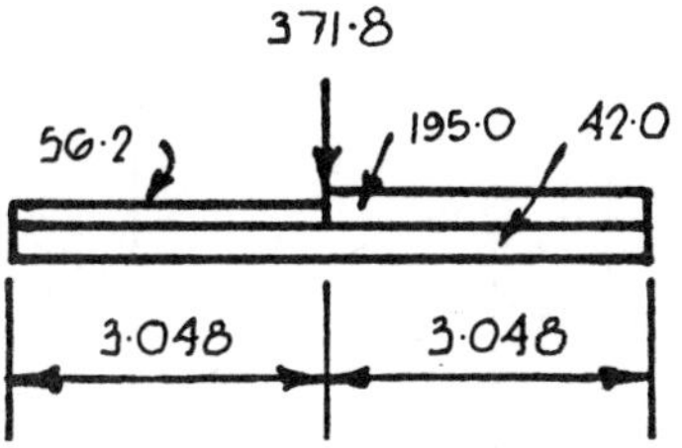

Fig. A.7.

O.W. & C. = 42·0 56·2	42·2	14·0		
Floor 371·8	185·9	185·9		
3·048 × 3·048 × 21·0 = 195·0 195·0	48·8	146·2		
Landing = 24·0 42·0	21·0	21·0		
666·3	297·9	367·1		
Wall				
3·048 × 2·4 × 4·4 = 32·2				
56·2 kN				

Point load at centre
    from S9× = 184·2
    from D54 = 82·5
    from D96 = 105·1
                371·8 kN

$M = 297 \cdot 9 \times 3 \cdot 048 - 56 \cdot 2 \times 1 \cdot 542 - 21 \cdot 0$
$\quad \times 1 \cdot 542 = 905 \cdot 6 - 86 \cdot 7 - 32 \cdot 4$
$\quad = 786 \cdot 5 \text{ kN m}$

$Z \text{ reqd} = \dfrac{786 \cdot 5 \times 10^3}{165} = 4766 \cdot 7 \text{ cm}^3$

Section: 686 × 254 × 170 kg UB (4902)

Beam ref.: 91

Fourth-floor steel (*continued*)

Loads and calculations	R/L kN	R/R kN	Section	Beam ref.

Span 6·096 m

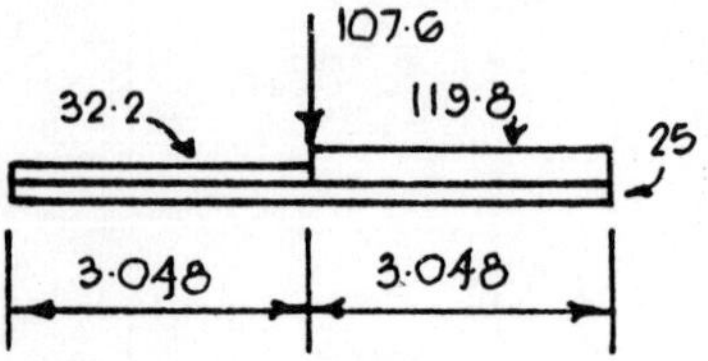

Fig. A.8

```
O.W. & C.        = 25·0    32·2      24·2    8·0
Floor              =        107·6    53·8   53·8
3·048×1·542×4·4 = 98·7    119·8    30·0   89·8
Wall 3·048×2·4 = 7·3       25·0    12·5   12·5
less             = 2·5    ─────   ─────  ─────
                          284·6    120·5  164·1
        4·4×4·8 = 21·1    ═════   ═════  ═════
                 ─────
                 119·8 kN
```

Wall
3·048 × 2·4 × 4·4 = 32·2 kN

$M = 120·5 \times 3·048 - 32·2 \times 1·524 - 12·5 \times 1·524 = 367·3 - 49·1 - 19·1$
$= 299·1 \text{ kNm}$

$Z \text{ reqd} = \dfrac{299·1 \times 10^3}{165} = 1812·7 \text{ cm}^3$

533 × 210 × 92 kg UB (2072)

131

$$f_{bxx} = \frac{141 \cdot 0 \times 10^3 \times 233}{2 \times 1313 \times 10^3} = 12 \cdot 5 \text{ N/mm}^2$$

$$f_{byy} = \frac{133 \cdot 0 \times 10^3 \times 107}{2 \times 456 \cdot 9 \times 10^3} = 15 \cdot 6 \text{ N/mm}^2$$

p_c for the section chosen for this stanchion length $= 125 \text{ N/mm}^2$, obtained as follows:

Effective length $= 1 \cdot 0 \times 395 \cdot 0$ cm ($1 \cdot 0$ is taken as this is a two-way connection on a bottom stanchion length, with a slab base at the lower end (see page 251)).

$$\frac{l}{r} = \frac{1 \cdot 0 \times 395 \cdot 0}{6 \cdot 57} = 61$$

p_c (see fig. 11.16) $= 125 \text{ N/mm}^2$

$$\frac{f_c}{p_c} = \frac{85 \cdot 8}{125} = 0 \cdot 686$$

$$\frac{f_{bc}}{p_{bc}} = \frac{12 \cdot 5 + 15 \cdot 6}{165} = 0 \cdot 170$$

$$\underline{0 \cdot 856}$$

which is less than unity and therefore the section is adequate.

The section of the stanchion is changed at the splice which is marked **X**. The symbol **X** (against the 'stanchion reference' column) indicates in the case of each stanchion that there is to be a splice at that point. These splices are required for erection and fabrication purposes, as it is inconvenient, and costly, to handle lengths of stanchions greater than about 12 m.

All the stanchions in the present example have been designed as uncased stanchions. As an exercise to illustrate the economy that may be achieved by the use of the casing, stanchion 1 will now be re-designed as a cased stanchion assuming a 50 mm casing all round the steel section (reference BS 449, Clause 30b).

Length C–B

Try $203 \times 203 \times 46$ kg UC $\qquad (r_{yy} = 0 \cdot 2(203 \cdot 2 + 100) = 60 \cdot 6 \text{ mm})$:

$$\frac{l}{r} = \frac{0 \cdot 85 \times 3350}{60 \cdot 6} = 47 \qquad p_c = 136 \text{ N/mm}^2$$

372 Structural Steelwork

$$\text{Allowable direct load} = \left(58{\cdot}8 \times 10^2 \times 136 + \frac{303{\cdot}2^2 \times 136}{0{\cdot}19 \times 165}\right) \text{N}$$

$$= 799{\cdot}7 + 398{\cdot}8 = 1198{\cdot}5 \text{ kN}$$

$$\frac{\text{Actual load}}{\text{Allowable load}} = \frac{602{\cdot}2}{1198{\cdot}5} = 0{\cdot}50$$

$$f_{bxx} = \frac{141{\cdot}0 \times 202}{2 \times 449{\cdot}2} = 31{\cdot}7$$

$$f_{byy} = \frac{133{\cdot}0 \times 103}{2 \times 151{\cdot}5} = \underline{45{\cdot}2}$$

$$\underline{\underline{76{\cdot}9 \text{ N/mm}^2}}$$

$$\frac{f_{bc}}{p_{bc}} = \frac{76{\cdot}9}{165} = 0{\cdot}47$$

$0{\cdot}5 + 0{\cdot}47 = 0{\cdot}97$, which is less than unity and therefore the section is adequate.

Length A–G

Try $254 \times 254 \times 73$ kg UC $\quad (r_{yy} = 0{\cdot}2(254{\cdot}0 + 100) = 70{\cdot}8 \text{ mm})$:

$$\frac{l}{-} = \frac{1{\cdot}0 \times 3950}{70{\cdot}8} = 56 \quad p_c = 129 \text{ N/mm}^2$$

$$\text{Allowable direct load} = \left(92{\cdot}9 \times 10^2 \times 129 + \frac{354^2 \times 129}{0{\cdot}19 \times 165}\right) \text{N}$$

$$= 1113{\cdot}3 + 515{\cdot}7 = 1629 \text{ kN}$$

$$\frac{\text{Actual load}}{\text{Allowable load}} = \frac{1172{\cdot}2}{1629{\cdot}0} = 0{\cdot}720$$

$$f_{bxx} = \frac{141{\cdot}0 \times 227}{2 \times 894{\cdot}5} = 17{\cdot}9$$

$$f_{byy} = \frac{133{\cdot}0 \times 104}{2 \times 305{\cdot}0} = \underline{22{\cdot}7}$$

$$\underline{\underline{40{\cdot}6 \text{ N/mm}^2}}$$

$$\frac{f_{bc}}{p_{bc}} = \frac{40{\cdot}6}{165} = 0{\cdot}246$$

$0.720 + 0.246 = 0.966$, therefore the section is adequate.

The economy in the main stanchion section is apparent.

It will be noticed, in the above calculation, that the bending stresses have been calculated on the *steel section alone*.

It is necessary to check that the axial load on the cased section does not exceed twice that permitted on the uncased section, and also that the slenderness ratio of the uncased section does not exceed 250 (see page 246).

Length C–B

203 × 203 × 46 kg UC, uncased:

$$\frac{l}{r} = \frac{0.85 \times 335}{5.11} = 56 \qquad p_c = 129 \text{ N/mm}^2$$

$$\text{Allowable direct load} = 58.8 \times 10^2 \times 129 \text{ N}$$
$$= 758.5 \text{ kN}$$
$$\text{Allowable direct load on cased section} = 1198.5 \text{ kN}$$

Length A–G

254 × 254 × 73 kg UC, uncased:

$$\frac{l}{r} = \frac{1.0 \times 395}{6.46} = 61 \qquad p_c = 125 \text{ N/mm}^2$$

$$\text{Allowable direct load} = 92.9 \times 10^2 \times 125 \text{ N}$$
$$= 1161.3 \text{ kN}$$
$$\text{Allowable direct load on cased section} = 1629 \text{ kN}$$

In both cases it will be seen that the conditions are satisfied.

Foundation calculations (typical only)

Allowable pressures:

i) on soil $= 300 \text{ kN/m}^2$
ii) on concrete $= 4200 \text{ kN/m}^2$.

Slab bases will be used throughout in the example, the BS 449 formula for determining the thickness of slabs being:

$$t = \sqrt{\frac{3w}{p_{bt}}\left(A^2 - \frac{B^2}{4}\right)}$$

Sketch	$\frac{l}{r}$	Calculations			Total kN	Reactions		Portion and Length	Section	Stan. Ref.
		p_c	$f_c + f_{bc}$	$\frac{f_c}{p_c} + \frac{f_{bc}}{p_{bc}}$		Amount kN	Beam ref.			
![sketch T3 51/54]						Tie 28·0 15·7 4·0	MR51 MR54 MR54× O.W. & C.	M.R. 2·400 m	As below	5× 9×
					47·7	47·7				
	$0.7 \times 275.0 \over 3.68$ $= 53$	131	$f_c = \dfrac{184 \cdot 2 \times 10^3}{29 \cdot 8 \times 10^2} = 61 \cdot 8$ $f_{bxx} = \dfrac{72 \cdot 6 \times 10^3 \times 176}{2 \times 165 \cdot 7 \times 10^3} = 38 \cdot 6$ $f_{byy} = \dfrac{32 \cdot 5 \times 10^3}{2 \times 52 \cdot 95} = 31 \cdot 6$ $\overline{70 \cdot 2}$	$\dfrac{61 \cdot 8}{131} = 0 \cdot 472$ $\dfrac{70 \cdot 2}{165} = 0 \cdot 425$ $\overline{0 \cdot 897}$	136·5	45·7 13·2 72·6 5·0	R51 T3 R54 O.W. & C.	Roof 2·750 m	$152 \times 152 \times 23$ kg UC	
					184·2	136·5				
Fig. A.10										
![sketch T4 22/62 T3]			As Stan. 5×			13·2 19·8 25·6 22·3 5·0	T3 T4 R22 R62 O.W. & C.	Roof 2·750 m	$152 \times 152 \times 23$ kg UC	6× 10×
					85·9	85·9				
Fig. A.11										

		1 4 16	As below	Roof 2·750 m	R12 R11 O.W. & C.	15·7 33·0 5·0
						53·7
		1 4 16 cont.	As below	4th 2·750 m	D12 D11 O.W. & C.	133·5 125·0 7·0
						265·5
						319·2
0.85×335.0 $\dfrac{5 \cdot 19}{= 55}$	130	$f_c = \dfrac{602 \cdot 2 \times 10^3}{75 \cdot 8 \times 10^2} = 79 \cdot 4$ $f_{bxx} = \dfrac{141 \cdot 0 \times 205}{2 \times 581 \cdot 1} = 24 \cdot 9$ $f_{byy} = \dfrac{133 \cdot 0 \times 107}{2 \times 199 \cdot 0} = 35 \cdot 8$ $\overline{60 \cdot 7}$	$\dfrac{79 \cdot 4}{130} = 0 \cdot 611$ $\dfrac{60 \cdot 7}{165} = 0 \cdot 368$ $\overline{0 \cdot 979}$			
				3rd 3·350 m	C12 C11 O.W. & C.	141·0 133·0 9·0
			$203 \times 203 \times 60$ kg UC			283·0
						602·2
			——X——			
			As below	2nd 3·350 m	B12 B11 O.W. & C.	141·0 133·0 10·0
						284·0
						886·2
$1 \cdot 0 \times 395 \cdot 0$ $\dfrac{6 \cdot 57}{= 61}$	125	$f_c = \dfrac{1172 \cdot 2 \times 10^3}{136 \cdot 6 \times 10^2} = 85 \cdot 8$ $f_{bxx} = \dfrac{141 \cdot 0 \times 233}{2 \times 1313} = 12 \cdot 5$ $f_{byy} = \dfrac{133 \cdot 0 \times 107}{2 \times 456 \cdot 9} = 15 \cdot 6$ $\overline{28 \cdot 1}$	$\dfrac{85 \cdot 8}{125} = 0 \cdot 686$ $\dfrac{28 \cdot 1}{165} = 0 \cdot 170$ $\overline{0 \cdot 856}$			
			$254 \times 254 \times 107$ kg UC	1st 3·950 m	A12 A11 O.W. & C.	141·0 133·0 12·0
						286·0
						1172·2

Fig. A.12

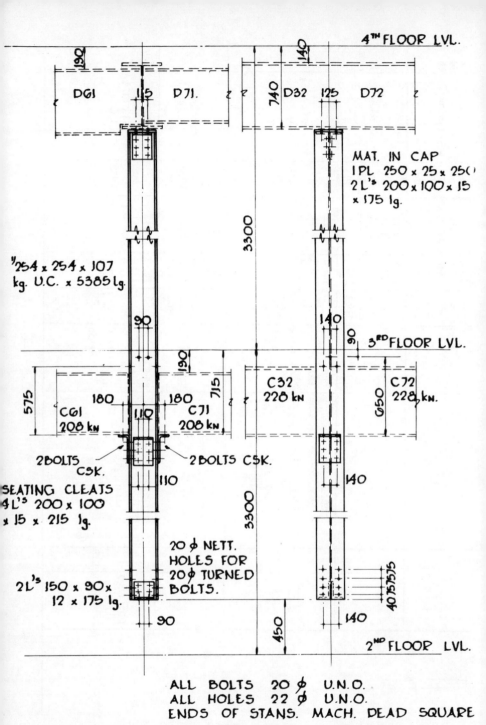

Fig. A.13   Stanchion detail.

For explanation of symbols, see Chapter 12. Where the stanchion base loads are approximately the same, they would be grouped together in the calculations.

Stanchions 1, 4 *and* 16

$$\text{Load} = 1180 \text{ kN}$$
$$\text{Weight of concrete base} = (\text{say}) \ \underline{120} \text{ kN } (10\% \text{ of stanchion load})$$
$$\text{Total} = \underline{\underline{1300}} \text{ kN}$$

$$\text{Concrete area required} = \frac{1300}{300} = 4\cdot 33 \text{ m}^2$$

Make base 2·1 m × 2·1 m $= 4\cdot 41 \text{ m}^2$

$$\text{Steel area required} = \frac{1180}{4200} = 0\cdot 281 \text{ m}^2$$

Make slab base 550 mm × 550 mm
$$= 0\cdot 303 \text{ m}^2$$

$$t = \sqrt{\frac{3 \times 1180 \times 10^3}{180 \times 550 \times 550}\left(146^2 - \frac{142^2}{4}\right)}$$
$$= \sqrt{0\cdot 065(21316 - 5041)}$$
$$= \sqrt{1058} = 32\cdot 6 \text{ mm, say } 40 \text{ mm}$$

$$\text{Maximum overhang} = \frac{2\cdot 100 - 0\cdot 550}{2} = 0\cdot 775 \text{ m}$$

Depth of concrete base (if unreinforced) should be about 150 mm more than the maximum overhang of the concrete beyond the slab base.

∴ Make concrete base 1·000 m deep.
$$\text{Concrete base} = 2\cdot 1 \text{ m} \times 2\cdot 1 \text{ m} \times 1\cdot 0 \text{ m deep}$$
$$\text{Slab base} = 550 \text{ mm} \times 550 \text{ mm} \times 40 \text{ mm thick}$$

Schedule of Foundations.

BASE NUMBER	SIZE		DEPTH OF UNDERSIDE BELOW GRD. FLR. LEVEL
	PLAN	DEPTH	
13	1900 SQ.	0.900	1.500
1. 4. 16.	2100 SQ	1.000	1.600
9	2750 SQ.	1.100	1.700
5	2600 SQ	1.200	1.800
2.3.8.12.14.15.	2700 SQ	1.200	1.800
6.7.10.11.	3500 SQ.	1.450	2.050

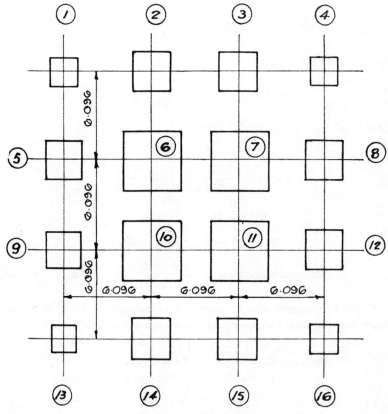

Fig. A.14 *Foundation plan.*

Appendix 2

British Standards and Codes of Practice

These lists are reproduced by kind permission of the British Standards Institution, and are up-to-date at the time of publishing this book. Information on British Standards Institution publications may be obtained from the BSI Yearbook and direct from the Institution.

Amendments to the British Standards and Codes of Practice are issued from time to time, and care should be taken to ensure that the publications used incorporate any such Amendments.

General

648:1964 Schedule of weights of building materials
1991: Letter symbols, signs and abbreviations
       Part 1: 1967 General
       Part 4: 1961 Structures, materials and soil mechanics
4466:1969 Bending dimensions and scheduling of bars for the reinforcement of concrete
Handbook 3: Summaries of British Standards including Codes of Practice, Drafts for Development and other publications for building (Issued annually)

BS materials for bridges and buildings, etc.

4: Structural steel sections
    Part 1: 1972 Hot-rolled sections
    Part 2: 1969 Hot-rolled hollow sections
153: Steel girder bridges
    Parts 1 & 2: 1972
    1. Materials and workmanship; 2. Weighing, shipping and erection
    Part 3A: 1972 Loads

380 Structural Steelwork

	Parts 3B & 4: 1972
	3B: Stresses; 4: Design and construction
405:1945	Expanded metal (steel) for general purposes
449:	The use of structural steel in building
	Part 1: 1970 Imperial units
	Part 2: 1969 Metric units
2853:1957	The design and testing of steel overhead runway beams
2994:1958	Cold rolled steel sections
4360:1972	Weldable structural steels
4449:1969	Hot rolled steel bars for the reinforcement of concrete
4461:1969	Cold worked bars for the reinforcement of concrete
4482:1969	Hard drawn mild steel wire for the reinforcement of concrete
4483:1969	Steel fabric for the reinforcement of concrete
4486:1969	Cold worked high tensile alloy bars for prestressed concrete
4848:	Hot-rolled structural steel sections
	Part 4: 1972 Equal and unequal angles

BS bolting

325:1947	Black cup and countersunk bolts and nuts
916:1953	Black bolts, screws and nuts
1768:1963	Unified precision hexagon bolts, screws and nuts (UNC and UNF threads). Normal series
1769:1951	Unified black hexagon bolts, screws and nuts (UNC and UNF threads). Heavy series
2708:1956	Unified black square and hexagon bolts, screws and nuts (UNC and UNF threads). Normal series
3139:	High strength friction grip bolts for structural engineering.
	Part 1: 1959 General grade bolts
3294:	The use of high strength friction grip bolts in structural steelwork
	Part 1: 1960 General grade bolts
3692:1967	ISO metric precision hexagon bolts, screws and nuts
4190:1967	ISO metric hexagon bolts, screws and nuts
4395:	High strength friction grip bolts and associated nuts and washers for structural engineering
	Part 1: 1969 General grade
	Part 2: 1969 Higher grade bolts and nuts and general grade washers
	Part 3: 1973 Higher grade bolts (waisted shank), nuts and general grade washers

4604: The use of high strength friction grip bolts in structural steelwork. Metric series
Part 1: 1970 General grade
Part 2: 1970 Higher grade (parallel shank)
Part 3: 1973 Higher grade (waisted shank)
4929: Steel hexagon prevailing-torque type nuts
Part 1: 1973 Metric series
Part 2: 1973 Unified (inch) series.

BS welding

938:1962 General requirements for the metal-arc welding of structural steel tubes to BS 1775
1856:1964 General requirements for the metal-arc welding of mild steel
2642:1965 General requirements for the arc welding of carbon manganese steels

BS cement lime and plasters

12: Portland cement (ordinary and rapid-hardening)
Part 1: 1958 Imperial units
Part 2: 1971 Metric units
146: Portland-blastfurnace cement
Part 1: 1958 Imperial units
Part 2: 1973 Metric units
890:1972 Building lime
915: High alumina cement
Part 1: 1947 Imperial units
Part 2: 1972 Metric units
1014:1961 Pigments for cement, magnesium oxychloride and concrete
1191: Gypsum building plasters
Part 1: 1973 Excluding premixed lightweight plasters
Part 2: 1973 Premixed lightweight plasters
1370: Low heat Portland cement
Part 1: 1958 Imperial units
Part 2: 1974 Metric units
4027: Sulphate-resisting Portland cement
Part 1: 1966 Imperial units
Part 2: 1972 Metric units
4246: Low heat Portland-blastfurnace cement
Part 1: 1968 Imperial units
Part 2: 1974 Metric units

4248:1974 Supersulphated cement
4550: Methods of testing cement
Part 2: 1970 Chemical tests
4627:1970 Glossary of terms relating to types of cements, their properties and components
4721:1971 Ready-mixed lime: sand for mortar

BS aggregates
812:1967 Methods for sampling and testing of mineral aggregates, sands and fillers
877: Foamed or expanded blastfurnace slag lightweight aggregate for concrete
Part 1: 1967 Imperial units
Part 2: 1973 Metric units
882, 1201: Aggregates from natural sources for concrete (including granolithic)
Part 1: 1965 Imperial units
Part 2: 1973 Metric units
1047:1952 Air-cooled blastfurnace slag coarse aggregate for concrete
Part 2: 1974 Metric units
1165:1966 Clinker aggregate for concrete
1198–1200:1955 Building sands from natural sources
3681: Methods for the sampling and testing of lightweight aggregates for concrete
Part 1: 1963 Imperial units
Part 2: 1973 Metric units
3797:1964 Lightweight aggregates for concrete
4619:1970 Heavy aggregates for concrete and gypsum plaster

BS timber
373:1957 Testing small clear specimens of timber
565:1972 Glossary of terms relating to timber and woodwork
881, 589:1955 Nomenclature of commercial timbers, including sources of supply
4169:1970 Glued-laminated timber structural members
4471: Dimensions for softwood
Part 1: 1969 Basic sections
Part 2: 1971 Small resawn sections
4978:1973 Timber grades for structural use

Codes of practice

CP 3   : Chapter V: Loading, Part 1: 1967 Dead and imposed loads
CP 3   : Chapter V: Loading, Part 2: 1972 Wind Loads
CP 110 : The structural use of concrete
        Part 1: 1972 Design, materials and workmanship
        Part 2: 1972 Design charts for singly reinforced beams, doubly reinforced beams and rectangular columns
        Part 3: 1972 Design charts for circular columns and prestressed beams
CP 112 : The structural use of timber
        Part 2: 1971 Metric units
        Part 3: 1973 Trussed rafters for roofs of dwellings
CP 114 : Structural use of reinforced concrete in buildings
        Part 2: 1969 Metric units
CP 2004: 1972   Foundations

Appendix 3

Spacing of Holes in Standard Sections

SPACING OF HOLES IN COLUMNS, BEAMS AND TEES

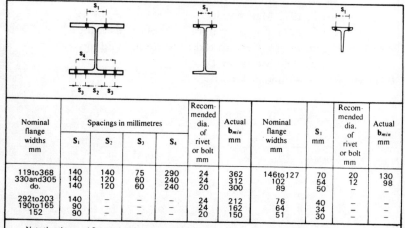

Nominal flange widths mm	Spacings in millimetres				Recommended dia. of rivet or bolt mm	Actual b_{min} mm	Nominal flange widths mm	S_1 mm	Recommended dia. of rivet or bolt mm	Actual b_{min} mm
	S_1	S_2	S_3	S_4						
119 to 368	140	140	75	290	24	362	146 to 127	70	20	130
330 and 305	140	120	60	240	24	312	102	54	12	98
do.	140	120	60	240	20	300	89	50	–	–
292 to 203	140	–	–	–	24	212	76	40	–	–
190 to 165	90	–	–	–	24	162	64	34	–	–
152	90	–	–	–	20	150	51	30	–	–

Note that the actual flange width for a universal section may be less than the nominal size and that the difference may be significant in determining the maximum diameter. The column headed b_{min} gives the actual minimum width of flange required to comply with Table 21 of BS 449:Part 2:1969.
The dimensions S_1 and S_2 have been selected for normal conditions but adjustments may be necessary for relatively large diameter fasteners or for particularly heavy weights of serial size.

SPACING OF HOLES IN CHANNELS

Nominal flange width mm	S_1 mm	Recommended dia. of rivet or bolt mm
102	55	24
89	55	20
76	45	20
64	35	16
51	30	10
38	22	–

SPACING OF HOLES IN ANGLES

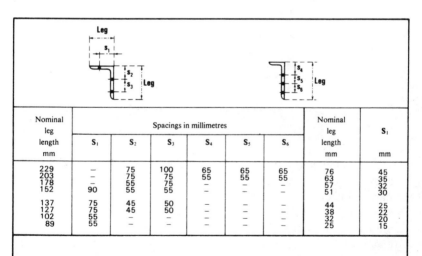

Nominal leg length mm	Spacings in millimetres						Nominal leg length mm	S_1 mm
	S_1	S_2	S_3	S_4	S_5	S_6		
229	–	75	100	65	65	65	76	45
203	–	75	75	55	55	55	63	35
178	–	55	75	–	–	–	57	32
152	90	55	55	–	–	–	51	30
137	75	45	50	–	–	–	44	25
127	75	45	50	–	–	–	38	22
102	55	–	–	–	–	–	32	20
89	55	–	–	–	–	–	25	15

Inner gauge lines are selected for normal conditions and may require adjustment for specially large diameters of fasteners or thick members. Outer gauge lines may require consideration in relation to a specified edge distance.

Appendix 4

Geometrical Properties of a Parabola

The properties given below will be found useful in bending moment and deflection problems.

Geometrical construction of a parabola

Fig. A.15 shows a simple construction, given the base and the maximum central ordinate (as in a B.M. diagram for a simply supported beam with uniformly distributed load).

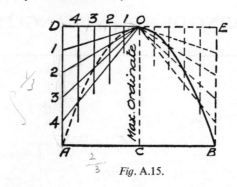

Fig. A.15.

The rectangle ADEB is drawn of height CO. OD is divided into a chosen number of equal parts (say 5), and DA is also divided into the same number of equal parts. The point of intersection of the vertical through 1 and the radial line O1 will be a true point on the parabola. Similarly the intersection points of the other corresponding verticals and radials will give further points through which to draw in the required curve. The right half of the diagram may be drawn in, if desired, by the principle of symmetry.

Area of a parabola

Area of parabola AOB (fig. A.15)

$= \tfrac{2}{3}$ of the base × height
$= \tfrac{2}{3}$ AB × CO

Area of semi-parabola AOC

$= \tfrac{2}{3}$ of the base × height
$= \tfrac{2}{3}$ AC × CO

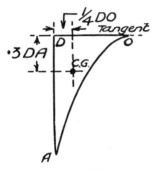

Fig. A.16.

Area of parabolic diagram ADO (fig. A.16)

$= \tfrac{1}{3}$ of the base × height
$= \tfrac{1}{3}$ OD × DA

Position of centre of gravity

In the parabolic diagram AOC (fig. A.17), the c.g. is at a point $\tfrac{2}{5}$ of CO, measured from CA, and $\tfrac{3}{8}$ of CA, measured from CO.

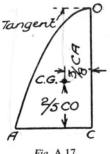

Fig. A.17.

In the case of diagram AOD (fig. A.16), the c.g. is $\tfrac{1}{4}$ DO from DA, and 0·3 DA from DO.

Answers to Numerical Questions

Exercises 1. Page 14
1 100 N/mm^2.   2 4.
3 92·5 mm.   4 1·25 m square. 1·5 ?
5 (a) 70·7 kN.   (b) 186 kN.
6 100 N/mm^2.   Necessary thickness = 8 mm.
7 Strain = 0·0003.   Contraction in length = 1·6 mm.
8 200 kN/mm^2.   10 0·64 mm.
11 240 kN.   12 60 kN/mm^2.

Exercises 2. Page 35
1 Ult. stress = 500 N/mm^2.   Working stress = 100 N/mm^2.
2 Y.P. stress = 350 N/mm^2.   Ult. stress = 640 N/mm^2.   Percentage elongation = 30·0.   Percentage contraction in area = 54.
3 Load at Y.P. = 120 kN.   Maximum load = 200 kN.   Elongation = 60 mm.   Diameter at fracture = 14·6 mm.
4 Working stress = 125 N/mm^2.   Thickness = 20 mm.
6 110 kN.   7 4.   8 Stress exceeds elastic limit.

Exercises 4. Page 79
1 20·4 kN; 60·8 kN.   2 76 kN.
3 Section 1: 195·3 kN.   Section 2: 192·7 kN.
   Rivet strength = 242·2 kN.
   Safe load (L) = 192·7 kN.
   Percentage efficiency = 92·9.
4 167·2 kN.
5 No. of rivets = 5.
   Suitable plate width = 160 mm.
   Cover thickness = 8 mm.
   Percentage efficiency = 88·5.
6 29·03 kN.
7 2 bolts.
8 7 bolts.

Exercises 5. Page 100
2 Centre: 1·5 kNm (negative) and 2 kN.
   Maximum: 5·4 kNm (negative) and 6 kN.
3 Maximum values: 6 kNm (negative) and 10 kN.
   Given section 3·375 kNm (negative) and 7·5 kN.
4 Given section: 24 kNm and 4 kN.
   B.M.s: 16 kNm; 28 kNm; 12 kNm.
5 (a) 1·62 kNm; (b) 66·96 kNm.

Answers to Numerical Questions   389

6 1·67 m.
7 From left end of beam (in kNm):
   0; 12 (neg.); 7·5; 10·5; 12 (neg); 0.
8 Left end support: 216 Nm (neg.).
   Right end support: 337·5 Nm (neg.).
   Maximum B.M. (in central portion) = 864 Nm (treated as simple beam between the supports for purpose of diagram construction).

EXERCISES 6.   PAGE 143

1 241 kN.   2 $406 \times 152 \times 60$ kg U.B.
3 121 N/mm².   4 330 cm³.
5 105·4 kN.
6 Timber beams 50×175.
   Steel beams $178 \times 102 \times 21\cdot54$ kg I ($f = 165$ N/mm²).

EXERCISES 7.   PAGE 158

1 $I_{xx} = 1324\cdot5$ cm⁴; $I_{yy} = 1127$ cm⁴.   2 85·8 mm.
4 1165 cm⁴.   5 $I_{maximum} = 83\,048$ cm⁴; $Z_{maximum} = 3322$ cm³.
6 555 kN.   7 $I_{max.} = 14\,063$ cm⁴; $Z_{max.} = 1012$ cm³.
8 1030 kN.

EXERCISES 8.   PAGE 183

1 15 mm.   2 (a) 159·3 N/mm²; (b) 2·85 mm.
3 0·0036 radians. 4 6610 N/mm².
5 10·0 m; 165 N/mm²; 104 kN.   6 11·5 mm.
8 11·5 mm.   10 $457 \times 191 \times 74$ kg. U.B.

EXERCISES 9.   PAGE 200

1 S.F. diagram crosses base line at 5 m from left end.
2 At 9 m from left end. B.M. max = 209 kNm.
3 Loads from left: 2 kN; 6 kN; 5 kN.
4 45·56 kNm at 5·25 m from left end.
5 42·7 N/mm²,   Maximum = 48 N/mm².
6 2·32 N/mm²; 46·4 N/mm²; 56·2 N/mm².
7 Two rows 80 mm pitch in the vertical legs of the angles.

EXERCISES 10.   PAGE 225

1 (a) 545 cm³   (b) 364 cm³   (c) 491 cm³.
   B.M.s maximum (a) 90 kNm; (b) 60 kNm; (c) 81 kNm.
   Fixing moment (b) 60 kNm; (c) 9 kNm.
2 28·8 kNm; 125 N/mm².
3 Left end: 80 kNm.   Right end: 40 kNm.
5 210 kNm.
   Reactions (from left): 67·5 kN; 387·5 kN; 145 kN.
6 $\tfrac{3}{8}$W.
7 B.M.s in kNm (from left):
   0; 4·14 (neg.); 3·78 (neg.); 0.
   Reactions in kN (from left):
   3·06; 25·92; 22·32; 0·54.

8 Fixing moments: (left end) 52·5 kNm.
  (right end) 37·5 kNm.
 Reactions: (left end) 73·75 kN.
  (right end) 46·25 kN.

Exercises 11. Page 261

2 (a) 21·65 mm; (b) 37·5 mm; (c) 51·1 mm; (d) 66·2 mm.
3 920 kN. 4 65·7 mm; 1697·3 kN. 5 382 kN
6 (i) 70·8 mm.
  (ii) 63·56.
  (iii) 1640 kN.
7 Load = 132 kN. 8 $\dfrac{94\cdot 7}{135}+\dfrac{38\cdot 5}{165} = 0\cdot 94$. (i.e. less than 1·0).

Exercises 12. Page 298

1 38 mm.
2 Upper tier: 3 No. 305 × 127 × 37 kg U.B.s.
  Lower tier: 12 No. 152 × 89 × 17·09 kg Joists.
3 0·15 N/mm². 4 451 kN/m².
5 60 mm. 6 457 × 191 × 74 kg U.B.s will be suitable.

Exercises 13. Page 305

3 (a) 80·5 mm. 5 6·3 kN/m².
  (b) 27 517 cm⁴ (concrete units). Yes. Maximum spacing
  (c) 17·0 kN/m². = 7 × thickness.
6 (b) 50% = 7 × 126 = 882 mm

Exercises 14. Page 324

2 115 N/mm² for grade 43 steel.
  160 N/mm² for grade 50 steel.
  195 N/mm² for grade 55 steel.
3 (i) 7 mm. 4 (i) 0·67 kN.
  (ii) 1680 mm². (ii) 0·90 kN.
  (iii) 126 mm. (iii) 3·23 kN.
5 346 kN. 6 279 kN.
7 Safe end reaction = 115 kN.
  Safe uniformly distributed load = 230 kN.
8 Max. horizontal shear = 1·29 kN per mm.

Index

Allowance for holes, 15, 65, 154, 158, 327, 329
Analysis of steel, 2
Angles:
 – as struts, 248
 – Commercial data, 19, 41
 – Equal, 41, 278–280
 – Rivet-hole positions in, 385
 – Unequal, 41
Area, Effective, in flanges, 327, 332
Area-moment method for deflection, 169
Area of parabola, 386
Assumptions in theory of bending, 103

Bars, commercial data, 20, 39
Base plate, 285
Bases:
 – Column, 281, 285
 – Slab, 281–285, 377
Beams:
 – British Standard, 38
 – Commercial data for, 39
 – Connection to beam, 35, 72, 74
 – Connection to column, 52, 71, 261
 – Continuous, 215
 – Deflection of, 103, 107, 161, 197, 213, 221, 338
 – Design in warehouse, 356
 – Fixed, 202
 – Moment of resistance of, 82, 105, 326
 – Overhanging, 97, 203
 – Relationship of depth to span in, 161, 176, 329
 – Hole positions in, 385
 – Simple bending in, 107
 – Simply supported, 90–100, 169, 181, 185
 – Slope of, 161, 165, 185
 – Supporting brick walls, 118, 161
 – Theory of, 103
 – Universal, 39, 123–135
Bearing pressure on subsoils, 297

Bearing stress:
 – Definition of, 55
 – Working values of, 31, 33
Bearing values of rivets/bolts: 59–63
Bending moment:
 – Convention of sign for, 85
 – Definition of, 82
 – Diagrams, 85–99, 203–224, 337
 – Eccentric, 238, 252
 – Maximum, 85–99, 187, 203–224
 – Relation between deflection and, 169
 – Relation between shear force and, 185
Bending stress, 26, 300
Bend test, 19
Bolted joints (*see* Riveted)
Bolts (*see also* Rivets), 44
 – Black, 33, 47
 – Clearance for, 47
 – Close tolerance, 47
 – Connections using, 35, 52, 53, 67
 – High strength friction grip, 48, 64, 76
 – In tension, 32, 50
 – Proportions of, 47
 – Turned and fitted, 33, 47
 – Working stress in, 33
Bracing in warehouse roof, 353
Brick walls, Beams supporting, 118–161
British Standards Institution, 1
British Standards:
 – Selected list of, 379
Buckling:
 – in columns, 6, 232
 – in webs, 198, 287, 296
Butt joint, 67–69

Calculation sheets for warehouse, 357–375
Camber, Allowance for, 338
Cantilevers:
 – B.M. and S.F. in, 85–90
 – Deflection of, 165
 – Use in beam deflection, 165, 203

Cap, Regulations for thickness in columns, 282
Carbon, Effect of, on steel, 1, 20
Cased struts, 246, 371
Casings, 289, 300
Centre of curvature, 107, 165
Centre of gravity:
 – For parabola, 387
 – Importance for N.A. position, 110
 – Use in fixed beams, 204
Channels:
 – Commercial data, 42
 – Rivet-hole positions in, 385
Characteristic points, 222
Circular bending, 163
Circular columns:
 – Base slab for, 282
 – Radius of gyration for, 229
Clearance:
 – for bolts, 47
 – for rivets, 47
Cleat connections, 35, 52, 71, 261
Coefficient of eccentricity (*see* Eccentricity coefficient)
Cold bend test, 19
Columns (*see also* Stanchions)
Combined bending and direct stress, 238, 253
Combined grillage, 287
Combined load systems, 86–100, 183
Commercial:
 data, 39
 – elastic limit, 17
 – testing of steel, 19
Complementary stresses, 190
Compound girders:
 – Allowance for rivet holes in, 157, 158
 – Bolt pitch in, 199
 – Section properties for, 154
Compressive strain, 9
Compressive stress:
 – Definition of, 5
 – in columns, 227
 – in flanges, 26, 329
 – in webs, 190, 296
 – Working values of, 27

392 Index

Concrete, Safe bearing pressure, 285, 373
Connections:
- Typical riveted and bolted, 51–53, 58
Continuous beams:
- Characteristic points for, 222
- Deflection of, 221
- Support reactions in, 218
- Theorem of three moments for, 215
- with concentrated loads, 220
- with U.D. load, 217
Contraction in area in test, 20
Contraflexure, Points of, 202, 213, 259
Curtailment of flange plates:
- in plate girder, 336
Curvature in beams, 85, 107, 163

Data sheet for warehouse, 347
Deflection:
- Allowances, 161
- Coefficients of, 162
- Due to shear, 197
- Graphical method for, 173
- in circular bending, 152, 163
- in general case, 164
- Mathematical treatment, 179
- Mohr's theorem for, 164
- of cantilevers, 164
- of continuous beams, 221
- of fixed beams, 213
- of plate girder, 338
- of simply supported beams, 169, 181
- Relationship with span, 176
- Secondary B.M. method for, 169
- with several load systems, 183
Depth:
- Importance in deflection of beams, 177
- in plate girders, 327, 329
- Relationship to span, 177, 329
Diameter of rivet (see also Rivets)
- Nominal, 58
- Relationship to plate thickness, 58
- Usual values of, 58
Distribution of stress:
- in beam section, 104
- in webs, 192
Double covered joint, 67
Double shear, 56
Drifts, Use of, 46
Ductility, Importance of, 1

Eccentricity:
- Coefficients of, 241
- Effect on columns of, 238
- Equivalent concentric load, 241
- in continuing columns, 253
- on rivet groups, 72
- with respect to both axes, 255
Edge distance of rivets, 66
Effective area (see Rivet holes), 65
Effective column length, 233, 251
Effective span, 119
Efficiency of joint, 70
Elasticity:
- Nature of, 10–14
- Shear modulus of, 12, 197
- Young's modulus of, 11, 103, 107, 164, 235
Elastic law, 10
Elastic limit, 11, 17
Electrodes, 308
Elongation in steel test, 18
End restraint in columns, 232, 251
Equal angle:
- Commercial data, 41
- Rivet-hole position in, 385
Equivalent length (see Effective column length)
Equivalent U.D. load, 119, 346, 357
Euler's theory, 234
Extensometer, 14
Extras (see Commercial data)

Fabrication of steelwork, 38
Factor of safety, 24
Field rivets, 33, 60
Filler floor beams, 302
Fire-resistance regulations, 302
Fire-resisting floors, 302
Fish-plate connections, 68
Fixed beams:
- Characteristic points in, 222
- Deflection of, 213
- Equivalent overhanging beam, 203
- Fixing B.M. diagram for, 204
- Points of contraflexure in, 203, 213
- Standard cases of, 205
- with non-central loads, 208
Fixed ends, Importance with columns, 232, 251
Fixing-moment diagram, 204
Flanges:
- Allowance for rivet holes in, 154, 157
- Curtailment of plates in, 337
- Design of, 328
- Joints in, 336
- Riveting/bolting detail for, 58
- Rivet/bolt pitch in, 198
- Unsupported length of, 329
- Welding, 334

- Width of plates in, 66
- Working stresses in, 332
Flat bars, 43
Floor loads:
- Alternative concentrated loads, 301
- Reduction permissible in, 302
- Typical values in, 301
Floor steel in warehouse, 361
Force diagrams:
- for lattice girder, 339
- for warehouse truss, 351
Forms of stress, 3
Foundations:
- Allowable pressure on, 284
- Allowable pressure under, 297
- Combined grillage, 292
- for warehouse, 377
- Single grillage, 287

Gauge length, 23, 24
Girder (see Compound, Lattice and Plate)
Graph, stress-strain, 11, 17
Graphical method for deflection, 173
Grey process for rolling, 44
Grillage, B.M. and S.F. for, 291, 295
- Buckling in web of, 296
- Combined, 292
- Shear in webs of, 291, 296
Gusseted base, 285
Gyration, Radius of, 228

High yield stress steel, 2
Hooke's Law, 10–14, 18, 25, 103
Horizontal reaction in trusses, 350
Horizontal shear in beams, 190
Horizontal wind pressure, 257
Hydraulic riveting, 46

I-beams (see Universal beams)
Inertia, Moment of, 110, 147, 154, 236, 304
Influence of carbon on steel, 1, 20
Intensity of pressure:
- permissible on subsoils, 297
- Wind on roof trusses, 348
Intensity of stress, 3

Joints:
- in columns, 251, 256
- in flange, 336
- Riveted (see Riveted joints)
- Welded, 306
Joists:
- Filler, 302
- Rolled steel, 136
- Timber, 106

Laboratory steel tests, 20
Lap joint, 67

Index 393

Lateral strain, 18
Lattice girder:
– Design of members of, 338
– Method of sections for, 344
– Stress diagram for, 339
Least radius of gyration, 228
Limit of elasticity, 11, 17
Limit of proportionality, 17
Link polygon in deflection, 175
Lintols, 118
Loads:
– Concentrated, 84–87, 90–93, 208–214
– Eccentric, 238, 252
– Equivalent, uniformly distributed, 119, 346, 357
– Floor, 301
– Reduction in floor, 302
– Safe, on foundations, 284
– Uniformly distributed (*see* U.D. load)
– Wind, 257
Long columns, 227

Maximum:
– allowable deflection, 161
– and minimum moment of inertia, 112
– bending moment and shear force, 84–100, 187, 206
– flange compressive stress, 326
– section modulus, 112
– shear stress in beams, 192
Metal arc welding, 306
Method of sections, 344
Mill:
– Plate rolling: facing 22
– Roll forms for, 38, 44, 45
Modulus:
– Section (*see* Section modulus)
– Shear, 12, 197
– Young's (*see* Young's modulus)
Mohr's theorem, 165
Moment of inertia:
– Addition of values of, 110
– Definition of, 110
– Maximum and minimum values of, 112
– of compound section, 154
– of plate girder section, 157, 326
– of rectangular section, 111
– of U.B. section, 114, 123, 125
– Principle of parallel axes in, 146
– Principle of translation in, 145
Moment of resistance:
– Definition of, 84
– General expression for, 109
– Value of rectangular beams, 104
– Value in plate girders, 326

Multiple load systems:
– for B.M. and S.F., 100
– for deflections, 183

Net areas, 15, 69, 329, 333 (*see also* Rivet/Bolt holes)
Neutral axis, 104, 108, 111, 194
Neutral layer, 104, 108
Nominal rivet diameter, 58
Nominal stress, 17
Nuts and washers, 47

Overhanging beams:
– B.M. and S.F. diagrams for, 97
– compared with fixed beams, 203
– Points of contraflexure in, 203

Packing:
– in column joint, 257
Parabola, Properties of, 386
Parallel axes, Principle of, 146
Parallel translation of areas, 145
Percentage:
– contraction in area, 20
– elongation, 18
Pitch of bolts:
– in compounds, 58, 199
– in plate girders, 199
– Maximum value of, 58
– Minimum value of, 58
– Straight-line, 66
Plate girder:
– Camber of, 338
– curtailments, 336
– Deflection of, 164, 338
– Depth of, 327, 329
– Design of flanges, 327, 332
– Moment of inertia, 157, 328
– Practical proportions of, 329
– Bolt hole allowance in, 158, 327, 332
– Bolt pitch in, 199
– Stiffeners, use in, 197, 334
– Web design in, 331
– Weight of, 331
– Width of flange plates in, 66, 329
– Working stresses for, 330
Plates, Commercial data, 43
Pneumatic hammer, 47
Points:
– Characteristic, 222
– of contraflexure, 202, 213, 259
Pressure:
– Wind, 257
– on foundations, 284, 297
Purlins, in warehouse roof, 354

Radius of curvature, 107, 163
Radius of gyration, 228
– for B.S. sections, 123–142
– for cased struts, 246

Rankine's formula, 236
Reactions for truss, 350
Rectangular beam section, 104, 193
Reduction in area in test, 20
Resistance moment (*see* Moment of resistance)
Rivet bars, Testing of, 20
Riveted/bolted joints:
– Beam to beam, 35, 74
– Beam to column, 52, 71, 261
– Column base, 281
– Column splice, 256
– Efficiency of, 70
– Fish-plated, 68
– in lattice girder, 340
– Methods of failure in, 68
– Tie-bar example of, 67
Rivet groups, Eccentric load on, 72
Rivet/bolt holes:
– Allowance for, 15, 65, 155, 158
– Clearance in, 47
Riveting, Methods of, 46
Rivets/Bolts:
– Bearing in, 59–63
– Clearance for, 47
– Edge distances of, 66
– Forms of heads for, 46
– in B.S. sections, 385
– Leading, 68
– Pitch of (*see* Pitch)
– Proportions of, 46
– Shear in, 7, 33, 55, 59–63
– Shop, 33, 45, 59
– Site, 33, 60
– Table of strength of, 59–63
– Tension, in 32
– Working stresses for, 33
Rolled sections, 38, 122–127, 136–142
Rolling mill, 38
Rolls, 38, 44, 45
Roof steel in warehouse, 356–360
Roof truss in warehouse, 348
Round bars:
– Testing of, 19
Round columns, Base slab for, 282

Safety, Factor of, 24
Second moment (*see* Moment of inertia)
Section, British Standard, 38
Section modulus:
– Definition of, 115
– Maximum and minimum values of, 116
– B.S. sections, 122–142
– Values in tension and compression, 116
Sections, Method of, 344
Separators, 292
Shear:
– Convention of signs for, 85
– Definition of, 83

Index

Shear—cont.
- Deflection due to, 197
- diagrams, 86–99, 291, 295
- in continuous beams, 220
- in fixed beams, 207
- in rivets and bolts, 7, 33, 55, 59–63
- in webs, 192, 291
- Maximum value of, 86–99
- Modulus, 12, 197
- Relationship to B.M., 86–99, 186
- Rules for constructing diagrams of, 189
- stress distribution, 195
- Use in finding B.M. maximum, 187

Shear strain, 9, 197
Shear stress, 6, 190 (see also Shear)
Sheets, Commercial data, 43
Shop rivets, 33, 45, 59
Simple bending, 107
Simply supported beams, 90–100, 169, 181, 185
Site riveting, 33, 45, 60, 61
Slab bases, 281–285, 377
Slag, in welding operation, 308
Slenderness ratio (or l/r), 242
Slope, 161, 165
Soils, Safe pressure on, 297
Span:
- Effective, 119
- Minimum for end connections, 71
- Relationship of beam depth to, 176, 329

Specification, British Standard, 1, 379
Specimen, Standard test, 21, 22
Splices:
- in column, 256
- in flange plates, 336
- in web, 338

Stanchions or columns:
- Base design in, 281, 284, 285
- Bending moment in, 238–256
- Cased, 246, 371
- Continuing, 238
- Design in warehouse, 369
- Double eccentricity in, 254
- Eccentric loading of, 238–256
- Effective length of, 233, 251
- End restraint in (see Chapter XI)
- Equivalent concentric load for, 240
- Euler's theory for, 234
- Joints in, 251, 256
- Long and short, 227
- Practical end fixture rules for, 238
- Radius of gyration in, 228
- Rankine's formula for, 236
- Solid round, 229, 282
- Stiffness, 255

- Table of safe loads for, 266–280
- Wind loads on, 257

Steel:
- Commercial testing of, 19
- Laboratory testing of, 20
- Nature of steel, 1
- Rolling sections of, 38, 44, 45

Stiffeners, 332
Stiffness, 151, 243
Strain:
- Compressive, 9, 104, 108
- Nature of, 7
- Relationship to stress, 10, 17, 104, 108
- Shear, 9, 197
- Tensile, 7, 104, 108

Stress:
- Bearing, 31, 33, 55
- Bending, 26, 300
- Compressive, 5, 26, 27, 190, 227, 329
- diagrams, 339
- Forms of, 3
- Intensity of, 3
- Nature of, 2
- Real, 18
- Relationship to strain, 10, 17, 104, 108
- Shear (see also Shear)
- Tensile, 3, 13, 17, 20, 190, 240
- Ultimate, 17
- Varying, 3, 103, 190
- Working values of, 27–33, 243–245

Strut, 227, 246, 269–280

Tabular method for moment of inertia, 152
Tees, Commercial data, 42
Tensile strain, 7
Tensile stress:
- Commercial test limits, 19
- Graph of test values of, 17
- in rivets and bolts, 33
- in sections, 32
- Nature of, 3
- Working stress values for, 32, 33

Tensile test, 18–22
Test, Cold bend, 19
- Standard specimens for, 21, 22

Testing machine, Fig. 1.13
Theorem of three moments, 215
Theory of bending, 107
Ties, 4, 5, 15, 67, 341, 343
Timber floor beams, 105
Tolerances, Rolling, 39
Truss for warehouse, 352

Ultimate stress, 17
Unequal angles, Commercial data, 41
- Rivet-hole positions in, 385
Uniformly distributed load, 88, 93, 167, 170, 213

Units for:
- bending moment, 84
- moment of inertia, 112
- moment of resistance, 83
- section modulus, 112
- shear force, 84
- strain, 8
- stress, 3
- Young's modulus, 12

Universal beams, 39, 123–135
Universal columns, 40, 265–276
Unsymmetrical beam, section, 116, 152

Value of one rivet or bolt, 56, 59–64
Varying stress, 3, 104, 192
Vertical reaction in truss, 350
Vertical shear, relation to horizontal, 191

Walls, beams supporting, 118
Warehouse, Design of, 346
Web:
- Buckling of, 197, 292
- Complementary stresses in, 190
- joint, 338
- Stiffening of, 332
- Thickness of, 332
- Varying stress in, 190
- Working stresses in, 31

Weight of steel sections, 39
Welding:
- Metal arc, 307
- Methods of, 307
- Practical considerations in, 309
- Typical calculations in, 316

Welds:
- Butt, 310
- Fillet, 315
- Forms of, 308
- Nomenclature of, 311
- Return, 315
- Typical dimensions of, 312–314, 319

Width of flange plates, 66
Wind:
- Bracing in warehouse roof, 353
- C.P.3 requirements, 257
- on columns, 257
- on roof truss, 348

Working stress:
- Definition, 24
- in bearing, 33
- in bending, 27–30
- in filler beams, 300
- in grillage beams, 289
- in shear, 31
- in tension, 32

Yield point, 17, 18
Young's modulus, 11, 103, 162

Z-symbol for section modulus, 116